Biorational Tree-Fruit Pest Management

Biorational Tree-Fruit Pest Management

Edited by

Martín Aluja

Tracy C. Leskey

and

Charles Vincent

www.cabi.org

CABI is a trading name of CAB International

CABI Head Office
Nosworthy Way
Wallingford
Oxfordshire OX10 8DE
UK

Tel: +44 (0)1491 832111
Fax: +44 (0)1491 833508
E-mail: cabi@cabi.org
Website: www.cabi.org

CABI North American Office
875 Massachusetts Avenue
7th Floor
Cambridge, MA 02139
USA

Tel: +1 617 395 4056
Fax: +1 617 354 6875
E-mail: cabi-nao@cabi.org

A catalogue record for this book is available from the British Library, London, UK.

Library of Congress Cataloging-in-Publication Data

Biorational tree fruit pest management / edited by Martín Aluja, Tracy C. Leskey, and Charles Vincent.
p. cm.
Includes bibliographical references and index.
ISBN 978-1-84593-484-2 (alk. paper)
1. Pests–Integrated control. 2. Fruit–Diseases and pests. I. Aluja, Martin. II. Leskey, Tracy C. III. Vincent, Charles, 1953- IV. Title.

SB951.B48 2009
634′.049–dc22

2008045145

ISBN-13: 978 1 84593 484 2

Typeset by AMA Dataset, Preston.
Printed and bound in the UK by the MPG Books Group, Bodmin.

The paper used for the text pages in this book is FSC certified.
The FSC (Forest Stewardship Council) is an international network to promote responsible management of the world's forests.

Contents

Contributors

Aluja, Martín, *Unidad de Entomología Aplicada, Instituto de Ecología, AC, Km 2.5 Carretera Antigua a Coatepec No 351, 91070 Xalapa, Veracruz, Mexico. Email: martin.aluja@inecol.edu.mx*

Berlocher, Stewart H., *Department of Entomology, University of Illinois at Urbana-Champaign, 320 Morrill Hall, 505 S. Goodwin Avenue, Urbana, IL 61801, USA. Email: stewartb@life.uiuc.edu*

Chouinard, Gérald, *Institut de Recherche et de Développement en Agroenvironnement (IRDA), 3300 Sicotte, CP 480, Saint-Hyacinthe, QC J2S 7B8, Canada. Email: gerald.chouinard@irda.qc.ca*

Coli, William M., *Department of Plant, Soil and Insect Sciences, Fernald Hall, University of Massachusetts, Amherst, MA 01003, USA. Email: wcoli@umext.umass.edu*

Cooley, Daniel R., *Department of Plant, Soil and Insect Sciences, Fernald Hall, University of Massachusetts, Amherst, MA 01003, USA. Email: dcooley@microbio.umass.edu*

Dorn, Silvia, *ETH Zurich, Institute of Plant Sciences/Applied Entomology, Schmelzbergstrasse 9/LFO, 8092 Zurich, Switzerland. Email: silvia.dorn@ipw.agrl.ethz.ch*

Feder, Jeffrey L., *Department of Biological Sciences, PO Box 369, Galvin Life Science Center, University of Notre Dame, Notre Dame, IN 46556-0369, USA. Email: jfeder@nd.edu*

Green, Thomas A., *IPM Institute of North America, Inc., 4510 Regent St., Madison, WI 53705, USA. Email: ipmworks@ipminstitute.org*

Hilton, Richard J., *Southern Oregon Research and Extension Center, Oregon State University, 569 Hanley Road, Central Point, OR 97502, USA. Email: richard.hilton@oregonstate.edu*

Kogan, Marcos, *Integrated Plant Protection Center and Department of Horticulture, 2040 Cordley Hall, Oregon State University, Corvallis, OR 97331, USA. Email: koganm@science.oregonstate.edu*

Lauzon, Carol R., *Department of Biological Sciences, California State University, East Bay, Hayward, CA 94542, USA. Email: carol.lauzon@csueastbay.edu*

Leskey, Tracy C., *USDA-ARS Appalachian Fruit Research Station, 2217 Wiltshire Road, Kearneysville, WV 25430-2771, USA. Email: Tracy.Leskey@ars.usda.gov*

Opp, Susan B., *Department of Biological Sciences, California State University, East Bay, Hayward, CA 94542, USA. Email: susan.opp@csueastbay.edu*

Papaj, Dan R., *Department of Ecology and Evolutionary Biology, University of Arizona, Tucson, AZ 85721, USA. Email: papaj@email.arizona.edu*

Piñero, Jaime C., *College of Tropical Agriculture and Human Resources, University of Hawaii at Manoa, Honolulu, HI 96822, USA. Email: jpinero@ctahr.hawaii.edu*

Roitberg, Bernard D., *Department of Biological Sciences, Simon Fraser University, Burnaby, BC, V5A 1S6, Canada. Email: roitberg@sfu.ca*

Rull, Juan, *Unidad de Entomología Aplicada, Instituto de Ecología, AC, Km 2.5 Carretera Antigua a Coatepec No 351, 91070 Xalapa, Veracruz, Mexico. Email: juan.rull@inecol.edu.mx*

Vincent, Charles, *Horticultural Research and Development Centre, Agriculture and Agri-Food Canada, 430 Gouin Blvd., Saint-Jean-sur-Richelieu, QC J3B 3E6, Canada. Email: Charles.Vincent@agr.gc.ca*

Wright, Starker E., *USDA-ARS, Appalachian Fruit Research Station, 2217 Wiltshire Road, Kearneysville, WV 25430-2771, USA. Email: starker.wright@ars.usda.gov*

Preface

The inception of this book is tied closely to the legacy of Ronald J. Prokopy (1935–2004). Ron was not only one of the most highly regarded applied behavioural ecologists worldwide, but also a pioneer in the development of one of the first successful integrated pest and disease management (IPM) programmes for apple orchards. His research programme was aimed at studying the behavioural ecology and management of tree-fruit pests, particularly fruit-fly species. His work on the behaviour of the apple maggot fly, *Rhagoletis pomonella* (Walsh), and many other tephritid species, has led not only to an archetypal behavioural model for those interested in studying host finding and selection, but also to practical behaviourally based management strategies that fit into IPM, integrated fruit production and organic production systems. His keen interest in basic behaviour also led to novel research in areas such as learning and social facilitation. Ron was also a strong and tireless proponent of IPM strategies, developing theoretical approaches, working extensively with growers and conducting numerous large-scale experiments throughout the important apple-growing region of New England, USA.

Our goal as editors is to produce a book that will represent a compendium of current theory, philosophy and innovative research in the area of biorational tree-fruit pest management, an area that Ron was so instrumental in nurturing. The inception of this book began at the 2005 Entomological Society of America National Meeting held in Fort Lauderdale, Florida. There, a special Programme Symposium was held to honour Ron, who unfortunately passed away before the symposium took place. It was entitled 'Bridging the Gap between Basic Behavioral Research and Crop Protection Applications for Tree-Fruit Agroecosystems: Honoring the Life and Work of Ronald J. Prokopy'. Presentations were made by former graduate students, postdoctoral students and close colleagues. The overall themes presented during that symposium were the trigger for creating this volume. After the symposium was held, it was decided that a long-lasting tribute to Ron would be to produce a book in his honour. Our role as editors has been to envision the theme of this book and to unify, as much as possible, the written material produced by a collection of invited authors, so that the final product would serve as a lasting tribute to the late Ron Prokopy and foster environmentally friendly tree pest management schemes. So, three of us took on the task of preparing a book proposal that incorporated some of the themes presented during the symposium, but decided to expand the scope of the book by inviting other former colleagues of Ron to share their vast experience in the area of tree-fruit pest management (for example, Marcos Kogan). We, together with the authors, wanted to demonstrate how one could utilize and apply biological information gathered from basic research concerning pest species to develop meaningful crop protection strategies. Importantly, we wanted to carry this theme not just to an experimental level but also to an applied one that highlighted grower involvement and successful business owners supplying pest management information and tools. In other words, we wanted to convey a complete story of biorational pest management from how pests evolved in nature, to selling environmentally friendly products for their control. The topics that are addressed in the book begin with the most fundamental ecological, evolutionary and behavioural questions related to insect pests and subsequently build from there, until reaching the market phase.

With this goal in mind, we invited one of the most influential thinkers in the field of IPM, Marcos Kogan, to write an introductory chapter and provide a fundamental framework for the rest of the volume. We also invited Stewart Berlocher and Jeffrey Feder, two of the most successful students of Guy Bush, one of the leading insect evolutionary biologists of our times, to answer the question of how pest species evolve. Guy was Ron's mentor, and so this book represents a beautiful example of what academia is all about: leaving intellectual progeny. In this case, the grandfather, father, sons and grandchildren are all represented. The introductory chapter by Marcos Kogan and Richard Hilton is followed by a chapter on the evolution of key tree-fruit pests written by Stewart Berlocher and Jeffrey Feder. Next is a chapter about the functional and behavioural ecology of tree-fruit pests written by two of Prokopy's former graduate students, Bernard Roitberg and Susan Opp, and postdoctoral research associates, Daniel Papaj and Carol Lauzon (an innovative microbiologist). Silvia Dorn, in collaboration with Prokopy's final graduate student, Jaime Piñero (who, in turn, was a former undergraduate student of Martín Aluja, one of Ron's graduate students), address the topic of how coleopteran pests of fruit trees detect their hosts and how this information can be exploited to develop effective monitoring tools for pests. Tracy Leskey (another former graduate student of Ron's), Charles Vincent and Gérald Chouinard honour Prokopy's research legacy with a chapter describing behaviourally based tactics for managing apple maggot fly and plum curculio. Two of Ron's closest colleagues, Daniel Cooley and William Coli, describe one of the most successful IPM stories worldwide: the New England tree-fruit IPM programme. Then, two former graduate students, Martín Aluja and Juan Rull (both from Mexico), tackle a topic that is currently in the forefront of insect pest management: environmental manipulation. Finally, Thomas Green (a former graduate student of Ron's) and Starker Wright (former field coordinator for the Massachusetts Tree-Fruit IPM team), successful IPM-related business owners and advocates, discuss a topic often overlooked in most scientific books: turning basic and applied science into products and programmes that improve health, environmental and economic impacts in the marketplace. By providing growers with tools to reduce pesticide use and consumers with tools to recognize and support these producers, private businesses can flourish, improve production practices and simultaneously educate the public, and in particular children, about the many benefits of IPM.

This wide range of topics and distinguished list of contributors is designed to produce a unique contribution to the field of IPM. We are not aware of any similar effort to date or of any other book that has treated, in one volume, the topic with such breadth and depth, evaluating pest management from its very roots, the evolution of a pest, to the most applied of angles, including market considerations. The group of authors not only represent international perspectives on biorational pest management, with four countries represented (Canada, Mexico, Switzerland and the USA), but also their areas of expertise span topics that include basic behavioural, ecological and evolutionary research, as well as implementation of applied pest management programmes. Furthermore, the authors represent diverse professional backgrounds, including university faculty, government scientists, extension personnel and business owners. It is unusual to have such a combination of authors working together with the aim of synergizing a deeper understanding of IPM and biorational pest management principles.

Based on the above, we are confident that this book will interest a wide audience. We hope it will be used in courses on IPM and sustainable agriculture throughout the world. We hope that a wide array of specialists in both the academic world and more applied realms also will enjoy reading the book. At times, we disagreed with the ideas, theories and concepts put forward by the contributing authors. However, we believe strongly in the virtue of academic freedom and have allowed the authors to express their own perspectives accordingly. Therefore, the authors themselves ultimately are responsible for the ideas they present within this text.

Finally, and as noted at the very beginning of this Preface, a special feature of the book is the fact that it is dedicated to Ronald J. Prokopy, a highly respected pioneer in the field of biorational pest management. We hope that many people and libraries throughout the world will want to place this significant title on their bookshelves.

Ron was a remarkable observer, educator, researcher, writer and apple grower. His accomplishments include uncovering many of the secrets that the apple maggot fly; Mediterranean fruit fly, *Ceratitis capitata* (Weidemann); Oriental fruit fly, *Bactrocera dorsalis* (Hendel); plum curculio, *Conotrachelus nenuphar* (Herbst); and other fruit and vegetable pests have held tightly for many years. And, as anyone who spent time with him soon realized, Ron was also a wonderful role model for all who marched to the beat of their own drum. His numerous eccentricities have generated a wealth of humorous anecdotes that have enlivened many a discussion between colleagues and friends.

Among his many skills, Ron was a 'supernormal' observer of nature, verging on supernatural, as apple grower, Michael Phillips (2004) reports:

> Learning to 'project your consciousness' into the mind of an insect . . . is a powerful way to understand the interconnectedness of nature. The late Ron

> Prokopy . . . did this time and time again in developing numerous strategies to deter orchard pests over the course of his lifetime. Any and all apple growers need to hone their observation skills and then literally listen to the teachings that the orchard willingly shares. Ron's influence will live on for a long time to come in each of us who seeks to understand deeply.

Cardé and Stoffolano (2004) cited a 'clear impression that Ron, while closing his eyes, would try to envisage himself as the insect perceiving the very situation you just described to him' and that 'Ron's keen sense of observation could lead to previously unimagined behaviours'.

Ron worked hard to convey to Massachusetts's apple growers the insights he gained from his unique perspective. According to Wes Autio (2004), horticulturist and colleague at the University of Massachusetts, 'At a mid-June twilight meeting, Ron spent a good portion of his 20 minutes . . . describing how plum curculios found an apple tree . . . mostly given from the orchard floor, that is Ron was lying down and crawling along the ground (and speaking) showing the growers how the curculio saw the world.' Ron very much regarded these growers as his employers and made sure that his research and extension priorities addressed their concerns, including maximizing fruit quality, minimizing input costs and reducing hazards to themselves, their workers, consumers and the environment.

Ron was a consummate grant rainmaker and researcher, as Dan Papaj (2004), collaborator and former postdoctoral research associate tallied from Ron's 46-page curriculum vitae:

> With more than US$5.25 million in competitive grant awards, despite no award greater than US$250,000 and most in the US$50–100,000 range . . . every aspect of the insect's biology – oviposition, mating and feeding, chemical cues and visual cues, larval and adult stages, experience and physiological state – was subject to scrutiny . . . taken together, the studies are a tour de force . . . to tell a complete story of an insect's behavior in nature.

Ron's prolific publication rate was unmatched by any of his contemporaries, despite his lifelong refusal to use a computer or word processor. He authored or co-authored more than 450 publications with more than two dozen technicians and undergraduate, graduate and postdoctoral students he mentored, many of who remain in science or education. Dan Cooley (2004), plant pathologist and colleague, relates, '. . .the number of publications a person writes gives other academics a quick read on the stature and impact a scientist carries. It's a career batting average, and Ron was a Ted Williams . . . Ron used time much like he used research dollars, squeezing the minutes.' On local trips, for example, if he had a driver, he would 'retreat to the back seat to fully devote attention to his papers. . . For longer trips, every plane ride, every hotel room served as an office and study for the constant, nearly undecipherable pencil scribbling on page after page of yellow-lined paper.' Even sunny, sandy beaches served as a writing environment for Ron.

Without a hint of malice, Ron effectively applied the knowledge he gained about his insect friends against them – to reduce production costs and reliance on pesticides for his growers. As reported in the posthumous presentation of the L.O. Howard Award by the Eastern Branch of the Entomological Society of

America in 2005, 'He pioneered the use of plant visual and odor cues for developing control tactics. His studies of foraging behavior and investigations of genetic and environmental components of behavior were used as the basis for creating new, biologically based approaches for pest management.' His work (and passing) was recognized in the New York Times (Gilbert, 1989; Pearce, 2004), which reported that 'in a five-year experiment begun in 1978 by Ronald J. Prokopy . . . 40 orchard owners were able to reduce their use of insecticides by 40% and bring down their costs without sacrificing yields or quality.' Of course, in his own orchard, Ron far surpassed this accomplishment, growing disease-resistant cultivars with only two applications of insecticide and, in some years, a fungicide (Prokopy and Cooley, 1991; Earles *et al.*, 1999; Prokopy, 2001; Cooley and Coli, Chapter 6, this volume).

Ron greatly appreciated his enduring marriage of nearly 40 years to Linda Olson Prokopy, often remarking how fortunate he was to have such a wonderful partner who granted him so much latitude to pursue his passion, let alone sleep with the window open throughout the dead of winter, memorialized in a poem presented at a symposium in his honour. Together, Ron and Linda successfully parented and launched two fine sons. Joshua Prokopy, 37, is currently taking time off from his work in support of public housing to write and to be a full-time father to daughters, Annabel, aged 5, and Margaret, 4, and son David, 2, with Linda Stalker Prokopy, an assistant professor in natural resource policy and planning at Purdue. After teaching chemistry to secondary school students for several years, Max Prokopy, 34, recently completed his Masters in exercise science and physiology and has accepted a position as Strength and Conditioning Coach for athletics at Amherst College. Ron's family enjoyed his many hobbies with him, including singing, gardening, swimming, skating, cross-country skiing, hiking and golf. At Ron's memorial service, his son, Josh, remarked that he never heard his father say that he was tired, an incredible claim for such an active life.

Agnello and Kain (2004) provide a fitting closing synopsis to an amazing life:

> Always a source of startling inquisitiveness and ingenuous encouragement, Ron enlightened even as he challenged us to constantly think and re-think our assumptions about fruit insects and the part they played in the natural world that consumed his professional life. An optimist who motivated, inspired, amused, and at times exasperated those he came in contact with, he was an individualist whose memory will provide numerous stories and a high standard of professionalism as a continuing influence on his peers. We are sure that, although he will be missed for many reasons, Ron would regret having to depart this world with so much yet left to do.

This last statement may be subject to debate – regret possibly not being a word in Ron's vocabulary. In fact, at Ron's memorial service at his home in Conway, Massachusetts, Dan Papaj related how when he and Ron and a planeload of passengers were asked to assume the crash position during a (fortunately successful) emergency landing in Hawaii, Ron turned to Dan and remarked, 'Well, I've had a very good life.'

In sum, we are happy to have edited a book that pays tribute to a scientist of great influence by offering a *tour de force* of intellectual effort in an area that has become of seminal importance to humanity: managing pests by minimizing damage to the environment. With global warming and an international food crisis lingering on the horizon, nothing better can be done than offering insight into solutions that will help this world become better. This is the legacy of the human being we honour in this book.

Martín Aluja, *Xalapa, Veracruz, Mexico*
Tracy C. Leskey, *Kearneysville, West Virginia, USA*
Charles Vincent, *Saint-Jean-sur-Richelieu, Quebec, Canada*
Thomas A. Green, *Madison, Wisconsin, USA* (guest)

References

Agnello, A. and Kain, D. (eds) (2004) Semper fly; remembering Ron. *Scaffolds Fruit Journal* 13 (http://www.nysaes.cornell.edu/ent/scaffolds/2004/040524.html#general, accessed 12 July 2008).

Autio, W.R. (2004) A *Fruit Notes* issue in honor of Ronald J. Prokopy. *Fruit Notes* 69 (2), 1–2.

Cardé, R.T. and Stoffolano, J.G. Jr (2004) Ron Prokopy 1935–2004. *Physiological Entomology* 29, 489–490.

Cooley, D.R. (2004) Ron's papers. *Fruit Notes* 69 (2), 3–4 (http://www.umass.edu/fruitadvisor/fruitnotes/v69n2/692RonsPapers.pdf, accessed 12 July 2008).

Earles, R., Ames, G., Balasubrahmanyam, R. and Born, H. (1999) *Organic and Low-Spray Apple Production. Appropriate Technology Transfer for Rural Areas.* Publication #IP020 (http://attra.ncat.org/new_pubs/attra-pub/PDF/apple.pdf, accessed 12 July 2008).

Gilbert, S. (1989) America tackles the pesticide crisis. *New York Times*, October 8 (http://query.nytimes.com/gst/fullpage.html?res=950DE4D9123EF93BA35753C1A96F948260&sec=&spon=&pagewanted=print, accessed 12 July 2008).

Papaj, D.R. (2004) In praise of a good colleague: Ronald John Prokopy. *Journal of Insect Behavior* 17, 569–577.

Pearce, J. (2004) R.J. Prokopy, 68, entomologist who focused on pest controls. *New York Times* (http://query.nytimes.com/gst/fullpage.html?res=9E05E7DA113EF93AA15756C0A9629C8B63&n=Top/News/Science/Topics/Science%20and%20Technology, accessed 29 May 2004).

Phillips, M. (2004) *Holistic Consciousness. Lost Nation Orchard: Grower Network Research* (http://www.herbsandapples.com/orchard/research.php, accessed 12 July 2008).

Prokopy, R.J. (2001) Twenty years of apple production under an ecological approach to pest management. *Fruit Notes* 66, 3–10.

Prokopy, R. and Cooley, D. (1991) Very low-spray apple growing. *Fine Gardening* 22, 46–49.

Ronald J. Prokopy

Ron's orchard in Conway, Massachusetts (see Chapter 6)

Acknowledgements

All chapters in this book were peer-refereed by two or more outside reviewers. We are very grateful to the following colleagues for sharing their valuable time and expertise: Dietmar Schwarz, Norman Barr, John Sivinski, Lukasz L. Stelinski, David C. Robacker, Larry Gut, Mark Whalon, Art Agnello, Alberto Anzures-Dadda, Elizabeth Henderson, Michael Phillips, John Mumford, James R. Miller, Charlie Pickett, John Banks, Francisco Díaz-Fleischer, David Nestel, Harvey Reissig, David Rosenberger, Charles Pickett and Odile Carisse.

We gratefully acknowledge the important support of Nigel Farrar and particularly Meredith Carroll, Kate Hill and Chris McEnnerney at CAB International. We also sincerely thank Alberto Anzures-Dadda and Pierre Lemoyne for their invaluable support throughout the editorial process.

Martín Aluja was supported financially by the Instituto de Ecología, AC (INECOL) and the Mexican Campaña Nacional Contra Moscas de la Fruta (Secretaría de Agricultura, Ganadería, Desarrollo Rural y Pesca – Instituto Interamericano de Cooperación para la Agricultura (SAGARPA-IICA)). He also acknowledges support from the Mexican Consejo Nacional de Ciencia y Tecnología (CONACyT) through a Sabbatical Year Fellowship (Ref. 79449) and thanks Benno Graf and Jörg Samietz (Forschungsanstalt Agroscope Changins-Wädenswil ACW) for providing ideal working conditions to finish editing this book.

1 Conceptual Framework for Integrated Pest Management (IPM) of Tree-Fruit Pests

MARCOS KOGAN[1] AND RICHARD J. HILTON[2]

[1]Professor of Entomology and Director Emeritus, Integrated Plant Protection Center and Department of Horticulture, Oregon State University, Corvallis, Oregon, USA; [2]Senior Research Assistant/Entomologist, Southern Oregon Research and Extension Center, Oregon State University, Central Point, Oregon, USA

When is it IPM?

At the peak of the popularity of the co-evolutionary theory, Daniel Janzen wrote a paper titled 'When is it co-evolution?' (Janzen, 1980). The theory had been proposed by Ehrlich and Raven (1964) and immediately caught the interest of ecologists, evolutionary biologists, entomologists and other biologists who studied interspecies interactions. The paper has been cited in over 800 books and scientific publications. It soon became apparent that the theory was being misapplied to many species interactions that were not necessarily co-evolved. Hence the need for Janzen's attempt to define the criteria whereby a plant/herbivore system was considered co-evolved.

Paraphrasing Janzen, it seems pertinent to question 'When is it IPM?'. The expression 'integrated pest management' has been used in the English language technical and popular literature since at least 1968 (Kogan, 1998). Over time, the expression slowly seems to have lost the unique conceptual content intended by its early proponents. Often, the term 'integrated' was dropped and 'pest management' began to be used as synonymous with 'pest control'. Consequently, any action taken to reduce a pest problem was called 'pest management', even if it meant simply the application of a pesticide on a preset schedule. Two main criticisms arose regarding the IPM expression. First, plant pathologists and weed scientists resented the entomological bias in most of the early IPM literature. To these critics, IPM came to mean 'insect pest management'. Second, the increasing use of techniques developed under the aegis of IPM, such as weather-driven models for predicting plant disease and insect population events to refine and optimize the use of pesticides, led to an impression that pesticides were central to IPM; in this context, IPM became 'integrated pesticide management'. Aiming to restore the integrity of the IPM concept, new terms were coined, or new expressions proposed. Thus, IPM was 'biointensive' according to Frisbie and Smith (1991), apparently in contrast to another IPM that, supposedly, would be 'chemically intensive' or, as in the title of this volume, 'biorational' (see also Hall and Berry, 1995), as if there were another kind of IPM that was bio-irrational. A panel of the National Academy of Sciences, largely at the insistence of its plant pathologist members, proposed what, in their view, was a new paradigm – 'ecologically based pest management' (NRC, 1996).

Since the late 1960s, IPM has been the fundamental paradigm in plant protection. A major contribution of IPM to agriculture, beyond the limited scope of plant protection, has been the demonstration of the need to base all phases of the production system on sound ecological principles, with the ultimate goal of 'designing' economically and ecologically sustainable agroecosystems (Gliessman, 1990; Glen *et al.*, 1995; Vandermeer, 1995; Gurr *et al.*, 2004; Nicholls and Altieri, 2007; Wratten *et al.*, 2007). Thus, by definition, a pest control programme that is 'chemically intensive', 'bio-irrational', or is not 'ecologically based' should not be called IPM, because one of the basic principles of IPM is that decision makers must adopt control strategies that are most suitable to the agroecological and socio-economic conditions of the region and, therefore, they must be ecologically based, biologically rational and make as intensive use of biological control agents (biointensive) as technologically and economically feasible.

Regrettably, the attempt of the NRC panel (NRC, 1996) to redefine IPM replaced the term 'integrated' with 'ecologically based' in the IPM expression. However, 'integration' is the real conceptual foundation of IPM, with the understanding that integrated in IPM are: (i) the set of available control tactics; (ii) the considered impact of the various pest categories (arthropod and vertebrate pests, plant pathogens and weeds); and (iii) the scales of agricultural units targeted for pest management. Thus, IPM is a dynamic and flexible concept that exists as a continuum within which control tactics may be applied singly or in various combinations that may involve only two or multiple control tactics. Consequently, integration of control tactics range from rather low to very high levels. Similarly, within the continuum, target pests range from single species to the total pest complex affecting the crop; and agricultural scales range from single fields to multiple fields of multiple crops and the surrounding vegetation (Kogan, 1988; Prokopy and Kogan, 2003).

Criteria that must be met to label a pest control programme as IPM have been defined in terms of the number of control tactics employed in the programme (e.g. Benbrook *et al.*, 1996). So, if both chemical and biological control were used, then it would be correct to label the programme as IPM. If even more tactics were included, the more robust would be the label. Although easy to understand and apply, this criterion is not necessarily universally valid. If, for example, the effectiveness of naturally occurring pest mortality factors is sufficient to keep pest populations below the economic injury level, there would be no need for human intervention. This would be an ideal IPM system. 'Do nothing' would be the best IPM decision. If, however, two or more control tactics were not applied intentionally, the programme would not fit the number of tactics criterion.

Figure 1.1 attempts to capture the notion of the multiple levels of pest management integration discussed above and to suggest an IPM definition that

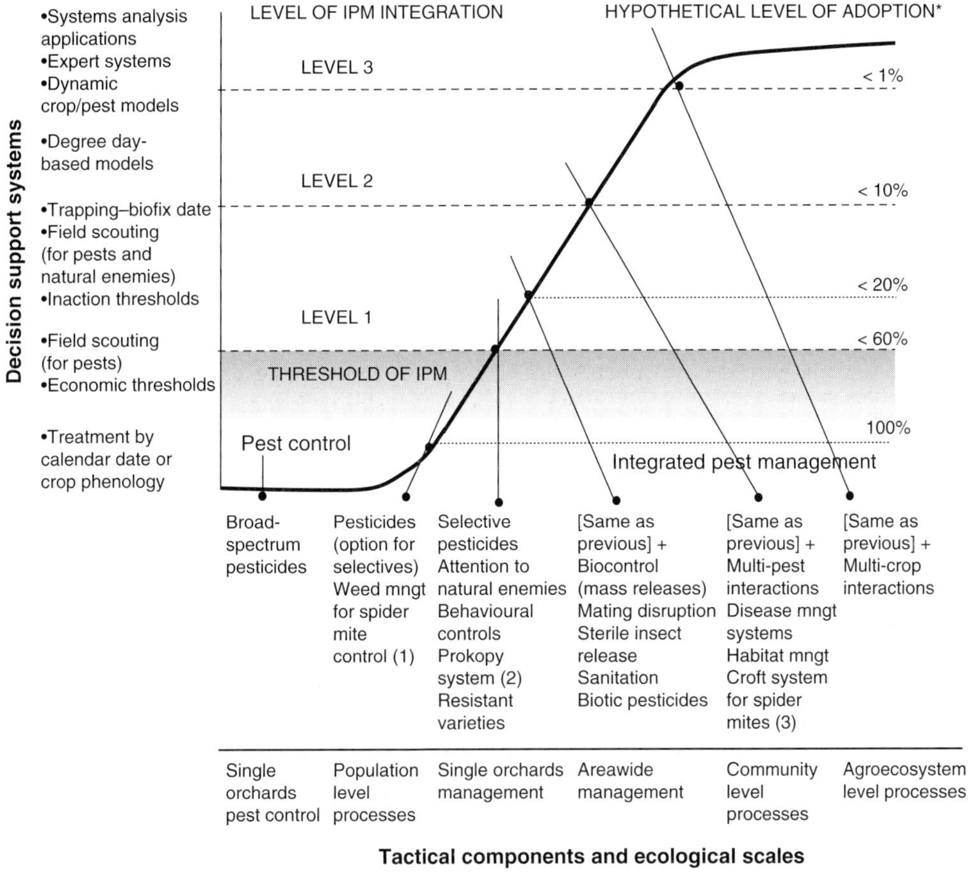

Fig. 1.1. Graph of the continuum from conventional pest control to level 3 IPM in apple/pear orchards in major commercial production regions of the USA. The graph suggests that a pest control programme becomes IPM when a certain combination of control decisions (decision support system) guides the selection and use of the control tactics best suited to the operational ecological scale. Use of the most basic decision support system, i.e. field scouting and economic thresholds, and careful consideration of the role of natural control agents combine to determine the 'IPM threshold'. Pest control programmes above this threshold are rightfully considered at level 1 of pest management integration.
(1) Weed management for spider mite control – model based on the interaction of the European red mite, *Panonychus ulmi*, its main predator, *Amblyseius fallacis*, and the need for adequate ground vegetation for conservation of the predaceous mite population (Tanigoshi *et al.*, 1983).
(2) Prokopy system – use of visual and chemical attractants for control of apple maggot flies (Duan and Prokopy, 1992).
(3) Croft system for spider mites – a complex community level management for enhancing and augmenting predatory mites in a multi-crop system that includes pome fruit orchards (Kogan *et al.*, 1999).
*Hypothetical levels of adoption, on the right, are assumed for fruit crops in the USA. Percentages of adoption add up to more than 100% because, in these gross estimates, the numbers below incorporate all those above.

considers IPM, primarily, a decision support system (Kogan, 1998; Norris *et al.*, 2003; Prokopy and Kogan, 2003). The figure proposes an 'IPM threshold', i.e. a fuzzy band within the pest control continuum above which a pest control programme rightfully may be called IPM. In Fig. 1.1 we reference both basic and advanced tactics in pome fruit IPM (on the *x*-axis) and decision support systems of increasing complexity (on the *y*-axis). The levels of adoption on the right vertical axis are hypothetical values, but partial surveys and expert opinions support these numbers.

Consequently, to be IPM, a pest control programme must apply the best possible ecological information at the population, community and ecosystem levels. Pome fruit crops offer excellent examples of IPM and ample opportunities to explore the highest levels of pest management integration. Indeed, Ron Prokopy, whom we honour in this volume, pioneered some of the most advanced IPM systems for apples (Prokopy *et al.*, 1990, 1996; Prokopy, 1994; Prokopy and Croft, 1994). In this chapter, we offer a brief overview of the ecological features of tree-fruit production in North America that have made this cropping system particularly suitable for advancing IPM to higher levels of integration. Our focus is the orchard community, particularly the arthropod community, based mainly on the work of Liss *et al.* (1982, 1986), Croft and Hull (1983) and Gut *et al.* (1988, 1991). The focus on the orchard community is justified by the importance of the biotic component of the ecosystem in the formulation of advanced IPM strategies. Most pest problems result from disturbances of community level processes. By stressing the biotic community, we do not underestimate the importance of the abiotic factors, but these are generally integral elements in models of population dynamics, phenological models and models of trophic dynamics, and thus are incorporated into research on the agronomic principles that guide orchard establishment, development and management.

The Biotic Orchard Community

In the late 1970s and early 1980s, few areas of ecology generated as much controversy among theoreticians as community ecology. There was, of course, full agreement about what fell under the purview of community ecology: species interactions such as competition, predation, mutualism, tritrophic interactions, food webs and their influence on populations and community regulation; issues related to diversity, stability and models that addressed species packing and community function, development and structure. The controversy centred mainly on the role of competition in community structure (Lewin, 1983; Roughgarden, 1983; Strong *et al.*, 1984). The conflicting ideas were brought to focus and reconciled, to some extent, in a round table on 'Research in Ecology and Evolutionary Biology' (*The American Naturalist*, 1983, Vol. 122, No. 5). Since then, much progress has been made in the analysis of both natural and crop-associated communities.

Caswell (1976), in a seminal paper on community structure, explained that 'a structural approach to the study of a system emphasizes the components, examining their nature, number, distribution, arrangement and pattern of interaction', whereas a functional approach 'focuses on the dynamics of behavior and development of the system or selected subsystems' (Caswell, 1976, p. 327). According to Caswell, species diversity, i.e. the number and relative abundance of the component species of a community, is one of the most commonly examined aspects of the structure of biological communities. The relationship between diversity and the functional dynamics of communities has been the object of a large body of ecological theory and, it should be added, the focus of much of the debate around the foundations of community ecology.

Caswell's analysis of the principal models used to describe the structure of natural communities suggested that the niche theory seemed to provide a reasonable approximation of the effect of biological interactions on species diversity and community structure (Caswell, 1976). This interpretation should be reassuring to applied ecologists (which include IPM specialists) who tend to use niche structure, with emphasis on the trophic components of the niche, to assess crop community diversity and dynamics (e.g. Kogan, 1981; Gut *et al.*, 1988; Imura, 2003; Ignacimutu and Jayaraj, 2006).

The niche concept, however, has not been immune to controversy. From its introduction into the ecologist's vocabulary in 1917 (Grinnell, 1917) to its present use in the modern ecological literature, the species niche has experienced a profound evolution. Chase and Leibold (2003) offered two new definitions of 'niche' as a synthesis of those proposed earlier by Grinnell (1917), who described the niche as the place in the environment that a species occupies; Elton (1927), who described the niche as the role of the species in or its impact on the environment; and Hutchinson (1957), who described the niche as an *n*-dimensional hypervolume. The first, less mathematically formal and discussed by Chase and Leibold (2003, p. 15), was that the niche was 'the joint description of the environmental conditions that allow a species to satisfy its minimum requirements so that the birth rate of a local population is equal to or greater than its death rate along with the set of per capita effects of that species on these environmental conditions'. For the purpose of our discussion, the specific food resource of the orchard arthropod community is the environmental condition or niche vector (in Hutchinsonian terms) of interest.

Gut *et al.* (1991) described the organization of the pear orchard community as a hierarchical system of subcommunities: systems of populations, guilds (or functional groups) and populations, each with their corresponding habitats (Fig. 1.2). The habitats may be equated to what we might call, in a more restrictive way, the specific food resource in the niche analysis of the orchard community.

A detailed niche analysis of the orchard community must take into account both spatial and temporal components. The spatial components relate to the geometry of the trees, depending on row spacing, pruning and training (use of trellises, for instance). The temporal components include the dynamic seasonal changes that occur in the orchard within 1 year and the changes that happen as the orchard ages from planting to full maturity over a period of 10–15 years. The following section provides a summary of features of the orchard plant matrix on which the arthropod community develops.

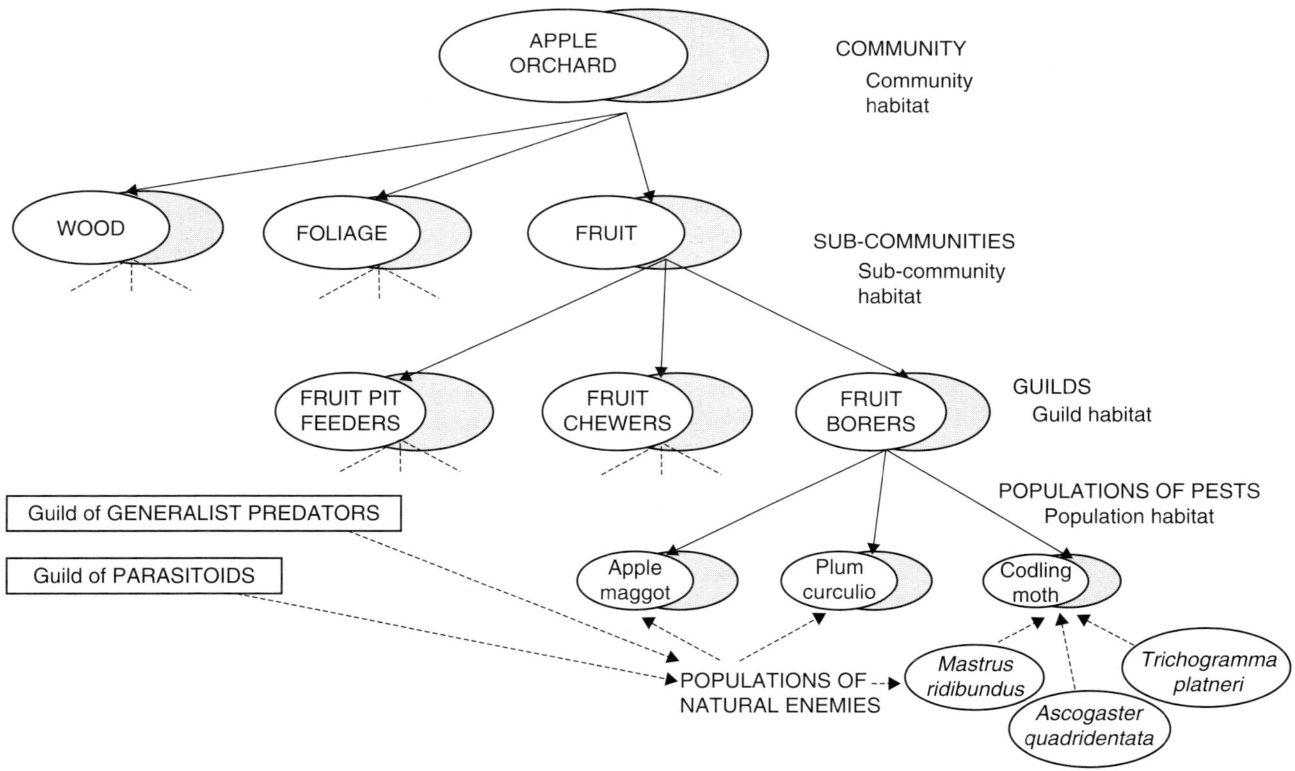

Fig. 1.2. Hierarchy of community structure showing examples of representative components of the fruit borer guild, their natural enemies and their corresponding habitats (represented by the shaded areas) (adapted from Liss *et al.*, 1982).

The orchard habitat

The domesticated apple appears to have originated in the Caucasus region near Asia Minor; pears and stone fruit are thought to be native to central Asia and China (Westwood, 1993). Apples had been cultivated in Europe for at least 2000 years before their introduction to North America. The transition from a marginal business to industrial fruit production in the USA occurred between 1850 and 1930 (Perkins, 1982). Deciduous fruit orchards were planted where there was sufficient chilling in the winter to break dormancy, and enough light, warmth and length of growing season to allow for full fruit growth and maturation. Such climatic restrictions largely limit these deciduous tree fruits to the temperate zone, although in lower latitudes, some of these tree fruits may be grown at higher altitudes. Selection of the first orchard sites was, presumably, decided based on water and nutrient availability. With modern intensive management, however, these factors can be supplied artificially when they are not present in the environment to support tree growth.

The orchard site, with its distinctive slope, aspect and soil, is a prime determinant of orchard productivity. With the domestication of the various types of deciduous tree fruits, site requirements for successful production have become well known. Constraints that were originally determined by trial and error are now delineated and, in some cases, quantified so that new orchards are established on sites with appropriate soil depth and type and trees are planted in a manner to optimize light interception. In addition to irrigation and fertilization, orchard management includes tree training, pruning, frost control, pollination and groundcover vegetation management, which ranges from clean cultivation to planted cover crops with herbicides applied along the tree row to maintain weed-free strips. Each of these orchard management techniques exerts an influence on the arthropod complex to some degree. For example, pruning, which affects partitioning of tree resources between fruit and vegetative growth, rejuvenates the tree and stimulates proliferation of new shoots. Such shoots can promote populations of sucking insects. Conversely, removal of these vigorous shoots, or watersprouts, can deter increase of these same pest populations. Frost control, to the extent that survival of fruit buds is enhanced, will impact those insects that use the fruit as a resource.

Among crops, orchards have higher than average uniformity because trees of the same cultivar are clones produced via vegetative propagation, and grafted on to rootstocks that are now commonly produced vegetatively as well. While many orchards have mixed cultivars and even mixed crops, it is not uncommon for large blocks of trees to be planted to the same genetically identical cultivar. If, however, the crops are not self-fertile, such as apples and pears, then some proportion of the trees must be planted to pollinizer varieties to promote cross-pollination. The fact that this clonal resource persists over some period of years presents an environment of minimal genetic variability to herbivorous arthropods and plant pathogens. Most pome fruit and stone fruit cultivars that are commonly planted today were chosen originally for their horticultural and fruit quality characteristics. Resistance to insect pests and diseases generally has not been a focus for selection until recently, when apples resistant to scab, a fungal disease, and pears resistant to fire blight, a bacterial infection, have been selected by breeders and are now available commercially. Rootstock selection, again, has focused primarily on horticultural traits such as dwarfing and precocity, which leads to early fruit production, or the ability to withstand less than optimal soil conditions. However, tolerance to root-attacking pathogens and insects occasionally has been selected for in the breeding process. It is on this relatively genetically uniform, highly managed perennial vegetational matrix that the apple arthropod community develops. Within this community, some highly competitive and dominant species become the major apple pests.

Arthropod community structure and pest characteristics

The arthropod community that develops on the fruit tree can be divided into distinct arboreal habitat-based subcommunities: wood, leaf, fruit buds and flowers and fruit-feeding pests (Gut *et al.*, 1991). For the purpose of this discussion, we have omitted the very important belowground arthropod community. When the arthropod species inhabiting unsprayed fruit trees are inventoried, the total number of species present in a locale is often in the hundreds (Oatman *et al.*, 1964; Meszaros, 1984; Szentkirályi and Kozár, 1991). Southwood (1961), testing a host-abundance hypothesis to explain the number of insects on trees, mentions records of 93 insect species on apple in England and 77 in Russia. Reports referring to species in the hundreds certainly include both phytophagous species and natural enemies. Beers *et al.* (1993) described 76 species of interest for pest management in orchards of the Pacific Northwest states of the USA. Selected arthropod species that feed on apple and are found in North America are listed in Table 1.1, with their country or region of origin and host range, where known.

As an orchard matures and is colonized by arthropods, the species tend to pack into the available niches before reaching a climax community. In the fully mature and unmanaged fruit tree, which is typified by excessive fruit spurring, lack of foliar growth except in the tree periphery, and dead and dying limbs in the shaded centres, species diversity will decrease as the wood component of the habitat comes to predominate (Knisley and Swift, 1972). However, in a managed orchard, where high productivity is encouraged, this climax state is not allowed to develop.

It is a common observation that of the many arthropod species that inhabit fruit orchards, relatively few are considered pests and very few are persistent pests (Table 1.2). Species that cause direct injury to the fruit are most likely to be regarded as key pests. In North American apple orchards, most of the key pests are fruit borers. These key pests include two native species, apple maggot (*Rhagoletis pomonella*) and plum curculio (*Conotrachelus nenuphar*), along with two introduced tortricids, codling moth (*Cydia pomonella*) and, more recently, Oriental fruit moth (*C. (Grapholita) molesta*). When the fruit is attacked directly, the economic injury level becomes so low that naturally occurring biological control agents are often unable to reduce the pest population sufficiently and in time to avoid economic losses (Westigard, 1973).

Table 1.1. Selected apple pests in North America: origin, current distributions and host ranges based on published records in papers cited throughout the chapter.

Pest and species name	Origin	Current distribution	Host range
Apple maggot *Rhagoletis pomonella* (Walsh)	N. America	N. America	Apple, hawthorn, plum, apricot, pear, sweet cherry, sour cherry, wild rose, cotoneaster and pyracantha
Codling moth *Cydia pomonella* (Linnaeus)	Eurasia	Multi-continent (not found in Japan, eastern China and Western Australia)	Apple, pear, hawthorn, quince, walnut, prune, plum and apricot
European apple sawfly *Hoplocampa testudinea* (Klug)	Europe	Europe; north-eastern USA; Ontario, Quebec and Vancouver Island, Canada	Apple
Oriental fruit moth *Cydia molesta* (Busck)	China	Multi-continent	Peach, nectarine, quince, apricot, apple, plum, cherry, pear, rose, flowering cherry
Plum curculio *Conotrachelus nenuphar* (Herbst)	N. America	N. America (east of the Rockies and Utah)	Apple, nectarine, plum, cherry, peach, apricot, pear and quince, wild plum, hawthorn and native crabapple
Tarnished plant bug *Lygus lineolaris* (Palisot de Beauvois)	N. America	N. America	Polyphagous
Western tarnished plant bug *Lygus hesperus* (Knight)	N. America	Western N. America	Polyphagous
Brown stink bug *Euschistus servus* (Say)	N. America	Eastern N. America	Polyphagous
Consperse stink bug *Euschistus conspersus* Uhler	N. America	Western N. America	Polyphagous
Green stink bug *Acrosternum hilare* (Say)	N. America	N. America	Polyphagous

(*Continued*)

Table 1.1. *continued*

Pest and species name	Origin	Current distribution	Host range
Apple ermine moth *Yponomeuta malinella* (Zeller)	Eurasia	Eurasia, BC Canada, Washington and northern Oregon, USA	Apple and crabapple
Fruit tree leaf roller *Archips argyrosphila* (Walker)	N. America	N. America	Almond, apple, apricot, citrus, cherry, pear, plum, prune, quince, walnut, cane berries, currant, loganberry, gooseberry and some non-fruit trees
Light brown apple moth *Epiphyas postvittana* (Walker)	Australia	Aus., NZ, UK, New Caledonia, Hawaii and (recently) California, USA	Polyphagous
Oblique-banded leaf roller *Choristoneura rosaceana* (Harris)	N. America	N. America	Polyphagous
Red-banded leaf roller *Argyrotaenia velutinana* (Walker)	N. America	N. America	Polyphagous
Tufted apple bud moth *Platynota idaeusalis* (Walker)	N. America	Eastern N. America	Polyphagous
Lacanobia fruitworm *Lacanobia subjuncta* (Grote & Robinson)	N. America	N. America	Polyphagous
Spotted tentiform leaf miner *Phyllonorycter blancardella* (Fabricius)	Europe	Europe and N. America	Apple, crabapple, pear, cherry
Apple rust mite *Aculus schlechtendali* (Nalepa)	Unknown (closely associated with apple)	Multi-continent	Apple and pear
European red mite *Panonychus ulmi* (Koch)	Europe	Multi-continent	Polyphagous
McDaniel spider mite *Tetranychus mcdanieli* (McGregor)	N. America (first described in Michigan, USA)	N. America and France	Polyphagous

Two-spotted spider mite *Tetranychus urticae* (Koch)	Unknown	Multi-continent	Polyphagous
Apple aphid *Aphis pomi* (De Geer)	Europe (closely associated with apple)	Multi-continent	Apple, pear, hawthorn, loquat, pyracantha, quince
Spirea aphid *Aphis spireacola* (Patch)	N. America	Multi-continent	Polyphagous
White apple leaf hopper *Typhlocyba pomaria* (McAtee)	N. America	N. America	Apple, pears, cherry, prune
Mullein bug *Campylomma verbasci* (Meyer)	Eurasia	Eurasia and N. America	Polyphagous (also predacious)
Rosy apple aphid *Dysaphis plantaginea* (Passerini)	Europe	Multi-continent	Apple, pear and hawthorn (alternate hosts include herbaceous plants, such as narrow-leafed plantain and dock)
San José scale *Quadraspidiotus perniciosus* (Comstock)	China	Multi-continent	Polyphagous
Woolly apple aphid *Eriosoma lanigerum* (Hausman)	N. America	Multi-continent	Apple, pear, hawthorn, mountain ash, cotoneaster and pyracantha (in the eastern USA, American elm serves as an alternate host)
Dogwood borer *Synanthedon scitula* (Harris)	N. America	N. America (east of the Rockies)	Many trees and woody plants

Table 1.2. Pest status of selected arthropods occurring in commercial apple orchards in four U.S.A. states. (Based on Croft and Hull, 1983 and updated in 2008).

Insects/Mites	Pennsylvannia	Michigan	Washington	New York
Apple maggot	2	1	2	1
Codling moth	1 (2)	1	1	1 (2)
Plum curculio	1	1	---	1
Tarnished plant bug (*Lygus* spp.)	1 (2)	2	2	2
Fruit tree leaf roller	2	2	2	--- (2)
Oblique-banded leaf roller	2	1 (2)	1 (2)	1
Red-banded leaf roller	2	2 (1)	---	--- (2)
Tufted apple bud moth	2 (1)	2	---	--- (2)
Tentiform leaf miner	3	3	3 (2)	3
Apple rust mite	3	3	--- (3)	3
European red mite	3	3	3	3
McDaniel spider mite	---	---	3 (2)	---
Two-spotted spider mite	3	3	3	3
Apple aphid	see Spirea aphid (3)	3	3	3
White apple leafhopper	3	2 (3)	3	3
Rosy apple aphid	1 (2)	3	2	2
San José scale	3	2 (3)	2 (1)	2 (3)
Woolly apple aphid	3	3	3	2 (3)
*Oriental fruit moth	1	1	---	1
*Stink bugs	2	2	2	2
*Mullein bug	2	2	---	2
*Lacanobia fruitworm	---	---	2	---
*European apple sawfly	2	---	---	---
*Spirea aphid	3	---	---	---
*Dogwood borer	3	---	---	---

Symbols:
1 = annual key pest
2 = sporadic or occasional pest
3 = secondary or induced pest
--- species not ranked: either not present or considered to be of minimal importance

Note: numbers in parentheses are from the original ranking of 1983 and indicate a change in the pest ranking for that region.
*These pests were not included in the ranking done in 1983 but were added to the updated ranking as they were considered to be of some importance currently in that state.
New pest ratings were provided by:
Pennsylvannia = Larry Hull, Greg Krawczyk, and Dave Biddinger
Michigan = Larry Gut
Washington = Jay Brunner and Elizabeth Beers
New York = Harvey Reissig.

The species classified as sporadic pests (and rarely as key pests) include the leaf rollers that feed on both leaves and the fruit surface and some sucking insects, such as *Lygus* spp., San José scale (*Quadraspidiotus perniciosus*) and rosy apple aphid (*Dysaphis plantaginea*), which can cause direct injury to the fruit. These sporadic or occasional pests often have been kept at low levels by the insecticides used to control perennial key pests. However, when these sporadic

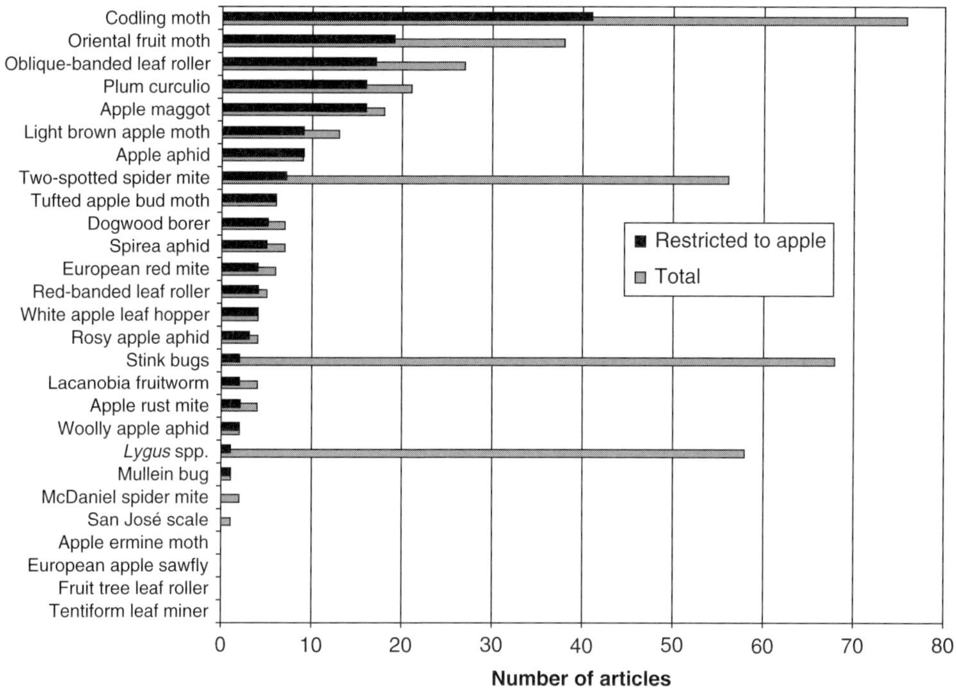

Fig. 1.3. Numbers of selected articles concerning apple pests published between 1999 and 2007 in publications of the Entomological Society of America: *Annals of the Entomological Society of America*, *Journal of Economic Entomology* and *Environmental Entomology*.

pests become resistant to currently used pesticides, or when selective controls are used for the key pests, the full extent of their impact becomes apparent. By contrast, those species in apple that are mainly leaf feeders, such as spider mites and leaf miners (*Phyllonorycter* spp.), are considered secondary pests, only exceeding economic injury levels when natural controls are disrupted. Outbreaks of these induced pests became quite common following the use of broad-spectrum insecticides in orchards that started in the late 1940s (Barnes, 1959). The subsequent development and adoption of IPM in tree fruit was, to a large degree, a response to this 'ecological backlash', characterized by resistance, resurgence and replacement (Norris *et al.*, 2003), which was observed repeatedly in orchards relying on chemical control methods. An indirect way of assessing the relative importance of these various pests to apple is the research effort dedicated to them and measured as the number of research papers published. Figure 1.3 uses the number of papers published in the three main journals of the Entomological Society of America from 1999 to 2007, to indicate the current level of importance of some common apple pests in North America.

Apples have been grown in the north-eastern USA for over 300 years. It seems important to stress that, as a perennial crop, the orchard is vulnerable to potential new invaders. These are either native species that evolve to adopt the orchard plant as a new host, as has been the case with the apple maggot fly, or exotic species well adapted to feeding on fruit crops in their regions of origin. Both apple maggot and plum curculio have moved successfully from their native rosaceous hosts into the managed apple orchards of eastern North America (Aluja and Mangan, 2008; see also Berlocher and Feder, Chapter 2, this volume). The chemical ecology of the orchard and the nature of the host-selection behaviour of apple-associated arthropods will play an increasingly important role in IPM as more emphasis is placed on behavioural methods of control. With regard to the introduced tortricids, which are key pests in apple, it is notable that the Oriental fruit moth was not even included when apple pests were ranked just 25 years ago (Croft and Hull, 1983; see Table 1.2) and has appeared only recently as a pest in pear orchards in California, where mating disruption had been used successfully for the control of the codling moth. To date, the Oriental fruit moth is not considered a pest in apples in the Pacific Northwest, although it is a frequent pest of peaches throughout the area. The codling moth, which has a close association with pome fruit, moved to walnuts at some point in the past; this host shift was observed again during the first half of the 20th century in California and the codling moth is currently a key pest in California walnuts (Barnes, 1991). Clearly, the shifting nature of host preference and attraction has been one of the primary determinants in shaping the current pest complex in tree fruit.

Sources of arthropod colonizers

There has been considerable interest and research on the dynamic interchange of arthropods among agricultural, ruderal and native vegetational patches within a regional landscape (Ekbom *et al.*, 2000). A detailed study of the movement of tortricid moths towards or away from a set of abandoned orchards and commercial orchards under effective IPM was conducted in Switzerland by Jeanneret (2000). The 2-year survey yielded a maximum of 44 species in the commercial orchards and 81 in the abandoned ones. Roughly one-third of the species was tortricids associated with apple. The landscape surrounding the orchards included forests, intensive crops (not specified), vineyards, other orchards, meadows and gardens. These sources of orchard arthropod colonizers also are sources of natural enemy species. It is critical for IPM purposes to identify the most desirable mix of crop and non-crop vegetation that will provide the most effective balance between colonizing herbivores and their potential natural enemies. Such has been the focus of research on habitat management as a new frontier in IPM (Gurr *et al.*, 2004; Nicholls and Altieri, 2007; Wratten *et al.*, 2007; see also Aluja and Rull, Chapter 7, this volume).

The pool of phytophagous arthropods and their predators and parasitoids potentially capable of colonizing an orchard derives from three main sources. The first source consists of species endemic to the orchard, i.e. those that spend their entire life cycle and overwinter in the orchard habitat. The second source is the fauna associated with habitats immediately surrounding the orchard, such as adjacent crop fields, hedges and surrounding woodlots. The third source includes highly mobile species dispersing from distant habitats. Species in this last group are an integral part of other communities from which they migrate as the orchard habitat develops and offers an attractive food resource. The landscape, of which the orchard is one component, is in fact a mosaic of all those community habitats. Jonson and Fahrig (1997), based on work by Merriam (1984), Fahrig and Merriam (1985, 1994) and Hanski (1994) suggested that:

> An important concept in landscape ecology is that environments are comprised of a heterogeneous mix of habitat patches, where habitat patches are discrete areas in which an organism obtains resources and/or breeds. The pattern of habitat patches on the landscape changes over time. The landscape has a dynamic spatial structure. This structure can influence the persistence of both local and regional populations of organisms. In order for regional populations to survive, local extinctions must be balanced by (re)colonizations of habitat patches via dispersal of organisms over the landscape. Changes in landscape structure, such as the reduction of the proportion of one or more patch types or the increase in patch isolation, will alter the ability of organisms to disperse. Species that cannot disperse effectively as a result of a change in structure will suffer reductions in regional population sizes.
>
> (Jonsen and Fahrig, 1997, p. 185)

Obviously, agricultural landscapes are highly variable when they include both annual and perennial crops and a variety of options exist for managing the non-crop vegetation. Recolonization of orchards, therefore, follows the patterns described by Jonsen and Fahrig (1997).

Food specialists and food generalists follow different colonization paths and probably display different colonization rates depending on the composition of the surrounding flora. The drive for individuals of any given species to migrate from one habitat to the next may have been determined over evolutionary time to increase fitness of the species as local conditions in the habitat of origin deteriorate, local intra- and interspecific competition increase and the sink habitat becomes more attractive, and, in the case of agriculturally dominated landscapes, offers a highly concentrated resource (Stinner *et al.*, 1983). Depending on the species, movement can occur over long distances, even from remote, unrelated landscapes, or it can be localized and of short duration, which has been referred to as trivial movement (Kareiva, 1986).

Thus, orchard colonizers are individuals of species associated with the diverse habitat patches that form the landscape in which the orchard community develops. Those species, in turn, differ in fundamental life-history traits, host-plant associations and geographic origin (Table 1.1). As a crop introduced into the USA from another continent, the fauna presently associated with apple has similar characteristics to those observed in association with other crops of foreign origin (Kogan, 1981; see also Berlocher and Feder, Chapter 2, this volume). These colonizers fall into four categories:

1. Native polyphagous species that expanded their host range to include apple in the diet – examples are: spirea aphid; *L. lineolaris* and *L. hesperus*; and certain pentatomid species.
2. Native oligophagous species well adapted to feeding on local *Rosaceae* or plants in related families that present similar phytochemical characteristics – examples are: apple maggot and plum curculio.

3. Native oligophagous species that were adapted to feeding on other plants but switched, in ecological time, to feeding on apple. Such switches are less common than those involving Rosaceae oligophagous species, but there are records of such occurrences in other crops (Kogan, 1981).
4. Exotic invaders that are well adapted to feeding on apple in their countries of origin and, as a group, are generally the most serious actual or potential pests in the invaded areas. Examples of exotic species that invaded USA orchards are the codling moth, the Oriental fruit moth and the San José scale. Two more recent additions to the North American apple fauna include the oligophagous apple ermine moth (*Yponomeuta malinellus*), first detected in western Washington in 1985, and the polyphagous light brown apple moth (*Epiphyas postvittana*), first detected in California in 2007. Both of these introduced species belong to the functional group of leaf feeders.

The apple orchard is host to hundreds of arthropod species. The characteristics of the arthropod community that finally becomes established in any given production area depend on the diversity of the local landscape, the richness of the local fauna and the level of anthropogenic disturbance of the native landscape. Of those hundreds of species that visit the orchard, only a fraction is capable of exploiting the resources available, and even fewer become pests. To understand the adaptive features of successful colonizers, it is useful to group species into guilds. The following section discusses the guild concept briefly and describes the most important phytophagous arthropod guilds associated with apple in North America.

Arthropod guilds in established orchards

The concept of 'guild' was introduced in ecology by Root (1967), although similar concepts had been advanced since the early 1900s (Simberloff and Dayan, 1991). A guild was defined originally as: '. . .a group of species that exploit the same class of environmental resources in a similar way'. To this basic definition, Root added an explanatory note:

> This term (guild) groups together species, without regard to taxonomic position, that overlap significantly in their niche requirements. The guild has a position comparable in the classification of exploitation patterns to the genus in phylogenetic schemes.
>
> (Root, 1967, p. 335)

Since first introduced, the guild concept has been applied, not always in Root's strict context, in hundreds of community level studies (Price, 1986; Root, 1986; Simberloff and Dayan, 1991). Many studies organized species assemblages into guilds based on the class of environmental resources used, but ignored Root's second criterion that required those resources to be used in a 'similar way'. According to Price (1986), niche theory coupled with the guild concept 'help us focus on approaches to community analysis, and the species on which critical tests for interactions should be made'. Root (1967) offered three main reasons for adopting the guild concept in the study of communities:

1. The guild concept focuses attention on all sympatric species involved in a competitive interaction, regardless of their taxonomic relationship.
2. The guild concept removes from the definition of the niche the aspect of the functional role of a species in the community, thus limiting that definition to the set of conditions that permit a species to exist in a particular ecosystem (the Hutchinsonian niche). By reserving the functional role of a species to the guild, it is possible to reconcile observations that seem to refute the generality of the competitive exclusion principle. Consequently, groups of species having very similar ecological roles within a community may coexist because they are members of the same guild, not occupants of the same niche.
3. The guild concept is useful in the comparative study of communities because it is usually impractical to consider at once all species living in an ecosystem. Guilds permit focus on specific groups with specific functional relationships rather than studying taxonomic groups, within which different species may perform unrelated roles (see also Simberloff and Dayan, 1991).

Grouping arthropods associated with apple orchards into guilds brings the advantages mentioned by Root and others to the comparative analysis of a fauna that contains significant numbers of economically important species in different production regions. It seems, however, that from the point of view of management decisions, it is more important to group pest species into trophic functional groups, i.e. focus on the resource utilization aspect of the guild definition regardless of the ways the species in the group use the same resource. Perhaps we could call these 'super-guilds' but we prefer to use the more established expression 'functional groups'. There is a precedent for this use of the concept (Simberloff and Dayan, 1991). The following list shows the main functional groups, with their more restrictively defined trophic guilds that are generally recognized in association with pome fruit ecosystems based mainly on Gut *et al.* (1991). For species lists in the guilds see Table 1.3.

1. Functional group of foliage feeders. Foliage feeders are indirect pests. By sucking the sap, chewing the leaf blade or mining the leaf parenchyma, they reduce the photosynthetic ability of the plant and impair optimal fruit development and total production. The specific guilds recognized in this functional group include: (i) strip feeders (also called leaf chewers by Gut *et al.*, 1991); (ii) pit feeders; (iii) sap feeders; (iv) leaf miners; and (v) leaf rollers.
2. Functional group of fruit bud and flower feeders. Although crop plants usually produce more flowers than the plant can bear finally in a commercial production, fruit bud and flower feeders may be considered direct pests as excessive damage to the reproductive organs may affect fruit set and development. There is no obvious subdivision of the group. Thus, the guild of fruit bud and flower feeders may be considered as a whole.
3. Functional group of fruit feeders. Fruit feeders are direct pests and include the most economically important arthropod pest species because even relatively minor injury to the fruit can result in a major loss of value in the fresh produce market. The following are the main guilds under this functional group: (i) fruit skin scrapers; (ii) fruit borers; and (iii) fruit juice feeders.

Table 1.3. Representative arthropod pest species on apple within primary functional groups and trophic guilds.

Functional groups	Guilds	Main representative species
Fruit feeders	Fruit borers	Apple maggot, codling moth, oriental fruit moth, European apple sawfly
	Fruit sap feeders	Tarnished plant bug, western tarnished plant bug, brown stink bug, consperse stink bug, green stink bug
Foliage feeders	Leaf rollers	Fruit tree leaf roller, light brown apple moth, oblique-banded leaf roller, red-banded leaf roller, tufted apple bud moth
	Leaf strip feeders	Lacanobia fruitworm
	Leaf miners	Spotted tentiform leaf miner
	Leaf pit feeders	Apple rust mite, European red mite, McDaniel spider mite, two-spotted spider mite
	Leaf sap feeders	Apple aphid, spirea aphid, white apple leaf hopper
Wood feeders	Wood borers	Dogwood borer, shothole borer
Mixed plant organ feeders	Leaf and fruit sap feeders	Mullein bug, rosy apple aphid
	Wood, leaf and fruit sap feeders	San José scale
	Leaf and root sap feeders	Woolly apple aphid

4. Functional group of wood feeders. This group includes the following guilds: (i) wood borers; and (ii) wood pit feeders.

Each of these guilds may include hundreds of species capable of colonizing the apple orchard. For instance, the leaf rollers (Lepidoptera: Tortricidae) are an extremely trophically diverse group of moths. The family has about 5000 species worldwide. About 80 Eurasian species have been recorded whose larvae feed on apple (Meijerman and Ulenberg, 2000). Another group that includes actual or potential pome fruit pests are the tephritid flies in the genus *Rhagoletis*. Boller and Prokopy (1976) mentioned about 50 species, mostly in the Holarctic and Neotropical regions; several of these have shifted hosts from economically unimportant plant species to major introduced crop plants. Many species can be part of multiple guilds; sap feeders, in particular, often feed on various plant parts. For instance, the San José scale feeds on new wood, leaves and fruit, while the woolly apple aphid (*Eriosoma lanigerum*) feeds on the roots as well as the aboveground parts of the tree.

Additional guilds include those species in higher trophic levels: predators, parasitoids and hyperparasitoids, which range from generalist to specialist. The generalist predators, in particular, may act to connect herbivore populations inhabiting different subcommunities. Other arboreal guilds include the saprotrophs and fungivores, which are not as well described as the herbivorous or natural enemy guilds. However these saprotrophic and/or fungivorous groups (e.g. psocids and tydieds) may serve as alternate prey for natural enemies, thereby sustaining populations of biological control agents when arboreal herbivorous pest populations are low. Distinct from the arboreal fauna are those arthropods inhabiting the orchard groundcover vegetation. While some orchard-dwelling species move readily between the tree canopy and the ground cover vegetation, other species do not (Flexner *et al.*, 1991; Alston, 1994; Horton *et al.*, 2003). Still, species that remain in the orchard groundcover can serve as an important food source for orchard-dwelling natural enemies, as was mentioned for predaceous mites (Fig.1.1 and also Aluja and Rull, Chapter 7, this volume).

As shown in Table 1.2, the relative importance of the major apple pest species varies considerably among the major apple-producing regions of the USA. It is conceivable that a better understanding of the community structure, particularly a more in-depth analysis of the intraguild interactions, may explain differences in dominance among these species. From an IPM point of view, however, the central issue remains the nature of the orchard's biodiversity and its effect on the stability of the orchard ecosystem.

The diversity/stability concept and application to agroecosystems management

The idea that as the species diversity increases so the stability of the ecosystem increases has been proposed and supported by some of the most influential ecologists of the mid-20th century (Odum, 1953; MacArthur, 1955; Elton, 1958). It was assumed that populations in complex communities that included many predators and parasites would be less likely to suffer explosive growth or catastrophic declines. Much of the early literature focused on the correlation between diversity and population stability. The absence of substantial experimental validation, however, raised doubts about the generality of the hypothesis and its applicability to practical problems of conservation biology and pest control. Critical assessments of the hypothesis were offered by, among others, Goodman (1975) and Murdoch (1975). In summarizing the evidence offered, mainly by Elton (1958) and Pimentel (1961), Goodman (1975) suggested that the hypothesis had been formulated based on a set of observational or experimental data as follows:

1. Experimental evidence: one-predator/one-prey systems undergo violent population oscillations and usually become extinct rather quickly.
2. Simple mathematical models: models of one-predator/one-prey systems have no locally stable equilibrium point and generally go into oscillations, the amplitude of which is determined by initial population numbers, or simply become extinct.
3. Observation of agricultural systems: man-made monocultures are more vulnerable to pest outbreaks and more susceptible to new invaders than natural vegetation.

4. Data from island biogeography: island biota are vulnerable to invasion.
5. Comparison of ecological regions: the relatively depauperate arctic and boreal faunas are subject to obvious population fluctuations, whereas the very rich and complex tropical rainforests look stable.

Goodman questioned this evidence stating that, 'Individually, some of these claims are seriously flawed; and, as a group, they do not comprise a tight persuasive logical structure.' The reasons offered were:

1. and **2.** Both experimental and mathematical models should be able to demonstrate that 'many-predator/many-prey' systems are more stable than simpler ones.
3. Experiments with monocrops may not have been run long enough for the pest populations to stabilize and equilibrate with their hosts, even if at economically unacceptable densities. The use of annual crops continually replanted at high densities or left fallow and replaced by early successional vegetation denies the possibility of reaching that equilibrium. (We must raise the counter-argument that long-lived orchards offer a much better system to test the hypothesis and, indeed, abandoned orchards do seem to reach an equilibrium with colonizing herbivores.)
4. According to Goodman, the argument of the vulnerability of islands would be valid only if there was evidence that the vulnerability was not 'a consequence of accidents of distribution or of islands . . . accumulating species that are especially susceptible to competitive displacement'.
5. The enormous diversity and complexity of tropical biota may cause population fluctuations to go largely unnoticed and even very diverse systems may be vulnerable to invasions. Thus, the argument that tropical ecosystems are more stable than boreal ones may require further scrutiny.

It seems that Goodman's (1975) critical review raising doubts about the validity of the hypothesis triggered a flurry of interest in testing the diversity/stability relationship more rigorously. It became apparent that both the terms 'diversity' and 'stability' required clearer definitions. There are several aspects to diversity: (i) number of different species (richness); (ii) relative abundance of different species (evenness); (iii) ecological distinctiveness of different species, such as their functional differentiation; and (iv) evolutionary distinctiveness of different species (van Emden and Williams, 1974; Holsinger, 2007). Just as numerous are the definitions of stability. Van Emden and Williams (1974) recognized the difference between species population stability and ecosystem stability and emphasized the time concept of persistence of the equilibrium state as the best measure of stability. Other definitions of stability involve the concepts of resiliency and dynamic stability. The ability of a community to return to its initial state after changes induced by a disturbance has been termed resiliency; whereas dynamic stability characterizes a system in which future states are determined primarily by internal biotic interactions minimizing the influence of external factors and resulting in a system relatively immune to perturbation (McCann, 2000; Holsinger, 2007).

McNaughton (1977), based on observations made in the Masai Mara Game Reservation in Kenya, concluded that:

> The diversity–stability hypothesis developed over the past 23 years appears widely misunderstood by ecologists, although it simply states that species

> diversity mediates community functional stability through compensating interactions to environmental fluctuations among co-occurring species. Fluctuations in the abundance of species with different adaptive modes may be a mechanism stabilizing community function in a varying environment.
>
> (McNaughton, 1977, p. 523)

Recent reviews of the subject suggest that diversity can be expected, 'on average', to result in greater ecosystem stability. The evidence, however, suggests that diversity is not the driver of the relationship; rather, ecosystem stability depends on the existence in the community of species, or functional groups, that are capable of differential response (McCann, 2000; Holsinger, 2007; Ives and Carpenter, 2007). Considerable support for this view came from long-term field experiments, such as those reported by Tilman *et al.* (2006).

How all this translates to application in IPM also has been the subject of extensive debate. A major consideration is how the theory developed for natural ecosystems could apply to agroecosystems. The measures of stability certainly needed adjustments. From the point of view of the producer, stability would be measured as the maintenance of a high mean attainable yield over time. From the point of view of the IPM specialist, it would be measured as the maintenance of pest population fluctuations below the economic injury level and those of natural enemies at the most effective functional response level. Murdoch (1975) argued that farmers had adopted 'instability' as a tool in pest control, with the use of insecticides to disrupt pest populations to near extinction. He called the approach 'catastrophe strategy'. IPM, on the other hand, by emphasizing the role of biological control, uses a 'stabilizing strategy' whereby both prey and natural enemy populations are retained in the community at a density-dependent regulation level. Although Murdoch was critical of the general tenets of the diversity/stability hypothesis, he recognized the need to increase complexity, if not diversity, of agroecosystems as a means of enhancing the effectiveness of natural enemies. Similarly, van Emden and Williams (1974) recommended a cautious approach to the increase of biodiversity in agroecosystems. Instead, they recommended the addition of effective natural enemies and selective management of weeds and hedgerows to increase habitat for natural enemies.

These ideas have evolved, regardless of the questionable causal relationship between diversity and stability. The inherent diversity of polycultures came to represent an obvious alternative to the extreme ecological simplification of monocultural ecosystems. Attention to the merits of polycultures in IPM increased markedly in the 1990s (Andow, 1991), a development that was followed closely by interest in habitat management to conserve natural enemies (Barbosa, 1998; Pickett and Bugg, 1998; Landis *et al.*, 2000; Altieri and Nicholls, 2004; Gurr *et al.*, 2004; Nicholls and Altieri, 2007; Wratten *et al.*, 2007).

The enhancing of biological control through habitat management requires information on source–sink relationships among pests and natural enemies across crop plants, neighbouring crops, natural vegetation and managed vegetation in the form of cover crops, beetle banks, flight barriers and field hedges. Theoretically, diversification of the crop ecosystem leads to an increase in natural enemies and to greater stability of the system. The analysis of intra- and interfield movement, the host-selection behaviour of phytophagous and entomophagous insects, multitrophic interactions among community members, the dynamics of populations and a deep understanding of community structure and function are only a few of the many components of the knowledge base necessary to advance IPM systems to community-level integration (Kogan and Prokopy, 2003). With these concepts providing a reasonable ecological backdrop, we can now proceed to consider the fundamentals of design and implementation of higher-level apple IPM systems.

Levels of Integration of Orchard IPM Systems: Concluding Remarks

Within the spectrum of ecosystem complexity, extending from a maximum in tropical rainforests to a minimum in extensive annual monocultures, a well-managed, mature apple orchard occupies a place somewhere below the middle of the spectrum. For a man-made ecosystem, this is not a bad position as it offers ample opportunities to enhance complexity aimed at promoting increased overall stability of the system. To achieve progress in advancing orchard IPM to higher levels of integration, it will be necessary to shift the focus from single pests in individual orchards to the orchard community and the regional landscape. As we will see, it seems that incipient steps have been taken in this direction in research and, more importantly, in commercial production in the past 20 years (Prokopy *et al.*, 1990, 1996; Prokopy, 1994; Prokopy and Croft, 1994; Ferree and Warrington, 2003; see also Cooley and Coli, Chapter 6, and Aluja and Rull, Chapter 7, this volume). The transition from traditional pest control to even level 1 IPM in apple, however, has been a gradual and, at times, difficult process. In 1994, Prokopy wrote:

> It is doubtful that more than a few commercial fruit orchards in the world today are employing second, third . . . level IPM practices. Indeed, most are probably practicing little more than chemically-based IPM that emphasizes use of selective pesticides against key pests to minimize harm to beneficial natural enemies.
>
> (Prokopy, 1994)

While there is no distinct line that marks the threshold of IPM (see Fig. 1.1), the movement from a system of pest control that relies primarily on chemical intervention to a system that incorporates biological and cultural methods and limits pesticide use is a concrete step to level 1 IPM. This movement starts by limiting prophylactic applications of pesticides through the use of phenology models, economic injury levels and pest monitoring. The use of selective pesticides to minimize the mortality of biological control agents is a key component of level 1 IPM. Pesticide selectivity can also be enhanced by applying less of physiologically non-selective materials to the orchard when the presence of natural enemies is limited. Such actions, which Metcalf (1994) called ecological selectivity, include, for instance, dormant spray applications or placing the pesticide in a way that minimizes contact with beneficial species, such as the use of baits and trunk sprays. The combination of information-based application timing with selective pesticides (Fig. 1.4) has been termed the first half-step of IPM (Prokopy and

Fig. 1.4. An insecticide matrix of application frequency and timing versus pesticide toxicity. Decision support for pesticide applications included as an additional element in promoting the system towards the initial steps to level 1 IPM – the 'threshold of IPM'.

		many------------------Application frequency and timing----------------fewer	
		Repeated applications made on a calendar or fixed-schedule basis	Fewer applications made, timed with pest monitoring or phenology models. Economic injury levels used when available.
low------Toxicity to non-targets------high	**Broad-spectrum usually with long-lasting residual, many non-target effects**	Traditional pesticide-based programmes	Traditional programmes modified to reduce number of applications, lower pesticide costs and improve efficacy (often referred to as 'integrated pesticide management')
	Selective and/or short-lived residual, few non-target effects	'Soft' programmes with few information inputs, e.g. many organic programmes using short-lived biological or botanical insecticides	**Initial step to Level 1 IPM 'Threshold of IPM'**

Croft, 1994). Ehler (2006) called this 'the other IPM' or 'integrated pesticide management', i.e. the 'discriminate use of pesticides'. In a review of IPM in European apple orchards, we find the statement that 'most urgent is the need for more, and different, selective control methods' (Blommers, 1994). We believe that this first half-step brings the pest control programme close to the threshold of IPM. Once the disruptive effect of broad-spectrum insecticides has been removed, it becomes possible to move on to employing biological or other non-pesticidal methods to greater effect.

Advancement of IPM in apples has been slowed by the need to control the key fruit-boring pests with pesticides that generally were not selective. In order to conserve and realize the full potential of biological control, new approaches were needed. A significant development in the 1990s was the widespread adoption of mating disruption in tree fruit (Brunner *et al.*, 2001), aided in part by an extensive areawide project in the western USA states. This areawide project demonstrated that the adoption of mating disruption to control the codling moth could result in reduced pesticide use and associated economic savings, with no diminution of the high level of pest control that was demanded by the fresh fruit market (Calkins, 1998; Knight, 2008). The recent availability of the codling moth granulovirus to North American growers on a commercial scale has provided another highly selective tactic for managing this key pest (Arthurs *et al.*, 2005; Lacey and Unruh, 2007). These developments have been important contributors to the rapid increase in tree-fruit acreage using certified organic methods of production (Granatstein and Kirby, 2008). With the advent of these selective techniques, it is now realistic to look at managing vegetation in and around the orchard to enhance biological control.

Presently, it appears, IPM in tree fruits stands at the threshold of a move to higher levels of integration. To realize this desirable move, much work will be required on multiple fronts. New behavioural methods of control will require further studies of the basic biology of host attraction in key orchard pests (see Roitberg *et al.*, Chapter 3, and Dorn and Piñero, Chapter 4, this volume). Engineering the orchard ecosystem to promote biological control while simultaneously suppressing pest species includes incorporation of groundcover and the careful assessment of the surrounding vegetation. To achieve the desirable outcomes with this kind of ecosystem level engineering, it will be necessary to have detailed information on host associations and a deep understanding of the effects of habitat management. The 'Functional Ecology' programmes at Michigan State University represent a significant step in unravelling the benefits of the increased complexity of orchard communities (Whalon *et al.*, 2006).

The highest levels of IPM relate to the regional production system and the extant landscape features and involve broad market economics, along with legal constraints and political realities. As tree-fruit IPM advances beyond the threshold and begins to reach levels 2 and 3 of integration (Figs 1.5 and 1.6), it will be essential to proceed with the goal of assisting growers to produce fruit in a manner that is both profitable and socially responsible. It is doubtful that market forces alone will push IPM to higher levels of integration. Intelligent governmental policies will be needed. These policies should be directed at limiting the negative externalities associated with agricultural production and promoting research

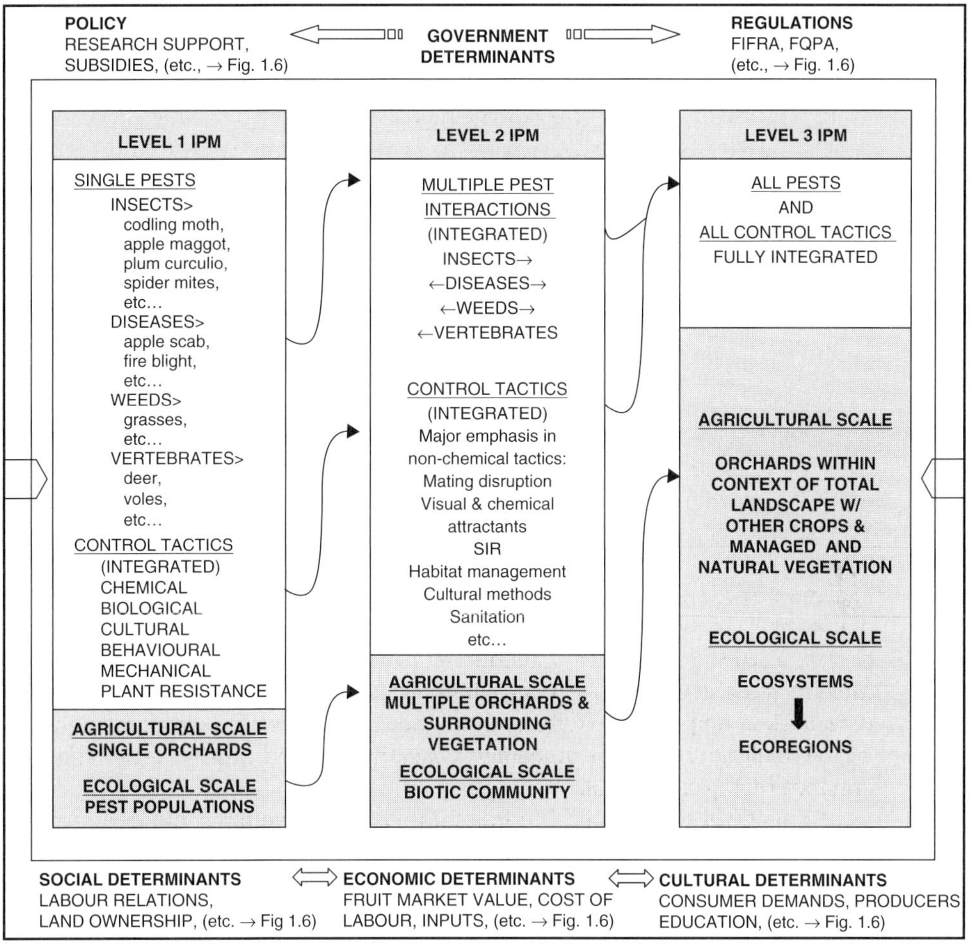

Fig. 1.5. Levels of IPM integration: main pest targets, ecological scales and levels of ecological complexity. Social, economic, cultural and governmental factors (determinants) impacting IPM design, development, implementation and level of adoption detailed in Fig. 1.6. Figures 1.5. and 1.6 are based on and expand on ideas presented graphically in Figure 1 in Prokopy and Kogan (2003).

that advances understanding of the complex interactions that occur at ecological scales greater than the population, and spatial scales greater than the individual crop field. A cadre of well-trained and properly supported extension specialists will be needed to help growers adopt concepts and approaches that certainly are more complex than the simple application of pesticides. Those government policies and socio-economic and cultural forces, which Prokopy and Croft (1994) called the '4th IPM level', affect IPM at all levels of integration and are identified as determinants and feedbacks in Figs 1.5 and 1.6, which are based on a diagram by Prokopy and Kogan (2003).

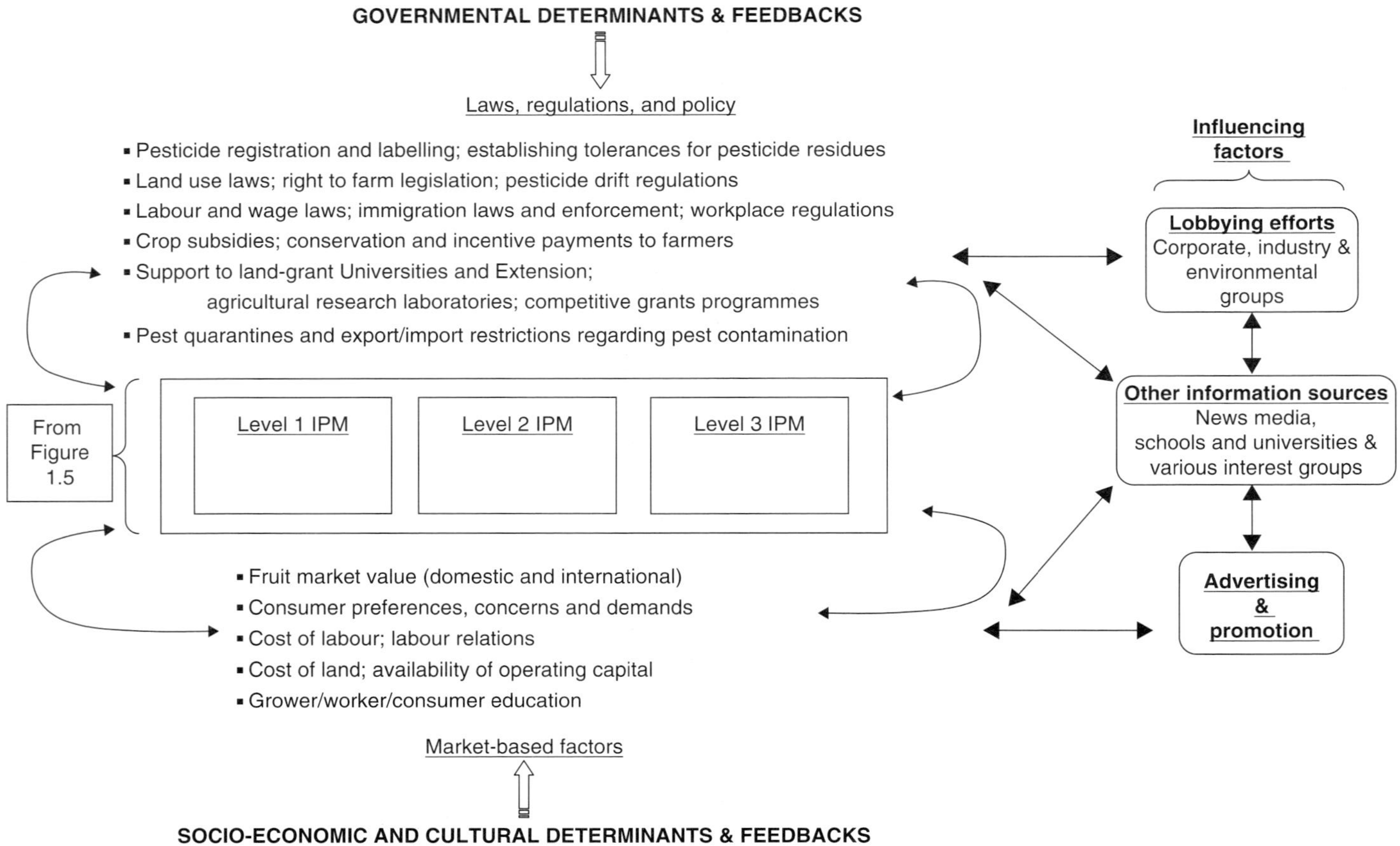

Fig. 1.6. Government, socio-economic and cultural factors that perform as determinants and provide feedbacks to all levels of IPM integration. The listings of factors are only a sample and do not presume to be exhaustive.

Although the ultimate goal of IPM practitioners should be to implement IPM at the highest possible levels of integration, the most advanced forms still are largely experimental. Emphasis is on integrating ecological and economic principles, information and decision support systems for all relevant control tactics, and for all pest categories (diseases, insect and vertebrate pests and weeds), in harmony with all other elements that affect long-term productivity and well-being of an ecosystem. For apple production, such elements include sound horticultural or husbandry practices. Level 3 IPM emphasizes attention to environmental and societal costs and benefits in making pest management decisions. The focal ecosystem may be an entity no larger than a small river valley, or it could be an entity as extensive as an entire watershed. For crops, level 3 IPM incorporates many elements of the concept of integrated crop management or, more specifically for apples, integrated fruit production. It is not, however, synonymous with organic agriculture, which disallows some materials and practices acceptable even under level 3 IPM.

There is growing international support for sustainable development rooted in the concept of ecosystem integrity, permeating all facets of human endeavour, whether economic, social or cultural (Kogan and Jepson, 2007). The concerns and practices of level 3 IPM correspond closely to those of sustainable development. Both sustainable development and level 3 IPM emphasize preservation of processes associated with natural ecosystems, long-term well-being of humans as members of social communities, economic viability, and deployment of exogenous resources only after careful consideration of possible negative impacts. For agriculture, concepts underlying level 3 IPM converge with the principal tenets of sustainable agriculture. We expect that sustainable agriculture will continue to evolve, and, as it does, it will require IPM systems that also have evolved to the highest and most effective levels of integration (Prokopy and Kogan, 2003).

Acknowledgements

We thank Peter Westigard, our colleague and pioneer of several ideas developed in this chapter, for his critical reading of an early draft of the paper. We thank also Charles Vincent and two anonymous reviewers for useful suggestions.

References

Alston, D.G. (1994) Effect of apple orchard floor vegetation on density and dispersal of phytophagous and predaceous mites in Utah. *Agriculture, Ecosystems and Environment* 50, 73–84.

Altieri, M.A. and Nicholls, C.I. (2004) *Biodiversity and Pest Management in Agroecosystems.* The Haworth Press Inc., Binghamton, New York.

Aluja, M. and Mangan, R.L. (2008) Fruit fly (Diptera: Tephritidae) host status determination: critical conceptual, methodological, and regulatory considerations. *Annual Review of Entomology* 53, 473–502.

Andow, D.A. (1991) Vegetational diversity and arthropod population response. *Annual Review of Entomology* 36, 561–586.

Arthurs, S.P., Lacey, L.A. and Fritts, R. (2005) Optimizing the use of the codling moth granulovirus: effects of application rate and spraying frequency on control of codling moth larvae in Pacific Northwest apple

orchards. *Journal of Economic Entomology* 98, 1459–1468.
Barbosa, P. (ed.) (1998) *Conservation Biological Control.* Academic Press, New York.
Barnes, M.M. (1959) Deciduous fruit insects and their control. *Annual Review of Entomology* 4, 343–362.
Barnes, M.M. (1991) Tortricids in pome and stone fruits, codling moth occurrence, host race formation and damage. In: Van der Geest, L.P.S. and Evenhuis, H.H. (eds) *Tortricid Pests, Their Biology, Natural Enemies and Control.* Elsevier Science, Amsterdam, pp. 313–327.
Beers, E.H., Brunner, J.F., Willett, M.W. and Warner, G.M. (eds) (1993) *Orchard Pest Management: A Resource Book for the Pacific Northwest.* Good Fruit Grower, Yakima, Washington.
Benbrook, C.M., Groth, E., Holloran, J.M., Hansen, M.K. and Marquardt, S.O. (1996) *Pest Management at the Crossroads*. Consumers Union, Yonkers, New York.
Blommers, H.M. (1994) Integrated pest management in European apple orchards. *Annual Review of Entomology* 39, 213–241.
Boller, E.F. and Prokopy, R.J. (1976) Bionomics and management of *Rhagoletis. Annual Review of Entomology* 21, 223–246.
Brunner, J., Welter, S., Calkins, C., Hilton, R., Beers, E., Dunley, J., Unruh, T., Knight, A., van Steenwyk, R. and Vanbuskirk, P. (2001) Mating disruption of codling moth: a perspective from the western United States. *IOBC/WPRS Bulletin* 25, 11–19.
Calkins, C.O. (1998) Review of the codling moth areawide suppression program in the western United States. *Journal of Agricultural Entomology* 15, 327–333.
Caswell, H. (1976) Community structure: a neutral model analysis. *Ecological Monographs* 46, 327–354.
Chase, J.M. and Leibold, M.A. (2003) *Ecological Niches: Linking Classical and Contemporary Approaches.* University of Chicago Press, Chicago, Illinois.
Croft, B.A. and Hull, L.A. (1983) The orchard as an ecosystem. In: Croft, B.A. and Hoyt, S.C. (eds) *Integrated Management of Insect Pests of Pome and Stone Fruits.* Wiley, New York.
Duan, J.J. and Prokopy, R.J. (1992) Visual and odor stimuli influencing effectiveness of sticky spheres for trapping apple maggot flies *Rhagoletis pomonella* (Walsh) (Dipt., Tephritidae). *Zeitschrift für Angewandte Entomologie* 113, 271–279.
Ehler, L.E. (2006) Integrated pest management (IPM): definition, historical development and implementation, and the other IPM. *Pest Management Science* 62, 787–789.
Ehrlich, P.R. and Raven, P.H. (1964) Butterflies and plants: a study in co-evolution. *Evolution* 18, 586–608.
Ekbom, B.S., Irwin, M.E. and Robert, Y. (eds) (2000) *Interchanges of Insects between Agricultural and Surrounding Landscapes.* Springer, New York.
Elton, C. (1927) *Animal Ecology.* Macmillan, New York.
Elton, C. (1958) *The Ecology of Invasions by Animals and Plants.* Methuen, London.
Fahrig, L. and Merriam, G. (1985) Habitat patch connectivity and population survival. *Ecology* 66, 1762–1768.
Fahrig, L. and Merriam, G. (1994) Conservation of fragmented populations. *Conservation Biology* 8, 50–59.
Ferree, D.C. and Warrington, I.J. (eds) (2003) *Apples: Botany, Production and Uses.* CAB International, Wallingford, Oxfordshire, UK.
Flexner, J.L., Westigard, P.H., Gonzalves, P. and Hilton, R.H. (1991) The effect of groundcover and herbicide treatment on two-spotted spider mite density and dispersal in southern Oregon pear orchards. *Entomologia Experimentalis et Applicata* 60, 111–123.
Frisbie, R.E. and Smith, J.W. Jr (1991) Biologically intensive integrated pest management: the future. In: Menn, J.J. and Steinhauer, A.L. (eds) *Progress and Perspectives for the 21st Century*. Entomological Society of America, Centennial Symposium, Lanham, Maryland, pp. 151–164.
Glen, D.M., Greaves, M.P. and Anderson, H.M. (1995) *Ecology and Integrated Farming Systems: Proceedings of the 13th Long Ashton International Symposium.* Wiley, Chichester, UK.
Gliessman, S.R. (ed.) (1990) *Agroecology: Research in the Ecological Basis for Sustainable*

Agriculture. Ecological Studies 7. Springer, New York.

Goodman, D. (1975) The theory of diversity–stability relationships in ecology. *The Quarterly Review of Biology* 50, 237–266.

Granatstein, D. and Kirby, E. (2008) *Washington State, National, and International Status of Organic Tree Fruit (2007).* Center for Sustaining Agriculture and Natural Resources, Washington State University, Wenatchee, Washington.

Grinnell, J. (1917) The niche-relationship of the California thrasher. *Auk* 34, 427–433.

Gurr, G., Wratten, S.D. and Altieri, M.A. (2004) *Ecological Engineering for Pest Management: Advances in Habitat Manipulation for Arthropods.* Cornell University Press, Ithaca, New York.

Gut, L.J., Westigard, P.H. and Liss, W.J. (1988) Arthropod colonization and community development on young pear trees in southern Oregon. *Melanderia* 46, 1–13.

Gut, L.J., Liss, W.J. and Westigard, P.H. (1991) Arthropod community organization and community development in pear. *Environmental Management* 15, 83–104.

Hall, F.R. and Berry, J.W. (eds) (1995) *Biorational Pest Control Agents: Formulation and Delivery.* ACS Symposium Series 595. American Chemical Society, Washington, DC.

Hanski, I. (1994) Spatial scale, patchiness and population dynamics on land. *Philosophical Transactions of the Royal Society of London B* 343, 19–25.

Holsinger, K.E. (2007) Diversity, stability and ecosystem function. University of Connecticut, published under Creative Commons Agreement (http://darwin.eeb.uconn.edu/eeb310/lecture-notes/diversity-stability/node6.html, accessed 26 May 2008).

Horton, D.R., Broers, D., Lewis, R., Granatstein, D., Zack, R., Unruh, T.R., Moldenke, A. and Brown, J. (2003) Frequency of mowing and effects on densities of natural enemies in the ground cover, in the tree canopy, and on the orchard floor at three Pacific Northwest pear orchards. *Entomologia Experimentalis et Applicata* 106, 135–145.

Hutchinson, G.E. (1957) Concluding remarks. *Cold Spring Harbor Symposia on Quantitative Biology* 22, 414–427.

Ignacimutu, S. and Jayaraj, S. (eds) (2006) *Biodiversity and Insect Pest Management.* Narosa Publishing House, New Delhi.

Imura, O. (2003) Herbivorous arthropod community of an alien weed *Solanum carolinense* L. *Applied Entomology and Zoology* 38, 293–300.

Ives, A.R. and Carpenter, S.R. (2007) Stability and diversity of ecosystems. *Science* 317, 58–62.

Janzen, D.H. (1980) When is it co-evolution? *Evolution* 34, 611–612.

Jeanneret, P. (2000) Interchanges of a common pest guild between orchards and the surrounding ecosystem: a multivariate analysis of landscape influence. In: Ekbom, B.S., Irwin, M.E. and Roberts, Y. (eds) *Interchanges of Insects between Agricultural and Surrounding Landscapes.* Kluwer Academic, Dordrecht, Germany, pp. 85–107.

Jonson, I.D. and Fahrig, L. (1997) Response of generalist and specialist insect herbivores to landscape spatial structure. *Landscape Ecology* 12, 185–197.

Kareiva, P. (1986) Trivial movement and foraging by crop colonizers. In: Kogan, M. (ed.) *Ecological Theory and Integrated Pest Management Practice.* Wiley, New York, pp. 59–82.

Knight, A.L. (2008) Codling moth areawide IPM. In: Koul, O., Cupernus, G. and Elliott, N. (eds) *Areawide Pest Management Theory and Implementation.* CAB International, Wallingford, UK.

Knisley, C.B. and Swift, F.C. (1972) Qualitative study of mite fauna associated with apple foliage in New Jersey. *Journal of Economic Entomology* 65, 445–448.

Kogan, M. (1981) Dynamics of insect adaptations to soybean: impact of integrated pest management. *Environmental Entomology* 10, 363–371.

Kogan, M. (1988) Integrated pest management theory and practice. *Entomologia Experimentalis et Applicata* 49, 59–70.

Kogan, M. (1998) Integrated pest management: historical perspectives and contemporary developments. *Annual Review of Entomology* 41, 243–270.

Kogan, M. and Jepson, P. (2007) Ecology, sustainable development, and IPM: the human factor. In: Kogan, M. and Jepson, P. (eds) *Perspectives in Ecological Theory and*

Integrated Pest Management. Cambridge University Press, Cambridge, UK, pp. 1–44.

Kogan, M. and Prokopy, R.J. (2003) Agricultural entomology. In: Resh, V.H. and Cardé, R.T. (eds) *Encyclopedia of Insects*. Elsevier Science, Amsterdam, pp. 4–9.

Kogan, M., Croft, B.A. and Sutherst, R.F. (1999) Applications of ecology for integrated pest management. In: Huffaker, C.B. and Gutierrez, A.P. (eds) *Ecological Entomology*. Wiley, New York, pp. 681–736.

Lacey, L.A. and Unruh, T.R. (2007) Biological control of codling moth (*Cydia pomonella*, Tortricidae: Lepidoptera) and its role in integrated pest management, with emphasis on entomopathogens. *Vedalia* 12, 33–60.

Landis, D.A., Wratten, S.D. and Gurr, G.M. (2000) Habitat management to conserve natural enemies of arthropod pests in agriculture. *Annual Review of Entomology* 45, 175–201.

Lewin, R. (1983) Santa Rosalia was a goat. *Science* 221, 636–639.

Liss, W.J., Westigard, P.H., Gut, L.J. and Warren, C.E. (1982) A perspective for understanding arthropod community organization and development in pear. *Acta Horticulturae (ISHS)* 124, 85–100.

Liss, W.J., Gut, L.J. and Westigard, P.H. (1986) Perspectives on arthropod community structure, organization, and development in agricultural crops. *Annual Review of Entomology* 31, 455–478.

MacArthur, R. (1955) Fluctuations of animal populations and a measure of community stability. *Ecology* 36, 533–536.

McCann, K.S. (2000) The diversity–stability debate. *Nature* 405, 228–233.

McNaughton, S.J. (1977) Diversity and stability of ecological communities: a comment on the role of empiricism in ecology. *The American Naturalist* 111, 515–525.

Meijerman, L. and Ulenberg, S.A. (2000) Eurasian Tortricidae. In: Ulenberg, S.A. (ed.) *Arthropods of Economic Importance*. Netherlands Biodiversity Information Facility (NLBIF), Amsterdam (http://nlbif.eti.uva.nl/bis/tortricidae.php?menuentry=inleiding, accessed 26 May 2008).

Merriam, G. (1984) Connectivity: a fundamental ecological characteristic of landscape patter. In: *Proceedings of the First International Seminar on Methodology in Landscape Ecological Research and Planning*. International Association for Landscape Ecology, Roskilde, Denmark, pp. 5–15.

Meszaros, Z. (ed.) (1984) Results of faunistical and floristical studies in Hungarian apple orchards (Apple Ecosystem Research No. 26). *Acta Phytopathologica Academiae Scientarum Hungarica* 19, 91–176.

Metcalf, R.L. (1994) Insecticides in pest management. In: Metcalf, R.L. and Luckman, W.H. (eds) *Introduction to Insect Pest Management*. Wiley, New York.

Murdoch, W.W. (1975) Diversity, complexity, stability and pest control. *Journal of Applied Ecology* 12, 795–807.

National Research Council (US) (1996) *Ecologically Based Pest Management: New Solutions for a New Century*. NRC, Washington, DC.

Nicholls, C.I. and Altieri, M.A. (2007) Agroecology: contributions toward a renewed ecological foundation for pest management. In: Kogan, M. and Jepson, P. (eds) *Perspectives in Ecological Theory and Integrated Pest Management*. Cambridge University Press, Cambridge, UK, pp. 431–468.

Norris, R.F., Caswell-Chen, E.P. and Kogan, M. (2003) *Concepts in Integrated Pest Management*. Prentice Hall, Upper Saddle River, New Jersey.

Oatman, E.R., Legner, E.F. and Brooks, R.F. (1964) An ecological study of arthropod populations on apple in northeastern Wisconsin: insect species present. *Journal of Economic Entomology* 57, 978–983.

Odum, E.P. (1953) *Fundamentals of Ecology*. Saunders, Philadelphia, Pennsylvania.

Perkins, J.H. (1982) *Insects, Experts, and the Insecticide Crisis: The Quest for New Pest Management Strategies*. Plenum Press, New York.

Pickett, C.H. and Bugg, R.L. (eds) (1998) *Enhancing Biological Control: Habitat Management to Promote Natural Enemies of Agricultural Pests*. University of California Press, Berkeley, California.

Pimentel, D. (1961) Species diversity and insect population outbreaks. *Annals of the Entomological Society of America* 54, 76–86.

Price, P.W. (1986) Niche theory and its use in the analysis of ecological systems.

In: Severinghous, W.D. and James, T.D. (eds) *Proceedings: Conference on Applications of the Guild Concept to Environmental Management.* US Army Corps of Engineers–CERL Technical Manuscript N-86/07. University of Illinois at Urbana-Champaign, Urbana, Illinois, pp. 16–27.

Prokopy, R.J. (1994) Integration in orchard pest and habitat management: A review. *Agriculture, Ecosystems and Environment,* 50, 1–10.

Prokopy, R.J. and Croft, B.A. (1994) Apple insect pest management. In: Metcalf, R.L. and Luckman, W.H. (eds) *Introduction to Insect Pest Management. 3rd Edition.* Wiley, New York.

Prokopy, R.J. and Kogan, M. (2003) Integrated pest management. In: Cardé, R.T. and Resh, V.H. (eds) *Encyclopedia of Insects.* Academic Press, San Diego, California, pp. 589–595.

Prokopy, R.J., Johnson, S.A. and O'Brien, M.T. (1990) Second-stage management of apple arthropod pests. *Entomologia Experimentalis et Applicata* 49, 59–70.

Prokopy, R.J., Mason, J.L., Christie, M. and Wright, S.E. (1996) Arthropod pest and natural enemy abundance under second-level versus first-level integrated pest management practices in apple orchards: a four-year study. *Agriculture, Ecosystems, and Environment* 57, 35–47.

Root, R.B. (1967) The niche exploitation pattern of the blue-gray gnatcatcher. *Ecological Monographs* 37, 317–350.

Root, R.B. (1986) Functional classification and the development of the guild concept. In: Severinghous, W.D. and James, T.D. (eds) *Applications of the Guild Concept to Environmental Management.* US Army Corps of Engineers–CERL Technical Manuscript N-86/07. University of Illinois at Urbana-Champaign, Urbana, Illinois, pp. 11–15.

Roughgarden, J. (1983) Competition and theory in community ecology. *The American Naturalist* 122, 583–601.

Simberloff, D. and Dayan, T. (1991) The guild concept and the structure of ecological communities. *Annual Review of Ecology and Systematics* 22, 115–143.

Southwood, T.R.E. (1961) The number of species of insect associated with various trees. *Journal of Animal Ecology* 30, 1–8.

Stinner, R.E., Barfield, C.S., Stimac, J.L. and Dohse, L. (1983) Dispersal and movement of insect pests. *Annual Review of Entomology* 28, 319–335.

Strong, D.R., Lawton, H. and Southwood, T.R.E. (1984) *Insects on Plants: Community Patterns and Mechanisms.* Harvard University Press, Cambridge, Massachusetts.

Szentkirályi, F. and Kozár, F. (1991) How many species are there in apple insect communities: testing the resource diversity and intermediate disturbance hypotheses. *Ecological Entomology* 16, 491–503.

Tanigoshi, L.K., Hoyt, S.C. and Croft, B.A. (1983) Basic biology and management components for mite pests and their natural enemies. In: Croft, B.A. and Hoyt, S.C. (eds) *Integrated Management of Insect Pests of Pome and Stone Fruits.* Wiley, New York, pp. 153–202.

Tilman, D., Reich, P.B. and Knops, J.M.H. (2006) Biodiversity and stability in a decade-long grassland experiment. *Nature* 441, 629–632.

Vandermeer, J. (1995) The ecological basis of alternative agriculture. *Annual Review of Ecology and Systematics* 26, 201–224.

Van Emden, H.F. and Williams, G.F. (1974) Insect stability and diversity of agro-ecosystems. *Annual Review of Entomology* 19, 455–475.

Westigard, P.H. (1973) Pest status of insects and mites on pear in southern Oregon. *Journal of Economic Entomology* 66, 227–232.

Westwood, M.N. (1993) *Temperate-zone Pomology: Physiology and Culture.* Timber Press, Portland, Oregon.

Whalon, M.E., Epstein, D., Bird, G., Gut, L. and Flore, J. (2006) Functional ecology: developing measures of sustainability from IPM. Electronic publication, Michigan State University (http://www.functional-eco.msu.edu/ team.htm, accessed 26 May 2008).

Wratten, S.D., Hochuli, D.F., Gurr, G.M., Tylianakis, J. and Scarratt, S.L. (2007) Conservation, biodiversity, and integrated pest management. In: Kogan, M. and Jepson, P. (eds) *Perspectives in Ecological Theory and Integrated Pest Management.* Cambridge University Press, Cambridge, UK, pp. 223–245.

2 The Evolution of Key Tree-Fruit Pests: Classical Cases

Stewart H. Berlocher[1] and Jeffrey L. Feder[2]

[1]Department of Entomology, University of Illinois at Urbana-Champaign, Illinois, USA; [2]Department of Biological Sciences, Galvin Life Science Center, University of Notre Dame, Notre Dame, Indiana, USA

Introduction

Ron Prokopy was an innovative leader and novel personality in insect behaviour, and a true pioneer of the field of pest biocontrol. Ron sought the fundamentals of how phytophagous insects recognize and specialize on new host plants, but always with an eye on using this information to manage insect populations with a minimal use of pesticides. Prokopy's work on his beloved apple maggot fly, *Rhagoletis pomonella* (Walsh) (Tephritidae), has become the gold standard of behavioural models for host finding in phytophagous insects. The *Rhagoletis* model has had important implications for integrated pest management, showing that flies shift from using fruit odour and tree shape at long ranges, to a combination of visual and chemosensory cues at short distances, to tactile, visual and chemical cues once they have alighted on fruit (Prokopy *et al.*, 1973, 1987; Prokopy and Roitberg, 1984). These discoveries led to a number of management strategies being devised by Ron to use the attractive nature of specific visual and olfactory cues to control *Rhagoletis* flies, methods that have become the common practice for insect biocontrol in general (see Leskey *et al.*, Chapter 5, Cooley and Coli, Chapter 6, and Aluja and Rull, Chapter 7, this volume).

But biocontrol was just one aspect of Ron Prokopy's curiosity and research interests. Ron's work also had a profound impact on evolutionary biology and, in particular, on speciation theory. In this chapter, we examine the legacy that Ron left for workers on sympatric host-race formation, organized in the form of 'Lessons from Ron'. We begin by exploring the formative years of Ron's research on *Rhagoletis* in Door County, Wisconsin, as a young postdoctoral associate with Guy Bush. We then discuss how these foundation studies by Ron and Guy and, later, follow-up experiments by Ron and his students established how the apple fly uses host-plant cues to distinguish between alternative host species. This research provided critical early empirical support for the theory of sympatric speciation, providing a mechanism for how host specialization and host-specific mating could generate pre-mating barriers to gene flow. We also briefly examine other important aspects of Ron's work, such as social facilitation of oviposition and performance differences between host races of *R. pomonella*, and his indirect sparking of recent fundamental discoveries about the role of ancient allopatric events in Mexico in facilitating much later sympatric host shifts in eastern North America. We conclude by discussing how Ron's initial research sparked new and exciting discoveries in host-plant discrimination and performance. In particular, we focus on the finding that *R. pomonella* flies not only orient preferentially to natal host-fruit volatiles, but also avoid non-natal odours. The latter discovery has important and previously unrecognized consequences for ecological divergence, especially in sympatry.

The Four Most Important Lessons from Ron: Love Your Insect, Work with the Best, Always Start with the Basics – and Have Fun

Although Ron's lifelong love affair with *Rhagoletis* had started in the mid-1960s, it was not until Ron met Guy Bush in the late 1960s that the future course of evolutionary research on *Rhagoletis* was set. In his ground-breaking thesis work, Guy observed that mating in *Rhagoletis* occurred on the host fruit (Bush, 1966, 1969), a key discovery that linked ecological adaptation and assortative mating – a connection that could allow speciation without the geographic isolation required by then current theory (Mayr, 1963). It was thus destiny fulfilled that Guy, with his passion for understanding speciation using *Rhagoletis* as a model, and Ron, with his passion for understanding host-seeking selection behaviour using *Rhagoletis* as a model, should work together. And 'together' is the operative word here – they worked side by side in the field on all of the work in the early papers. It is clear that Ron was happy to accept Guy's lead on the direction and evolutionary interpretation of many of the experiments, and that Guy frequently took a back seat to Ron when it came to the ultimate design of the experiments – but their early research truly was a joint effort. From their first meeting, Ron and Guy commenced not only a highly productive scientific relationship, but a deep and lifelong friendship as well.

With a 3-year postdoctoral appointment starting in Texas in 1969, the work could begin. But the question then became, where? Guy's standards were set high – the perfect research site had to have good access to orchards, with abundant flies of both the apple and hawthorn host races of *R. pomonella*, of course, but it also had to have a thriving art scene, lovely scenery and access to fly fishing and sailing. Guy found such a place in Door County, Wisconsin, a small sliver of rocky orchards and dairy farms surrounded by Lake Michigan, a destination for wealthy Chicagoans escaping the summer heat. Ron conceived of the perfect field site in different yet complementary terms: the weathered outbuildings nestled in an abandoned apple orchard outside Baily's Harbor were Ron's vision of heaven. Forget about the lack of running water and electricity; to Ron the essence of paradise was never the material but the ideal. Ron, his wife, Linda, and their young son, Josh, drew in scientists (Guy, of course, Jasper Loftus-Hills, Milton Huettel, Volker Moericke, Ernst Boller, Dale Norris) as well as guests, friends and assorted spiritual fellow travellers to participate in the grand experiment on flies and life that was the Prokopy orchard. One of us (Stewart H. Berlocher (SHB)) was a new graduate student of Guy Bush's during this time and spent most of the summer of 1972 working with Ron and Guy, living in a deluxe guest suite (that is, a tent) on the grounds. Wonderful memories remain of all of these people hunkered around a campfire, cooking dinner for 20 on a tiny camper stove, kids running everywhere and science being discussed endlessly.

And that was the remarkable thing about life on the farm – great science was being done. Both in the early 1970s, when Ron and family were renting the farm in the summer, and later that decade, when Ron and Linda bought the property (named, appropriately, the 'Prokopy Bioexperimental Farm') and eked out an existence through subsistence apple farming and pottery. Ron and Guy laid the foundations for almost all future work on *Rhagoletis* behaviour. Everything that we now know about behaviour related to speciation in *Rhagoletis* arose from the symbiotic relationship between these two creative individuals. Guy Bush (1974) summarized much of this early work lucidly in a flowchart of the steps in the life of a *Rhagoletis* fly searching for a host fruit (Fig. 2.1).

The remarkable feature of this behavioural flowchart is that essentially it outlines the programme of behavioural work on *R. pomonella* from the late 1960s (or in embryonic form even earlier, with ideas in a course paper of Guy's while at Harvard) through the next two decades and on into the future. As Ron and Guy had already deduced when they began work, an interplay of visual and chemical cues was necessary for the successful rendezvous, mating and oviposition of *R. pomonella* flies on the host fruit – behaviour that allowed sympatric host-race formation. As we will discuss, several steps in the chart have been explored in detail, while minor changes need to be made concerning other points based on research since Bush (1974).

Step 1 is for a *Rhagoletis* male or female to fly to the host tree. At this first step, a key bimodality of *Rhagoletis* host-selection behaviour becomes apparent: the flies use not only olfactory cues, like all other phytophagous insects, but also make significant use of visual cues. The purely visual aspect of orientation to host trees at moderate distances (8–50 m) was very clearly shown by Moericke *et al.* (1975). An array of 4 × 8 ft plywood panels were cut and rearranged into different shapes, painted a variety of different colours, covered with Tangle-trap® (by SHB, as his contribution to what would be his first publication) and then placed in an orchard to catch the attention of passing flies (and passing motorists). The outcome was that flies were most attracted to tall yellow panels with double sheets of a leaf-like pattern of alternating yellow and clear squares. In other words, *Rhagoletis* flies are visually attracted to tree-like shapes. This paper has not lead to any future work, but there seems little need for it, as the results were clear and sufficient. Moreover, visual response to tree shape, while an important step in general host recognition, probably is not involved in differential response to different host species (Bush, 1974).

Attraction to fruit odour at step 1, at the same distance as in the tree model experiments, was first shown experimentally by Prokopy *et al.* (1973). Although this paper was published earlier than the tree model work in Moericke *et al.* (1975), some of the work described was done concurrently. These first odour experiments were pure Prokopy simplicity: cloth bags of apples and same-sized bags of rocks were hung from one of the tree models (the solid yellow vertical rectangle) showing

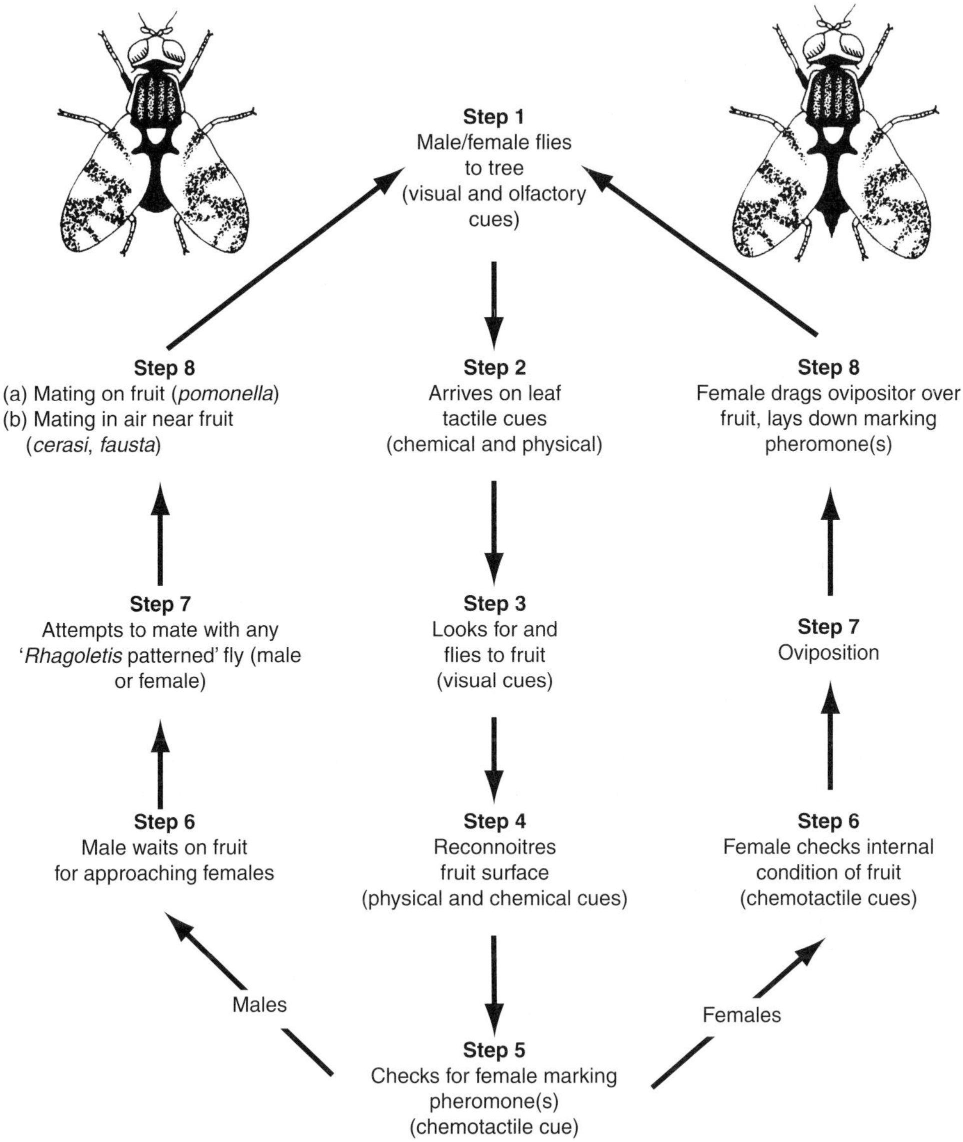

Fig. 2.1. The sequence of behaviours that males and females of all *Rhagoletis* spp. follow in mating and oviposition (pictured species is *R. fausta* (Osten Sacken), not *R. pomonella*). Modified and redrawn from Figure 1 of Bush (1974) – see text.

a good visual response. The results were unambiguous – fly capture on panels with fruit in cheesecloth bags hanging at the top and bottom was about double that on trees with bags of rocks hanging in them (Prokopy *et al.*, 1973). Similar experiments with actual trees also showed a key role for response to fruit odour (Prokopy *et al.*, 1973). We note that *Rhagoletis* species can be viewed as having replaced the long-distance sex pheromones of typical insects with plant kairomones. These experiments, as opposed to the tree model work, did lead to much further work by Ron and others, which we elaborate on later in this chapter.

Step 2, arriving on a leaf and evaluating tactile and chemical cues, has received little attention. Unpublished work by Ron and Guy, as well as observations by this chapter's authors and other *Rhagoletis* workers, have all suggested that this step in host-selection behaviour may be of some importance. But, because of a suspicion that this was a behaviour for coarsely identifying suitable hosts, and not important in fine discrimination of individual host species (Bush, 1974), Ron and Guy did not investigate step 2 in detail. But it is worth pointing out that later work shows that early season mating in several *Rhagoletis* species occurs on leaves, not on fruit (Smith and Prokopy, 1980). Thus, leaf cues could, in fact, be relevant to host-race formation and continued work on this step is warranted. We also note that if early season mating is found to be significant, then Fig. 2.1 will need to be redrawn with a 'shunt' from step 2 to step 8, and then back from 8 to 2, for females to continue to evaluate if they have the correct host plant for oviposition.

Step 3, orienting visually to fruit shape, was the focus of some of Ron's earliest *Rhagoletis* papers. Research on the visual cues used by the flies in finding fruit appealed greatly to Ron, in part because of his innate love of science that could be carried out with little equipment other than a perceptive eye and everyday objects. Ron was truly the commensurate field experimental behaviourist. He had a knack for designing simple field experiments that focused on just one or two aspects of behaviour at a time. His very earliest *Rhagoletis* papers (Prokopy, 1966, 1967, 1968) used hemispherical fruit models to show the importance of an unscented spherical shape in inducing an oviposition response from females, with the 1967 paper starting research into the effects of fruit colour as well as size. The Prokopy 1968 paper is a major work in the literature on *Rhagoletis*, exploring the effects of colour in flat panels and spherical fruit models, and also the size of the fruit. A later key work (Prokopy, 1977) on the effect of the size of artificial fruit showed that *R. pomonella* actually responded best to fruits somewhat larger than apples and much bigger than hawthorns – large size acts as a superstimulus. These papers were especially significant in that flies responded to fruit size purely on the basis of vision – in the field experiments, the artificial fruits were covered in Tangle-trap®, so that flies could not assess fruit size or other qualities further after landing.

After a fly is on a fruit (step 4), it appears to continue to assess fruit size as well as fruit chemistry by walking around the fruit's surface, probably sensing surface chemistry using the tarsal receptors, although flies may also lower their heads and apply their third antennal segments to the fruit surface. The role of fruit size following fly alightment is better understood than that of chemical cues. In what was to be one of the first quantitative behavioural studies of differences between members of the *R. pomonella* species group, Prokopy and Bush (1973c) showed that species preferred artificial fruits (dark cerasin wax domes) proportionately corresponding in size to their natural hosts. Hints of a possible role for fruit surface compounds in host selection can be found in early papers, but the first real experiments on the subject were carried out by Diehl and Prokopy (1986), who showed that apple flies oviposited more into artificial fruits on which surface extracts of apple had been placed than on artificial fruits lacking apple waxes.

No work on this subject has been done since, however, and the area is clearly ripe for further study.

Steps after 4 do not bear as directly on host choice as the early steps. Step 5 involves flies checking for the presence of oviposition deterrence pheromone, which tells females that an individual fruit is already 'occupied' by an egg or larva. This pheromone, a major discovery of Ron's (Prokopy, 1972), is discussed in detail elsewhere (see Aluja and Rull, Chapter 7, this volume).

After step 5, the behaviours split by sex. Step 6 for females involves attempted oviposition, at which point further assessment of fruit chemistry may be made by the sensillae at the tip of the ovipositor. However, this final assessment may be concerned more with detecting fruit quality (ripeness, presence of larvae, signs of decay) than with discriminating between host species. After oviposition (step 7), the female deposits the oviposition-deterrent pheromone on the surface of the fruit. Grouped females lay more eggs per female than do single females (Prokopy and Bush, 1973b).

Step 6 for males is to wait on the fruit for a female (perhaps arrested on the fruit by the oviposition deterrent or some other pheromone produced by females (Prokopy and Bush, 1972)) and then to attempt to mate with approaching females. On fruit, the common mating site, male waiting is followed by steps 7 and 8, in which males attempt to mate with females. Prokopy and Bush (1973a) concluded that males would attempt to mate with any '*Rhagoletis* patterned fly'. However, work since the early 1970s indicates that mating may involve more than just 'forced matings by males' on the host plant. First, Prokopy (1975) showed that virgin males produced a pheromone that acted either to attract females or as an aphrodisiac. If such a pheromone and response to it differed between taxa, potentially it could produce prezygotic reproductive isolation independent of host fruit, although this would most likely evolve after speciation was well under way and the populations had already evolved substantial host-related and other differences adversely affecting hybrid fitness. In any case, more work needs to be done on this aspect of mate choice. Second, in their work on early-season mating on leaves, Smith and Prokopy (1980) observed that many early-season matings were initiated by males approaching females from the front, while late-season matings on fruit were often initiated by males interrupting oviposition or attempted oviposition. Thus, there may be more opportunity for a role for pheromones and behaviour such as wing waving, which are not related directly to host specificity, in such early-season matings – in addition to the possible role for leaf chemicals already mentioned. On the other hand, late-season matings may contribute much more to the gene pool than early-season matings because of the second male sperm precedence of this multiple-mating insect (Opp *et al.*, 1990).

Lesson Five from Ron: Study Variability in Behaviour

Ron and Linda sold their farm in Door County and moved east in 1975, where Ron was to spend the remainder of his career at the University of Massachusetts. At Massachusetts, Ron trained a large number of students and postdoctoral researchers, who were part of a shift from the late 1970s through the 1980s from studies of host-selection behaviours common to all *Rhagoletis* toward more comparative studies. A particular focus was the evolution of the apple and hawthorn host races of *R. pomonella*. However, this shift was gradual and, even in the early years, some comparative work was done. Prokopy and Bush's (1973c) paper on differences in fruit size choice between *R. pomonella* host races and *R. pomonella* group species was the first quantitative behavioural work on reproductive isolation in the group.

Early steps in the comparative direction were the papers of Smith and Prokopy (1981, 1982) on oviposition behaviour in *R. mendax* Curran (Tephritidae), the blueberry maggot. These two species, although interfertile, do not interbreed when sympatric, even if the blueberry is growing under the apple (Feder and Bush, 1989a). The Smith and Prokopy papers represent the first detailed comparisons of mating behaviour in a close relative of *R. pomonella*. These papers reinforced the concept that mating behaviour in *R. mendax* was similar to that in *R. pomonella* – with the exception that blueberry flies preferred blueberry instead of apple or hawthorn in experiments involving real fruit, and smaller rather than larger models in tests involving artificial fruit, consistent with the reduced size of blueberries compared to hawthorn and apple. It is worth noting that even though early-season mating in *R. mendax* occurred on leaves, as with *R. pomonella* (Smith and Prokopy, 1980), almost all matings on leaves occurred on *blueberry* leaves, not surrounding non-host leaves (Smith and Prokopy, 1982), suggesting a role for leaf chemistry.

The two most explicitly comparative papers of the 1980s were Diehl and Prokopy (1986) on *R. mendax* and *R. pomonella* and Prokopy *et al.* (1988) on the apple and hawthorn host races of *R. pomonella*. Diehl and Prokopy (1986) demonstrated several important differences between the two sibling species. First, in laboratory experiments, the two species showed a significant preference to oviposit in their own natal fruit. Second, tests with similar-sized artificial fruits to which waxes and other fruit chemicals had been transferred (via simple rubbing with a dry swab) resulted in the same pattern of preference for the natal fruit. Third, field tests of several behaviours (time in tree, leaves visited, number of fruit visited and number of ovipositions) with potted hosts showed that each species consistently preferred its natal host for all behaviours (although not all comparisons were significant). Of notable interest here was the fact that the preference for the natal host extended to the leaf as well as the fruit, although the presence of fruit on the trees might well have confounded effects. Aside from showing unambiguously that these two interfertile species showed 'host fidelity' (reproductive isolation due to preference for and mating on a host; Feder *et al.*, 1994), these experiments were also the first comparative studies on the role of fruit surface chemicals in host choice. Throughout the late 1980s and early 1990s, more chemically sophisticated experiments (e.g. Bierbaum and Bush, 1990a) and the beginnings of electrophysiological experiments on chemoreception (e.g. Frey and Bush, 1990) would extend the classic, minimalistic but insightful Prokopy-style analysis of Diehl and Prokopy (1986) into new areas.

The experiments of Prokopy *et al.* (1988) on the host races should have received more attention than they did at the time. The key result was that the apple and hawthorn races did, in fact, differ consistently in host-choice behaviour, and were thus at least partially isolated reproductively. Prokopy *et al.* (1988) demonstrated that both apple- and hawthorn-origin females preferred to oviposit into hawthorn – but that apple-origin flies oviposited significantly more often in apples than did hawthorn flies. The preference for hawthorn makes sense in that this is the ancestral host. The paper also is notable for documenting better egg-to-larval survival for both host races in hawthorns than in apples, with no apparent difference between the races.

Lesson Six from Ron: Find New Research Areas that can be Explored with Your Favourite Organism

Ron had a deep interest in both exploiting and promoting *Rhagoletis* for basic research on insect behaviour – behaviour that might not be recognized immediately as related to host-race formation or insect control. Discussing the full range of all this work is beyond our scope here and is being covered elsewhere in this volume (see Roitberg *et al.*, Chapter 3, this volume), but we will discuss several of these behaviours briefly, chosen because they are, in fact, important in the process of speciation and host-race formation. A major area of research extending from the early 1980s to Ron's death was on learning (Prokopy *et al.*, 1982, 1994; Prokopy and Papaj, 1989). In essence, choice of host-fruit species is affected by the species of fruit into which an individual fly first oviposits. The connection to speciation is that this type of learning could reinforce an initially imperfect genetic polymorphism for host choice. Moreover, the ability to learn might itself have a genetic basis; hawthorn race flies seem to be poor at 'learning to like apple', whereas apple flies are much better at learning to prefer apples (Prokopy *et al.*, 1982, 1994). Another area that Ron developed extensively with students was the application of foraging theory to host-selection behaviour (Roitberg *et al.*, 1982; Averill *et al.*, 1996). Connections of optimal foraging theory with the theory of sympatric speciation have been discussed by Berlocher and Feder (2002). An excellent example of Ron's penchant for returning anew to subjects he had first explored years earlier is his work on social facilitation of oviposition (increased oviposition by grouped females) in *Rhagoletis*. The phenomenon first examined by Prokopy and Bush (1973b) was investigated much more thoroughly by Prokopy and Reynolds (1998), in which the potential explanations for facilitation were narrowed experimentally to just one, which was that females alighting on a fruit were more likely to oviposit if a female already on the fruit was ovipositing. Prokopy and Reynolds (1998) suggested that this behaviour could be adaptive if females were interpreting oviposition by another female as an indication of good fruit quality. If this is true, such behaviour could be important in a host shift by aiding females in the search for those individual host genotypes in which reproductive success is most likely.

Lesson Seven from Ron: Encourage Others to Expand on Your Work

Ron's work in the 1980s on host discrimination spawned, in the 1990s, several major, complementary research endeavours in the *Rhagoletis* community.

One was carried out by Ron's graduate student, Martín Aluja, who performed a novel and sophisticated set of experiments on the interaction of chemical and visual cues in *R. pomonella* (Aluja and Prokopy, 1993). Using potted apple trees in a large field enclosure, Aluja demonstrated that the flies relied more and more on fruit chemical cues as the strength of the visual stimulus was diminished, and vice versa. This finding is not only central to understanding host-finding behaviour in the flies, but essential knowledge for the design of the field odour response work, discussed later, that was carried out at the end of the 20th century (e.g. Linn *et al.*, 2003).

A second major undertaking was directed at quantifying the degree to which host fidelity due to behaviour, experience and plant phenology served as a gene flow barrier between *R. pomonella* host races and sibling species in the field. Field studies assessing host fidelity have used both mark–release–recapture and genetic techniques to assess the degree of interhost movement for flies. In one set of experiments involving marked flies released in an old field near Grant, Michigan (Feder *et al.*, 1994), it was determined that gene flow between hawthorn and apple fly populations was about 4–6% per generation. Host fidelity is therefore strong but not absolute for *R. pomonella*, consistent with the status of hawthorn and apple flies as partially reproductively isolated host races. Much as in Ron's earlier work, innate host preference was found to be a greater factor contributing to host fidelity for hawthorn than apple flies in the mark–recapture study (i.e. hawthorn flies showed a greater tendency than apple flies to avoid residing in non-natal trees they were released beneath). In contrast, possible experience associated with eclosing under the 'correct' or natal tree played a proportionally larger role in host fidelity for apple flies (Feder *et al.*, 1994). No evidence for non-host related pre-mating isolation was found in the mark–recapture data, as migrants from non-natal hosts did not mate at a statistically lower frequency with resident flies than natal flies did (Feder *et al.*, 1994, 1998).

In comparison to the results for the host races, genetic analysis of field-captured adults and larvae dissected from host fruit at a field site near Sawyer, Michigan, with sympatric apple trees and blueberry bushes revealed 100% host fidelity between the apple race of *R. pomonella* and *R. mendax*, within the limits of statistical detection (Feder and Bush, 1989a). It is therefore possible for host fidelity to evolve to the point were it may serve as a near complete barrier to gene flow between host-associated *Rhagoletis* populations – at which point they would be species.

Lesson Eight from Ron: Push Others to Push the Envelope

Moving from cheesecloth bag to solid-phase microextraction and GC-EAD

The work of Ron and his students and colleagues on the behavioural sequence that *Rhagoletis* flies use to recognize and find host trees, as summarized in Bush (1974) (Fig. 2.1), established a basis for testing in detail the roles that different olfactory, visual and tactile cues play in host discrimination. As noted previously, some important early steps were taken in the late 1980s and early 1990s by

Bush, Frey and Bierbaum to refine the role of chemicals and chemosensation in host selection. However, the techniques then available did not allow chemical cues to be identified rapidly from very tiny amounts of starting material and, in most of these initial studies, the extraction procedures produced a mixture of volatiles and relatively non-volatile fruit waxes, so that steps 1 and 4 potentially could be conflated.

Recent research on host discrimination in *Rhagoletis* has concentrated on the role that fruit volatiles play in medium-to-short range host-finding ability – the olfactory part of step 1. The exact range at which fruit odour is attractive has not been studied in this work, despite the fact that vision and olfaction are known to interact significantly (Aluja and Prokopy, 1993), because all the recent laboratory and field experiments thus far have had to combine odour response with visual response. A critical advance in the study of host-fruit odour discrimination in *Rhagoletis* was the development of synthetic fruit volatile blends for various *R. pomonella* host plants (Fein *et al.*, 1982). Recent work spearheaded by the Roelofs and Linn group at Cornell University has made extensive use of remarkable modern technological advances such as solid-phase microextraction (SPME) and gas chromatography coupled with electroantennographic detection (GC-EAD), which allows collection and storage of volatiles on unpicked fruit in the field and rapid identification of the key attractant molecules. The past few years have seen the development and refinement of numerous fruit volatile blends, including apple (Zhang *et al.*, 1999), hawthorn (Nojima *et al.*, 2003a), flowering dogwood (Nojima *et al.*, 2003b), as well as blueberry, snowberry and twinberry honeysuckle (C. Linn, Ithaca, 2006, personal communication). These blends have allowed for controlled testing of fly fruit odour behavioural discrimination in the laboratory and field, as well as for detailed study of its genetic basis. A key point is that all of the work described below involves step 1 from Fig. 2.1 – it all involves true response to volatiles, without a fruit (or leaf) contact response.

Laboratory flight tunnel assay studies have shown that apple, hawthorn and dogwood flies collected from several different field sites all oriented preferentially to their respective natal fruit blend (Linn *et al.*, 2003, 2004, 2005a). No apple, hawthorn or dogwood fly tested flew upwind toward a blank, odourless red sphere in the flight tunnel trails (Linn *et al.*, 2003, 2005a). However, essentially every apple fly tested took flight when the sphere was baited with 200 μg of the apple blend, with the majority (~75%, n = 255 flies tested) exhibiting upwind anemotactic flight tracing the odour plume in the tunnel to reach its source. In contrast, only 17% of apple flies flew upwind and reached the sphere when hawthorn volatiles were present, and even fewer (~11%) when the source sphere was baited with the dogwood blend. Hawthorn and dogwood flies displayed similar patterns. A total of 73% of hawthorn flies reached the sphere when it was seeded with the hawthorn blend (n = 277), compared to 5% for apple baited and 17% for dogwood blend spheres. Likewise, around 76% of dogwood flies reached spheres baited with their natal dogwood blend (n = 189), while 12% of flies reached apple spheres and 20% hawthorn spheres. One interesting finding was that a fair proportion of both hawthorn and dogwood flies initially took flight to each other's fruit volatiles in the tunnel, possibly due to chemical similarities in the hawthorn and dogwood blends (Linn *et al.*, 2005a). However, upwind flight was arrested quickly for hawthorn and flowering dogwood flies, with the majority of individuals not reaching non-natal odour spheres.

The flight tunnel results were nearly identical for paired apple and hawthorn fly populations tested from Grant and Fennville, Michigan, and Urbana, Illinois, as well as for a laboratory colony of Geneva, New York, apple flies established from the wild in the 1970s (Linn *et al.*, 2003). Thus, the host races showed a consistent preference for their natal blend, regardless of geographic location. Moreover, Urbana hawthorn flies reared for two generations in the laboratory on apple fruit displayed the same behavioural responses as hawthorn flies reared directly from field-collected haws, discounting an effect of the larval host-fruit environment on adult fly behaviour (Linn *et al.*, 2003).

In contrast to the results for apple, hawthorn and flowering dogwood flies, the current synthetic blends for blueberry, snowberry and twinberry failed to induce directed upwind flight from their native fly populations of *R. mendax*, *R. zephyria* Snow (Tephritidae) and the hybrid *Lonicera* fly (Schwarz *et al.*, 2005; D. Schwartz, Illinois, 2008, personal communication). It is conceivable that the lack of behavioural orientation displayed by these flies could be due to the synthetic blends lacking or having incorrect proportions of certain key, behaviourally critical compounds. However, it is also possible, if not probable, that these flies are relatively insensitive to fruit volatiles as a behavioural attractant at the distances (1 m) tested in the flight tunnel. If true, then this implies that not all *Rhagoletis* flies use fruit volatiles as intermediate range olfactory cues to identify host trees; in these cases, flies may mate and oviposit at times when fruit are not fully ripe and do not emit large amounts of volatile compounds. The finding also suggests that the responding apple, hawthorn and flowering dogwood flies may represent a clade of the most closely evolutionarily related races/species in the group, an inference supported by genetic data (Berlocher, 2000).

Analysis of the orientation behaviour of individual apple, hawthorn and flowering dogwood flies in flight tunnel tests to all three volatile blends presented in succession revealed the presence of population-level variation in response patterns (Linn *et al.*, 2005a). Apple, hawthorn and dogwood flies showed maximal response to their respective natal fruit blend, with significant decreases in their responses to non-natal blends. However, some flies of each taxon did respond to non-host volatiles. In almost every case, these non-natal responding flies also oriented to their natal blend. Indeed, in parental apple, hawthorn and dogwood fly populations, an average of about 60% of flies reached natal blend spheres only, 15–20% both natal and non-natal blend spheres, 3% non-natal spheres only and 15–20% did not respond to any blend. In each host population, a proportion of flies therefore exists with broad response specificity to multiple hosts. Linn *et al.* (2005a) proposed that these flies could be a source for host shifting, as they might not require a specific volatile mix to find a host to mate on or oviposit and be willing to accept a wider range of host fruit. However, it is not clear whether these broad responders represent the ancestral condition in hawthorn and dogwood populations (in which case, it could have served as a standing source of genetic variation triggering a host shift) or is due to ongoing gene flow between sympatric fly populations using different hosts (in which case, it would be a consequence of host shifting). Further testing of hawthorn fly populations from southern and northern USA, outside of their range of overlap with apple and dogwood flies, respectively, may help to resolve the nature of the observed olfactory response variation in *R. pomonella*.

Field trials conducted in the 2002 and 2003 seasons at study sites near Geneva, New York, Fennville and Cassopolis, Michigan, and Granger, Indiana, have confirmed the relevance of the laboratory flight tunnel results to nature. Trapping studies performed using sticky spheres baited with the apple, hawthorn or flowering dogwood volatile blend captured significantly higher numbers of the respective resident apple, hawthorn or dogwood flies than blank spheres or spheres baited with the alternative non-natal blends (Linn *et al.*, 2003; Forbes *et al.*, 2005). Most importantly, resident apple, hawthorn and flowering dogwood flies were captured significantly less often on their non-natal blends than on blank controls (from 25–39% capture on non-natal versus blank control spheres; Forbes *et al.*, 2005). Thus, not only do flies orient preferentially to their natal fruit odour, but they also appear to avoid non-natal volatiles. The field studies also showed that preference and avoidance behaviours for apple, hawthorn and dogwood flies extended to certain individual volatile compounds, as well as to entire fruit blends (Forbes *et al.*, 2005). However, not every natal or non-natal compound affected fly capture. For example, hawthorn flies did not appear to orient to or avoid butyl hexanoate, a minor component in the hawthorn blend and a major attractant to apple flies (Nojima *et al.*, 2003a). Combined mixtures of natal plus non-natal blends also resulted in significantly reduced captures of flies (Forbes *et al.*, 2005). This finding reiterates that apple, hawthorn and dogwood flies do not merely fail to recognize the odour of alternative host fruit, but also generally tend to avoid them. Moreover, avoidance of the non-natal elements of the mixed blends generally appears to be stronger than the inherent attraction invoked by components of the natal blend. Antagonistic responses of apple, hawthorn and dogwood flies to a mix of natal and non-natal blends have been confirmed in flight tunnel assays (Linn *et al.*, 2005b). The results for the combined blend tests in the field and laboratory imply that chemosensory changes accompanying *R. pomonella* host shifts appear to involve not just the derivation of a new preference for a novel fruit that overcomes an ancestral aversion behaviour, but also the rapid evolution of avoidance of ancestral fruit volatiles.

The genetic analysis of host-finding behaviour in *R. pomonella*

It is most important to understand the genetic basis for host discrimination in *Rhagoletis* as it provides insight into the kind of genetic changes responsible for initiating a shift to a new host and thus promoting host-race formation and speciation. To understand the genetic basis of the speciation process is thus the jewel in the crown of our evolutionary studies. Investigation of the genetic architecture underlying host-fruit odour discrimination has focused on the analysis of first and second generation hybrids constructed among apple, hawthorn and flowering dogwood flies. An unexpected finding from these crosses was that all combinations of F_1 hybrids between fly taxa were insensitive to host-fruit volatiles in flight tunnel assays (Linn *et al.*, 2004). The vast majority of F_1 hybrids did not respond to apple, hawthorn or dogwood fruit volatile blends in wind tunnel assays at doses of 200 μg that normally elicit maximal upwind flight of parental flies to source spheres (Linn *et al.*, 2004). A fraction of F_1 flies (30–57%) did respond when an order of magnitude higher blend dose (2000 μg) was used in the assays, a dose that resulted in arrestment of the upwind flight of parental flies to source spheres (Linn *et al.*, 2004). Almost every F_1 fly responding to 2000 μg doses of fruit volatiles (~97%) oriented to a 1:1 combination of their parent's blends, mixtures that tended to antagonize the upwind flight of parent flies at low doses (Linn *et al.*, 2005b). In addition, about 25% of F_1 hybrids also reached spheres baited with high doses of one or both of their parent's natal fruit blends (Linn *et al.*, 2004). Finally, F_1 apple × hawthorn hybrids responding at 2000 μg displayed a bias for orienting to their mother's natal blend (Linn *et al.*, 2004).

The reduced behavioural response pattern seen for F_1 flies implies that hybrids have an altered chemosensory system and are hampered in their ability to detect and orient to fruit odour. As a result, hybrids may suffer a fitness disadvantage finding host fruit in nature, a hypothesis requiring further testing in the field. If true, then host-specific mating would play a dual role in sympatric race formation and speciation, serving as an important post-zygotic, as well as pre-mating, reproductive isolating barrier (Linn *et al.*, 2004). The cause for the reduction in hybrid response remains to be determined definitively (see below), but could be due, alone or in combination, to F_1 flies being compromised behaviourally by possessing different alleles for avoidance to the volatiles of both parental fruit or to genetic incompatibilities between fly taxa disrupting normal development of the chemosensory system.

Analysis of the odour response patterns of F_2 and backcross hybrids in flight tunnel assays suggest that only a modest number of loci underlie the differences in fruit volatile discrimination among apple, hawthorn and dogwood flies (Dambroski *et al.*, 2005). In contrast to F_1 hybrids, many second hybrids (30–66%; n = 1115 total flies tested for all crosses) reached 200 μg baited source spheres in the flight tunnel (Dambroski *et al.*, 2005). A subset of the low-dose-responding F_2 flies was also tested at 2000 μg (n = 176 flies). Most (63–100%) of these 200 μg responders failed to reach the high dose source sphere, implying that they were antagonized by the elevated volatile level. A fair proportion of F_2 and backcross hybrids therefore had behavioural response patterns that mirrored those seen for parental apple, hawthorn and dogwood flies, implying that fruit odour discrimination variation among taxa was not overly polygenic.

F_2 flies also showed a grandmaternal effect suggestive of cytonuclear gene interactions (Dambroski *et al.*, 2005). The vast majority of F_2 flies responding to 200 μg doses of fruit volatiles reached spheres baited with the natal blend of their respective maternal grandmothers. The maternal grandmother effect was consistent across all of the different types and directions of F_2 and backcrosses conducted, as well as collecting sites (Dambroski *et al.*, 2005). Within a given cross type, responding F_2 and backcross males and females in 200 μg assays showed similar preferences for the maternal grandmother blend, discounting sex linkage as a major factor affecting host-fruit odour discrimination (Dambroski *et al.*, 2005). Taken together, the cross data for hybrids are consistent with a genetic model involving three epistatically interacting nuclear loci (one affecting odour preference and two influencing avoidance) coupled with a cytoplasmic maternal element underlying host-fruit volatile discrimination (Feder and Forbes, 2008). Subsequent studies directed at mapping quantitative trait loci (QTL) for fruit volatile preference and avoidance behaviours are in progress and have yet to reveal a statistically significant association between genetic markers and response (or lack of response) in flight tunnel assays.

The physiological basis for fruit odour discrimination

Olfactory discrimination in *Rhagoletis* could evolve via changes in the central processing centres of the brain and/or through variation in peripheral chemoreception. Based on single cell sensillum electrophysiology of fly antennae, Olsson *et al.* (2006a) showed that apple, hawthorn and dogwood flies, as well as *R. mendax*, possessed similar classes of olfactory receptor neurons (ORNs) responding to host and non-host volatiles. None of the taxa possessed significantly more ORNs tuned to one set of compounds than another, and topographical mapping indicated that ORN locations did not differ morphologically between apple, hawthorn, dogwood and blueberry flies (Olsson *et al.*, 2006a). Therefore, differences in host-plant preference among fly populations do not appear to be a function of altering receptor neuron specificity to host or non-host volatiles. Apple, hawthorn, dogwood and blueberry flies all have broad sensory palettes that can detect all the volatiles from the various different fruit (Olsson *et al.*, 2006a). Differences among fly populations were found, however, with respect to ORN response characteristics, including both sensitivity to specific volatiles and temporal firing patterns (Olsson *et al.*, 2006b). These peripheral sensory differences could contribute to fruit odour discrimination variation among apple, hawthorn and dogwood flies.

Single cell recordings were also made for F_1 hybrids between apple, hawthorn and dogwood host populations, flies that displayed a reduced behavioural response to fruit volatiles in flight tunnel assays. These recordings revealed distinct and diverse hybrid response profiles for certain ORNs not seen in parent populations (Olsson *et al.*, 2006b), consistent with compromised peripheral chemoreception in F_1 flies. Abnormalities in the peripheral system of F_1 hybrids imply some form of genetic and physiological breakdown in the development and function of the ORNs, possibly due to the misexpression and/or presence of novel combinations of receptors in hybrid neurons (Olsson *et al.*, 2006b). However, analysis of ORNs for fruit odour responding and non-responding second-generation hybrid flies in flight tunnel assays revealed that both behavioural classes of flies shared similar diverse sets of electroantennal response profiles (S.B. Olsson, New York, 2008, personal communication). Indeed, 61.2% (60/98) of the ORNs measured for behaviourally responding F_2 and backcross flies versus 64.8% (59/91) for non-responding flies displayed aberrant electroantennal profiles, a statistically non-significant difference. These data suggest that the observed differences in ORN response patterns in hybrid flies may not be the root cause for olfactory dysfunction and reduced chemosensory orientation in F_1 and F_2 progeny. However, it is still possible to develop models in which the total number or specific subtypes of ORNs must exceed a minimal threshold of functional, 'parent-like' units to result in behaviour. As shown by Olsson (S.B. Olsson, New York, 2008, personal communication), these threshold models can generate very similar predictions as the observed data. Consequently, the statistically non-significant difference in the proportions of parent-like ORNs in behaviourally responding versus non-responding hybrids may still be biologically relevant and indicate that disruption of the peripheral chemosensory system contributes to hybrid olfactory dysfunction.

Lesson Nine from Ron: Train and Encourage Students from Around the World (Fresh Insights from Mexico)

Although the heart of Ron's research was always with the apple maggot in eastern USA, he loved to travel and developed strong research partnerships in Switzerland, Poland, Hawaii, Mexico and other far-flung locales. The Mexican connection, in particular, runs deep for *Rhagoletis* research. Guy Bush worked for the USDA laboratory in Mexico City in the 1950s and, in the early 1970s, regularly regaled his laboratory with stories not only of life in 'La Capital', but of the many unsolved *Rhagoletis* mysteries to be found in Mexico. Therefore, it seems only appropriate that Ron went on, in the 1980s and 1990s, to train a new generation of students from Mexico. This partnership had an unexpected bearing on our understanding of the history of *R. pomonella* and led to new insights in speciation theory and fresh insights into the origins of genetic variation, contributing to sympatric host-race formation.

The biogeography of *R. pomonella* in Mexico differs from that of north-eastern North America. *Rhagoletis pomonella* in Mexico can live only at cool, high altitudes where their native hawthorn hosts (*Crataegus mexicana* DC. and several other *Crataegus* L. spp. (Rosaceae)) can thrive and this restriction to high altitudes means that the geographic range of the species is highly subdivided (Rull *et al.*, 2006). There are three suitable regions in which hawthorns are infested by *R. pomonella*: the Sierra Madre Oriental, the trans-Mexican volcanic belt and the Chiapan highlands (Rull *et al.*, 2006). Recent population genetic and phylogeographic work (Feder *et al.*, 2003a, 2005; Michel *et al.*, 2007; Xie *et al.*, 2007) unexpectedly has revealed that populations from the trans-Mexican volcanic belt are genetically very divergent from those in the Sierra Oriental and north-eastern North America, with mean sequence divergence >3.5% (Michel *et al.*, 2007; Xie *et al.*, 2007) for the mitochondrial cytochrome oxidase I (COI) locus, and with large mean pairwise F_{ST}s between volcanic belt flies and Sierra Madre Oriental and north-eastern North American flies of 0.314 and 0.323, respectively. These data, especially those for microsatellite loci (Michel *et al.*, 2007), suggest that current gene flow between the volcanic belt flies and those in the Sierra Madre Oriental is restricted. The Sierra Madre Oriental populations, by comparison, are relatively similar to populations in eastern North America, especially those in south-eastern North America (Michel *et al.*, 2007; Xie *et al.*, 2007), with a mean pairwise F_{ST} of 0.128 between them. The Chiapas population is probably introduced and not yet well understood.

The current hypothesis to explain these patterns of differentiation proposes three steps (Feder *et al.*, 2003a; Michel *et al.*, 2007; Xie *et al.*, 2007). First, the volcanic belt and Sierra Madre Oriental populations became isolated in the late Pleistocene, around 1.57 million years ago, and subsequently differentiated. This differentiation was not only at single loci but also for large blocks of chromosomes that are inverted in the different Mexican and North American populations (Feder *et al.*, 2003b; because the inversions are complex, likely to overlap and not cytologically suitable for analysis, the order of their origin in the different *R. pomonella* populations is not yet known). These inversions, or rather several alleles of linked loci which conveniently mark the different inversions, have been studied intensively in eastern North America, where they are associated closely with diapause phenology. Feder and colleagues (Feder *et al.*, 1997; Filchak *et al.*, 2000) have developed the 'slow/fast metabolism' model of diapause for *R. pomonella*, in which exhaustion of stored fat reserves in the pupa triggers adult development and in which inversion genotypes for slow metabolism alleles predominate in warm conditions and genotypes for fast metabolism predominate in cool conditions.

At a later point in the Pleistocene, some degree of gene flow occurred between these two populations, having the effect of enriching the amount of variation present in Sierra Madre Oriental populations for the inversions and for their contained single locus variation. Finally, flies from the Sierra Madre Oriental expanded north after the last glaciation, either mixing with flies from a North American refugium or, less likely, recolonizing an eastern North America in which *R. pomonella* had not survived the glacial maximum. Although *Crataegus* does not have a dense fossil record, other co-occurring plants well represented in the pollen record, such as *Quercus* L. (Fagaceae), were common in Florida at the last glacial maximum (Williams *et al.*, 2004), suggesting that *R. pomonella* could have survived in Florida with its host. In either event, very strong, linear, latitudinal clines of inversion polymorphism were established at this time (Feder and Bush, 1989b; Berlocher and McPheron, 1996), with genotypes for fast metabolism predominating in very warm southern North America and genotypes for fast metabolism predominating in cool northern North America (Dambroski and Feder, 2007). Taken together, all the recent work has led to the view that speciation in *R. pomonella* may be described better as a type of 'plural' speciation, with early allopatric events setting the stage for much later sympatric speciation (Xie *et al.*, 2007).

This complex history for *R. pomonella* has critical implications for host-race formation and sympatric speciation in *R. pomonella*, in other *Rhagoletis* and in other phytophagous insects. For *R. pomonella* itself, the clinal diapause metabolism variation provides the raw material for the temporal shift to the earlier fruiting time of apples. In what may seem paradoxical on first encounter, early emerging apple flies have slow metabolism inversions because the prediapause period is the critical stage for adaptation to a new thermal environment – and adapting to apples requires that the flies spend more time as a larva and pupa, in warmer weather, than do hawthorn flies. Thus, an adaptation to conditions in Mexico more than a million years ago may have produced the variation needed to colonize apple only 150 years ago.

Regardless of where the variation needed to colonize apple came from, it is clear that only part of the genome is involved in host adaptation. This fact emerges from the observation that larger differences between the apple and hawthorn host races are seen at loci on or next to inversions, compared to loci not on inversions. This pattern was first seen in the 1980s with allozyme data (Feder *et al.*, 1988; McPheron *et al.*, 1988) and has now become very clear with DNA polymorphisms and sequence data (Feder *et al.*, 2003b).

Speciation in the remainder of the *R. pomonella* species group also seems to be fuelled, to varying degrees, by ancient Mexican inversion polymorphism. The formally unnamed 'flowering dogwood fly', which infests *Cornus florida* L. (Cornaceae), has latitudinal clines for some of the same marker alleles, as seen in *R. pomonella*, albeit shallower ones (Berlocher, 1999). On the other hand, the blueberry maggot, *R. mendax*, has no clinal structure at all and some unique alleles (Berlocher, 1995), suggesting some different relationship between diapause and inversion polymorphism.

Looking more broadly, it would be of great interest to know if any members of the other North American *Rhagoletis* sibling species complexes, the *R. cingulata* (Loew) (Tephritidae) and *R. tabellaria* (Fitch) groups (Tephritidae) (Bush, 1966), have had their ability to shift to new hosts facilitated by variation from Mexico. This is feasible for the *R. cingulata* complex, in which the nominate species and *R. turpinae* Hernández-Ortiz (Tephritidae) occur in Mexico. Insufficient work has been done on this group to determine even if there are latitudinal clines or whether adaptation to domestic cherries represent host-race formation. In the *R. tabellaria* group, no members are known to exist in Mexico, but this does not mean that such members do not occur.

Lesson Ten from Ron: Provide a Benchmark and Beacon for Work Outside of *Rhagoletis*

What can we take from the apple maggot story and apply to host races or likely host races in other insects? (See Dorn and Piñero, Chapter 4, this volume.) From the standpoint of genetics, two key lessons stand out. First, genetic differences between host races will almost certainly not occur at all loci studied. This is because the barriers to gene flow between host races and close species are 'semi-permeable' – genes contributing to adaptation in one host race will be selected out of the other host race after flowing into it, and vice versa, while the rest of the genome will cross freely from one race to the other. The resulting heterogeneity of inter-race differentiation among loci is seen in all of the other host races for which there are enough data to test (Berlocher and Feder, 2002). Thus, while host races must be distinguished for some management issues, such as host activity patterns and parasitoids, for other issues, like the evolution of insecticide resistance, they probably should be treated as one population. In other words, from the standpoint of management, host races can be ignored or not, depending on the problem at hand.

Second, attention must be paid to the possibility that the variation making host shifts possible may be ancient. This is of great importance because, if all populations of the ancestral host race possess stores of long-standing genetic variation for adaptation to new hosts, then new derived host races are likely to arise on very short time scales, just tens of years, while derived host races could give rise to ancestral ones almost instantly, given a situation in which the ancestral host plant is uninfested. In fact, Berlocher, 1999, proposed that the concepts of host race and species could be distinguished logically on the criterion of time to recolonize a 'vacant' host plant. For example, if *R. pomonella*, after being introduced into north-western North America a few decades ago, had colonized introduced highbush blueberry in the same area almost instantly, then this would support the conclusion that *R. mendax* and *R. pomonella* differed only at the host race, not the species, level. The possible existence of ancient adaptive variation can be tested only by carrying out a thorough genetic study of patterns of geographic and host differentiation in all of the host-associated populations of the species in question. We note that while, of course, not all such patterns will be congruent with *R. pomonella* and involve Mexico, it is foolish to ignore the possibility of complex biogeography in native North American pest species whose range does extend into Mexico.

From a more ecological standpoint, one of the strongest lessons that Ron (and Guy Bush) can pass on to the study of other host races is that reproductive isolation must be studied on the host and in the field, if at all possible. The apple and hawthorn races of *R. pomonella* and, to varying lesser degrees, the species of the *R. pomonella* species group, display little reproductive isolation in cages with no fruit (see Berlocher and Feder, 2002).

Detailed comparison of the apple maggot with other cases of host-race formation in phytophagous insects is hindered by lack of exact correspondence in the life history of the pests. Perhaps the largest difference is that the simple historical pattern of a native ancestral race on a native non-crop plant and a derived host race on an introduced crop plant is found only in some cases. The consensus of opinion gradually has shifted to the view that the ancestral host plant of the cherry race of the European cherry fruit fly, *R. cerasi* (L.), is in fact honeysuckle, *Lonicera* L. (Caprifoliaceae) (Schwarz *et al.*, 2003). A variation of the *R. pomonella* pattern is seen in the recent (~10-year) host shift by a European weevil, introduced to control an introduced Asian milfoil (*Myriophyllum* L.) (Haloragidaceae), to a native North American milfoil (Sheldon and Jones, 2001). Other well-studied cases represent greater divergences from the *R. pomonella* pattern. In the goldenrod gallfly, *Eurosta solidaginis* (Fitch) (Tephritidae), all hosts are native plants, while in the pea aphid, *Acyrthosiphon pisum* (Harris) (Aphididae), both the insect and all of its hosts are introduced crops (Berlocher and Feder, 2002), so that it is impossible to know which is ancestral at this point. For the brown plant hopper, *Nilaparvata lugens* (Stål) (Delphacidae), it is plausible that *Leersia* Sw. (Poaceae) is the ancestral host and rice the derived one, although the role of wild rice ancestors has yet to be resolved (see review in Berlocher and Feder, 2002). Host races in the key tree-fruit pest, the codling moth *Cydia (Laspeyresia) pomonella* L. (Tortricidae), are likely (Witzgall *et al.*, 2005), but the insect is genetically too poorly known to allow strong inferences about which race is ancestral. Despite the fact that only a few host races fit the '*R. pomonella* model' precisely, much of what we have learned still can be applied successfully to these other cases. In particular, the 'divergence with gene flow' model of semi-permeable gene flow barriers will be found to be universal and must be considered in all research and control efforts.

Looking to the Future: What Would Ron Do?

As even our cursory tour through the work of Ron Prokopy shows, he initiated many significant areas of research on sympatric speciation in *R. pomonella*. For behavioural work, he was there at the beginning. And, as the discussion of our recent work on odour response shows, exploration of just one of Ron's research paths (step 1, response to fruit odour) has yielded a rich harvest of insights and new findings. What awaits us as we follow the many other roads down which Ron took a few steps and then left for us to follow and widen?

Many of the steps in Fig. 2.1 require further attention. For example, step 2 – the evaluation of leaves for tactile and chemical cues – is particularly lacking in detail. The fact that this step is relatively early in the host acceptance sequence, the reluctance of flies to spend much time on non-host foliage and the initiation of early-season mating on host leaves, all argue that a search for chemical cues on host leaf surfaces could lead to new insights for sympatric speciation. Another step that begs for attention is step 4 – the evaluation of fruit after landing using tactile and chemical cues. While this step is fairly deep into the flow chart of Fig. 2.1, the observations of Ron and others that flies spend time evaluating fruits once they alight argues that important decisions about host species are also being made at this stage of the sequence, especially with respect to ovipositing females.

Ron was also a pioneer in testing for fruit-related performance differences between apple and hawthorn flies. The results from these studies implied that the host races had not adapted differentially to any chemical or nutritional variation between hawthorns and apples (Prokopy *et al.*, 1988). Egg-to-larval survivorship was higher and similar for both races on the ancestral host hawthorn and lower on apple. However, there has not been adequate analysis testing for survivorship effects that may carry over from the larval feeding stage through pupal overwintering to adult eclosion. A recent study by Dambroski and Feder (2007) suggests that pupal-to-adult survival may be higher for apple than hawthorn flies reared as larvae under controlled conditions in apple. Thus, the initial characterization of the larval feeding environment as an aspect of the life history of *R. pomonella* not under differential selection requires revisiting. Larval feeding environment can be important at the species level (Bierbaum and Bush, 1990b).

There are also areas of *Rhagoletis* behaviour bearing on reproductive isolation and speciation that Ron did not investigate. Perhaps the most significant of these would be whether *Rhagoletis* flies differ in cuticular hydrocarbons and use them for differential mate recognition, as do many other insects (Ferveur, 2005). To date, there is no published research on cuticular hydrocarbons in *Rhagoletis*. The finding of a role for such compounds, apparently unrelated to host usage (though see Etges *et al.*, 2006), would require a revision of Fig. 2.1. The closest that Ron seems to have come to work that could have altered step 7 of Fig. 2.1

is in his little-noticed 1975 paper, in which he found that virgin males were attractive to females, presumably due to the production of a mating pheromone or aphrodisiac; if such a pheromone and the response to it varies among species, then this could also be a mechanism of reproductive isolation unrelated to host plant. As noted earlier, however, previous mark–release–recapture studies conducted at Grant, Michigan, showed no statistically significant non-host related pre-mating isolation (Feder *et al.*, 1994, 1998). Apple flies that migrated to hawthorn did not mate at statistically lower proportions with resident hawthorn flies than did natal hawthorn flies, and vice versa (Feder *et al.*, 1998). Although this does not rule out the possibility that cuticular hydrocarbons play a role in contributing to pre-zygotic isolation between apple and hawthorn flies, it does suggest that if they exist, their effects are likely to be subtle for the host races. The same may not be true for the sibling species, however, where more substantial hydrocarbon differences contributing to differential mate choice may have had more time and greater opportunity to accumulate.

But, whatever waits in the future of *Rhagoletis* research, Ron would have greeted it with great enthusiasm and plans for experiments – beautifully simple experiments that reveal the pure essence of what it means to think and act as a fly.

Acknowledgements

We thank Guy L. Bush for his thoughtful comments on this chapter, for many details about his early work with Ron Prokopy and for his many discoveries and insights about *Rhagoletis* and sympatric speciation. We thank Guy and Dorrie Bush and Ron and Linda Prokopy for many, many evenings of gracious, warm hospitality. Finally, we thank Ron Prokopy for his unqualified and unfailing support of both of us when we were but newly eclosed young scientists. This material is based on work supported by the National Science Foundation under Grant No. DEB 06-14528. Any opinions, findings and conclusions or recommendations expressed in this material are those of the author(s) and do not necessarily reflect the views of the National Science Foundation.

References

Aluja, M. and Prokopy, R.J. (1993) Host odor and visual stimulus interaction during intratree host finding behavior of *Rhagoletis pomonella* flies. *Journal of Chemical Ecology* 19, 2671–2696.

Averill, A.L., Prokopy, R.J., Sylvia, M.M., Connor, P.P. and Wong, T.Y. (1996) Effects of recent experience on foraging in tephritid fruit flies. *Journal of Insect Behavior* 9, 571–583.

Berlocher, S.H. (1995) Population structure of *Rhagoletis mendax*, the blueberry maggot. *Heredity* 74, 542–555.

Berlocher, S.H. (1999) Host race or species? Allozyme characterization of the 'flowering dogwood fly', a member of the *Rhagoletis pomonella* complex. *Heredity* 83, 652–662.

Berlocher, S.H. (2000) Radiation and divergence in the *Rhagoletis pomonella* species group: inferences from allozymes. *Evolution* 54, 543–557.

Berlocher, S.H. and Feder, J.L. (2002) Sympatric speciation in phytophagous insects: moving beyond controversy? *Annual Review of Entomology* 47, 773–815.

Berlocher, S.H. and McPheron, B.A. (1996) Population structure of *Rhagoletis pomonella*, the apple maggot fly. *Heredity* 77, 83–99.

Bierbaum, T.J. and Bush, G.L. (1990a) Host fruit chemical stimuli eliciting distinct ovipositional responses from sibling species of *Rhagoletis* fruit flies. *Entomologia Experimentalis et Applicata* 56, 165–177.

Bierbaum, T.J. and Bush, G.L. (1990b) Genetic differentiation in the viability of sibling species of *Rhagoletis* fruit flies on host plants, and the influence of reduced hybrid viability on reproductive isolation. *Entomologia Experimentalis et Applicata* 55, 105–118.

Bush, G.L. (1966) The taxonomy, cytology, and evolution of the genus *Rhagoletis* in North America (Diptera: Tephritidae). *Bulletin of the Museum of Comparative Zoology* 134, 431–562.

Bush, G.L. (1969) Sympatric host race formation and speciation in frugivorous flies of the genus *Rhagoletis* (Diptera: Tephritidae). *Evolution* 23, 237–251.

Bush, G.L. (1974) The mechanism of sympatric host race formation of the true fruit flies (Tephritidae). In: White, M.J.D. (ed.) *Genetic Mechanisms of Speciation in Insects.* Australia and New Zealand Book Company, Auckland, pp. 3–23.

Dambroski, H.R. and Feder, J.L. (2007) Host plant and latitude-related diapause variation in *Rhagoletis pomonella*: a test for multifaceted life history adaptation on different stages of diapause development. *Journal of Evolutionary Biology* 20, 2101–2112.

Dambroski, H.R., Linn, C. Jr, Berlocher, S.H., Forbes, A., Roelofs, W. and Feder, J.L. (2005) The genetic basis for fruit odor discrimination in *Rhagoletis* flies and its significance for sympatric host shifts. *Evolution* 59, 1953–1964.

Diehl, S.R. and Prokopy, R.J. (1986) Host-selection behavior differences between the fruit fly sibling species *Rhagoletis pomonella* and *R. mendax* (Diptera, Tephritidae). *Annals of the Entomological Society of America* 79, 266–271.

Etges, W.J., Veenstra, C.L. and Jackson, L.L. (2006) Premating isolation is determined by larval rearing substrates in cactophilic *Drosophila mojavensis*. VII. Effects of larval dietary fatty acids on adult epicuticular hydrocarbons. *Journal of Chemical Ecology* 32, 2629–2646.

Feder, J.L. and Bush, G.L. (1989a) A field test of differential host plant usage between two sibling species of *Rhagoletis pomonella* fruit flies (Diptera: Tephritidae) and its consequences for sympatric models of speciation. *Evolution* 43, 1813–1819.

Feder, J.L. and Bush, G.L. (1989b) Gene frequency clines in host races of *Rhagoletis pomonella* (Diptera: Tephritidae) in the midwestern United States. *Heredity* 63, 245–266.

Feder, J.L. and Forbes, A.A. (2008) Host fruit-odor discrimination and sympatric host-race formation. In: Tilmon, K.J. (ed.) *Specialization, Speciation, and Radiation: The Evolutionary Biology of Herbivorous Insects.* University of California Press, Berkeley, California, pp. 101–116.

Feder, J.L., Chilcote, C.A. and Bush, G.L. (1988) Genetic differentiation between sympatric host races of the apple maggot fly *Rhagoletis pomonella*. *Nature* 336, 61–64.

Feder, J.L., Opp, S.B., Wlazlo, B., Reynolds, K., Go, W. and Spisak, S. (1994) Host fidelity is an effective premating barrier between sympatric races of the apple maggot fly. *Proceedings of the National Academy of Sciences of the United States of America* 91, 7990–7994.

Feder, J.L., Roethele, J., Wlazlo, B. and Berlocher, S.H. (1997) Selective maintenance of allozyme differences between sympatric host races of the apple maggot fly. *Proceedings of the National Academy of Sciences of the United States of America* 94, 11417–11421.

Feder, J.L., Berlocher, S.H. and Opp, S.B. (1998) Sympatric host race formation and speciation in *Rhagoletis* (Diptera: Tephritidae): a tale of two species for Charles D. In: Mopper, S. and Strauss, S. (eds) *Genetic Structure in Natural Insect Populations: Effects of Host Plants and Life History*. Chapman and Hall, New York, pp. 408–441.

Feder, J.L., Berlocher, S.H, Roethele, J.B., Dambroski, H., Smith, J.J., Perry, W.L., Gavrilovic, V., Filchak, K.E., Rull, J. and Aluja, M. (2003a) Allopatric genetic origins for sympatric host-plant shifts and race formation in *Rhagoletis. Proceedings of the National Academy of Sciences of the United States of America* 100, 10314–10319.

Feder, J.L., Roethele, J.B., Filchak, K., Niedbalski, J. and Romero-Severson, J. (2003b) Evidence for inversion polymorphism related to sympatric host race formation in the apple maggot fly, *Rhagoletis pomonella. Genetics* 163, 939–953.

Feder, J.L., Xie, X., Rull, J., Velez, S., Forbes, A., Leung, B., Dambroski, H., Filchak, K.E. and Aluja, M. (2005) Mayr, Dobzhansky, and Bush and the complexities of sympatric speciation in *Rhagoletis. Proceedings of the National Academy of Sciences of the United States of America* 102 (Supplement), 6573–6580.

Fein, B.L., Reissig, W.H. and Roelofs, W.L. (1982) Identification of apple volatiles attractive to the apple maggot. *Journal of Chemical Ecology* 8, 1473–1487.

Ferveur, J.F. (2005) Cuticular hydrocarbons: their evolution and roles in *Drosophila* pheromonal communication. *Behavioral Genetics* 35, 279–295.

Filchak, K.E., Roethele, J.B. and Feder, J.L. (2000) Natural selection and sympatric divergence in the apple maggot, *Rhagoletis pomonella. Nature* 407, 739–742.

Forbes, A.A., Fisher, J. and Feder, J.L. (2005) Habitat avoidance: overlooking an important aspect of host specific mating and sympatric speciation? *Evolution* 59, 1552–1559.

Frey, J.E. and Bush, G.L. (1990) *Rhagoletis* sibling species and host races differ in host odor recognition. *Entomologia Experimentalis et Applicata* 57, 123–131.

Linn, C. Jr, Feder, J.L., Nojima, S., Dambroski, H., Berlocher, S.H. and Roelofs, W. (2003) Fruit odor discrimination and sympatric race formation in *Rhagoletis. Proceedings of the National Academy of Sciences of the United States of America* 100, 11490–11493.

Linn, C.E., Dambroski, H.R., Feder, J.L., Berlocher, S.H., Nojima, S. and Roelofs, W. (2004) Postzygotic isolating factor in sympatric speciation in *Rhagoletis* flies: reduced response of hybrids to parental host-fruit odors. *Proceedings of the National Academy of Sciences of the United States of America* 101, 17753–17758.

Linn, C.E. Jr, Dambroski, H., Nojima, S., Feder, J.L., Berlocher, S.H. and Roelofs, W. (2005a) Variability in response specificity of apple, hawthorn, and flowering dogwood-infesting *Rhagoletis* flies to host fruit volatile blends: implications for sympatric host shifts. *Entomologia Experimentalis et Applicata* 116, 55–64.

Linn, C. Jr, Nojima, S. and Roelofs, W. (2005b) Antagonist effects of non-host fruit volatiles on discrimination of host fruit by *Rhagoletis* flies infesting apple (*Malus pumila*), hawthorn (*Crataegus* spp.), and flowering dogwood (*Cornus florida*). *Entomologia Experimentalis et Applicata* 114, 97–105.

McPheron, B.A., Smith, D.C. and Berlocher, S.H. (1988) Genetic differences between host races of *Rhagoletis pomonella. Nature* 336, 64–66.

Mayr, E. (1963) *Animal Species and Evolution.* Harvard University Press, Cambridge, UK.

Michel, A.P., Rull, J., Aluja, M. and Feder, J.L. (2007) The genetic structure of hawthorn infesting *Rhagoletis pomonella* populations in Mexico: implications for sympatric host race formation. *Molecular Ecology* 16, 2867–2878.

Moericke, V., Prokopy, R.J., Berlocher, S.H. and Bush, G.L. (1975) Visual stimuli eliciting attraction of *Rhagoletis pomonella* (Diptera: Tephritidae) flies to trees. *Entomologia Experimentalis et Applicata* 18, 497–507.

Nojima, S., Linn, C. Jr, Morris, B., Zhang, A. and Roelofs, W. (2003a) Identification of host fruit volatiles from hawthorn (*Crataegus* spp.) attractive to hawthorn-origin *Rhagoletis pomonella* flies. *Journal of Chemical Ecology* 29, 321–336.

Nojima, S., Linn, C. Jr and Roelofs, W. (2003b) Identification of host fruit volatiles from flowering dogwood (*Cornus florida*) attractive to dogwood-origin *Rhagoletis pomonella* flies. *Journal of Chemical Ecology* 29, 2347–2357.

Olsson, S.B., Linn, C.E. Jr and Roelofs, W.L. (2006a) The chemosensory basis for behavioral divergence involved in sympatric host shifts. I. Characterizing olfactory receptor neuron classes responding to key volatiles.

Journal of Comparative Physiology A 192, 279–288.

Olsson, S.B., Linn, C.E. Jr and Roelofs, W.L. (2006b) The chemosensory basis for behavioral divergence involved in sympatric host shifts. II. Olfactory receptor neuron sensitivity and temporal firing pattern to individual key host volatiles. *Journal of Comparative Physiology A* 192, 289–300.

Opp, S., Ziegner, B.J., Bui, N. and Prokopy, R.J. (1990) Factors influencing estimates of sperm competition in *Rhagoletis pomonella* (Diptera: Tephritidae). *Annals of the Entomological Society of America* 83, 521–526.

Prokopy, R.J. (1966) Artificial oviposition devices for apple maggot *Rhagoletis pomonella*. *Journal of Economic Entomology* 59, 231–232.

Prokopy, R.J. (1967) Factors influencing effectiveness of artificial oviposition devices for apple maggot. *Journal of Economic Entomology* 60, 950–955.

Prokopy, R.J. (1968) Visual responses of apple maggot flies, *Rhagoletis pomonella* (Diptera, Tephritidae): orchard studies. *Entomologia Experimentalis et Applicata* 11, 403–422.

Prokopy, R.J. (1972) Evidence for a marking pheromone deterring repeated oviposition in apple maggot flies. *Environmental Entomology* 1, 326–332.

Prokopy, R.J. (1975) Mating behavior in *Rhagoletis pomonella* (Diptera: Tephritidae) Part V. Virgin female attraction to male odor. *Canadian Entomologist* 107, 905–908.

Prokopy, R.J. (1977) Attraction of *Rhagoletis* flies (Diptera: Tephritidae) to red spheres of different sizes. *Canadian Entomologist* 109, 593–596.

Prokopy, R.J. and Bush, G.L. (1972) Mating behavior in *Rhagoletis pomonella* (Diptera: Tephritidae). III. Male aggregation in response to an arrestant. *Canadian Entomologist* 104, 275–283.

Prokopy, R.J. and Bush, G.L. (1973a) Mating behavior in *Rhagoletis pomonella* (Diptera: Tephritidae). IV. Courtship. *Canadian Entomologist* 105, 873–891.

Prokopy, R.J. and Bush, G.L. (1973b) Oviposition by grouped and isolated apple maggot flies. *Annals of the Entomological Society of America* 66, 1197–1200.

Prokopy, R.J. and Bush, G.L. (1973c) Ovipositional responses to different sizes of artificial fruit by flies of the *Rhagoletis pomonella* species group. *Annals of the Entomological Society of America* 66, 927–929.

Prokopy, R.J. and Papaj, D.R. (1989) Can ovipositing *Rhagoletis pomonella* females (Diptera: Tephritidae) learn to discriminate among different ripeness stages of the same host biotype? *Florida Entomologist* 72, 489–494.

Prokopy, R.J. and Reynolds, A.H. (1998) Ovipositional enhancement through socially facilitated behavior in *Rhagoletis pomonella* flies. *Entomologia Experimentalis et Applicata* 86, 281–286.

Prokopy, R.J. and Roitberg, B.D. (1984) Foraging behavior of true fruit flies. *American Scientist* 72, 41–49.

Prokopy, R.J., Moericke, V. and Bush, G.L. (1973) Attraction of apple maggot flies to odor of apples. *Environmental Entomology* 2, 743–749.

Prokopy, R.J., Averill, A.L., Cooley, S.S. and Roitberg, C.A. (1982) Associative learning in egg laying site selection by apple maggot flies. *Science* 218, 76–77.

Prokopy, R.J., Aluja, M. and Green, T.A. (1987) Dynamics of host odor and visual stimulus interaction in host finding behavior of apple maggot flies. In: Labeyrie, V., Fabres, G. and Lachaise, D. (eds) *Insects–Plants: Proceedings of the Sixth International Symposium on Insect–Plant Relationships*. Junk, Hingham, Massachusetts, pp. 161–166.

Prokopy, R.J., Diehl, S.R. and Cooley, S.S. (1988) Behavioral evidence for host races in *Rhagoletis pomonella* flies. *Oecologia* 76, 138–147.

Prokopy, R.J., Bergweiler, C., Galarza, L. and Schwerin, J. (1994) Prior experience affects the visual ability of *Rhagoletis pomonella* flies (Diptera: Tephritidae) to find host fruit. *Journal of Insect Behavior* 7, 663–677.

Roitberg, B.D., van Lenteren, J.C., van Alphen, J.J. and Prokopy, R.J. (1982) Foraging behavior of *Rhagoletis pomonella* – a parasite of hawthorn (*Crataegus viridis*) in nature. *Journal of Animal Ecology* 51, 307–326.

Rull, J., Aluja, M., Feder, J.L. and Berlocher, S.H. (2006) Distribution and host range of hawthorn-infesting *Rhagoletis* (Diptera: Tephritidae) in Mexico. *Annals of the Entomological Society of America* 99, 662–672.

Schwarz, D., McPheron, B.A., Hartl, G.B., Boller, E.F. and Hoffmeister, T.H. (2003) A second case of genetic host races in *Rhagoletis*? – A population genetic comparison of sympatric host populations in the European cherry fruit fly, *Rhagoletis cerasi*. *Entomologia Experimentalis et Applicata* 108, 11–17.

Schwarz, D.B., Matta, M.N., Shakir-Botteri, L. and McPheron, B.A. (2005) Host shift to an invasive plant triggers rapid animal hybrid speciation. *Nature* 436, 546–549.

Sheldon, S. and Jones, K.N. (2001) Restricted gene flow according to host plant in an herbivore feeding on native and exotic watermilfoils (*Myriophyllum*: Haloragaceae). *Journal of Plant Science* 162, 793–799.

Smith, D.C. and Prokopy, R.J. (1980) Mating behavior of *Rhagoletis pomonella* (Diptera: Tephritidae). VI. Site of early season encounters. *Canadian Entomologist* 112, 585–590.

Smith, D.C. and Prokopy, R.J. (1981) Seasonal and diurnal activity of *Rhagoletis mendax* flies in nature. *Annals of the Entomological Society of America* 74, 462–466.

Smith, D.C. and Prokopy, R.J. (1982) Mating behavior of *Rhagoletis mendax* (Diptera: Tephritidae) flies in nature. *Annals of the Entomological Society of America* 75, 388–392.

Williams, J.W., Shuman, B.N., Webb, T., Bartle, P.J. and Leduc, P.L. (2004) Late-quaternary vegetation dynamics in North America: scaling from taxa to biomes. *Ecological Monographs* 74, 309–334.

Witzgall, P., Ansebo, L., Yang, Z., Angeli, G., Sauphanor, B. and Bengtsson, M. (2005) Plant volatiles affect oviposition by codling moths. *Chemoecology* 15, 77–83.

Xie, X., Michel, A.P., Schwarz, D., Rull, J., Velez, S., Forbes, A.A., Aluja, M. and Feder, J.L. (2007) Hawthorn-infesting populations of *Rhagoletis pomonella* in Mexico and speciation mode plurality. *Evolution* 61, 1091–1105.

Zhang, A., Linn, C. Jr, Wright, S., Prokopy, R., Reissig, W.H. and Roelofs, W.L. (1999) Identification of a new blend of apple volatiles attractive to the apple maggot, *Rhagoletis pomonella*. *Journal of Chemical Ecology* 25, 1221–1232.

3 Functional and Behavioural Ecology of Tree-Fruit Pests: The Four Fs of Fruit Flies (Diptera: Tephritidae)

Bernard D. Roitberg,[1] Carol R. Lauzon,[2]
Susan B. Opp[2] and Dan R. Papaj[3]

[1]*Department of Biological Sciences, Simon Fraser University, Burnaby, British Columbia, Canada;* [2]*Department of Biological Sciences, California State University, East Bay, Hayward, California, USA;* [3]*Department of Ecology and Evolutionary Biology, University of Arizona, Tucson, Arizona, USA*

Introduction – The Four Fs of Fruit Flies

Nearly all living animals must, on an almost daily basis, deal with the so-called 'Four Fs of Life': feeding, fighting, fleeing and reproducing. And so it is for the tephritid fruit fly, the second true love (Buss, 2006) of Ron Prokopy's life. Prokopy realized that true fruit flies could be excellent model systems for studying the four Fs for a variety of reasons that we will extol, not the least of which was the opportunity to study them under natural or semi-natural settings, a key criterion in Ron's research programme (Prokopy and Roitberg, 2007).

Each of the four Fs is a key contributor to Darwinian fitness and, as such, whole fields have developed to study them. Yet, they are linked so inextricably that they cannot be studied in isolation. Prokopy realized this fact and developed his research programmes in a multifaceted manner to take this into account. Our goal in this chapter is to show how fruit flies can be used to explore fundamental issues in evolutionary biology.

There are a number of ways by which one could study the four Fs, from molecular to population (or even community), proximate and ultimate or basic and applied. We have chosen to elucidate these issues as problems in functional and behavioural ecology and, in so doing, demonstrate the value of tephritid flies as model systems. We then offer some suggestions on how our approach might be applied to managing fruit flies in agroecosystems.

By functional ecology, we mean that we will emphasize the physiological, anatomical and life history characteristics of tephritids that determine their role in agroecosystems, and by behavioural ecology we mean that we will discuss the ecological and evolutionary basis of fly behaviour that allows them to adapt to their environment (both their internal and external states). Of course, there is great overlap in these two approaches and we will emphasize their congruence in understanding tephritid biology.

And what of the four Fs? It is well understood that as heterotrophs, tephritids must obtain energy and essential nutrients from external sources (Feeding). We will show how tephritids accomplish this feat through a symbiotic relationship with bacteria.

Tephritid fruit flies tend to live in close association with their fruit-bearing hosts wherein encounters with conspecifics occur frequently and competition for resources can be intense. We will discuss how female flies compete for these resources both directly (Fighting) and indirectly (via pheromones).

Individual fruit flies frequently find themselves at local sites where resources vary in quality and are often in short supply. Thus, a resource exploitation strategy for these individuals is necessary. In fact, a two-pronged strategy has evolved. First, we will consider how host selection (i.e. flies are finicky with regard to reproduction – the fourth F) varies as a function of ecological condition and physiological state and, second, we will elucidate the habitat-fleeing F through analysis of flight dynamics in fruit flies.

This work arises as a group effort from four scientists with broad yet diverse backgrounds in entomology, evolution and ecology. What we have in common and what we attempt to share here is our functional perspective on fruit fly life history and, furthermore, an interest in applying our perspective to agricultural problems.

F is for Feeding

An association with bacteria

As noted at the outset, tephritids must obtain essential resources to fuel somatic and gametic functions. Here, key resources are provided primarily by bacteria. Most of the work on tephritid–bacteria interactions historically has been the study of the attraction of certain pest tephritids to bacteria in culture. The catabolism of proteins in the culture medium by the bacteria was presumed to be the means by which attractive odours, such as ammonia (Gow, 1954), arose. Since that time, ammonia has been a useful component in several lure formulations. Other attractive odours from bacteria, alone or in combination with ammonia, have been shown to attract fruit flies (Drew and Faye, 1988; Robacker and Flath, 1995; Epsky *et al.*, 1997, 1998; Robacker and Bartelt, 1997). These findings have fostered continued interest in bacteria as potential sources of new odours that would improve the efficiency of lures and traps. This work continues today (i.e. Robacker and Lauzon, 2002; Robacker *et al.*, 2004).

The first functional study on the life history of frugivorous pest Tephritidae was conducted in the early 1930s, when Allen and colleagues (1932, 1934) described the association between *Phytomonas (Pseudomonas) melopthora* and *Rhagoletis pomonella* (Walsh) (Diptera: Tephritidae), the apple maggot fly.

The bacterium presumably was inoculated into the oviposition site, where it grew and caused rot of apple tissue. The identity of the bacterium was soon challenged (Luthy *et al.*, 1983) and this notion was supported later by the findings of others who explored tephritid–microbe interactions (i.e. Howard *et al.*, 1985; Lloyd *et al.*, 1986; Lauzon, 1991; MacCollom *et al.*, 1992). Although the bacterium was not identified correctly, its presence nevertheless caused fruit rot and loss.

Bacteria within two genera, *Enterobacter* and *Klebsiella*, have been isolated consistently from several tephritid species and are accepted fruit-fly symbionts. Bacteria within these genera are Gram-negative, facultative anaerobe rods commonly found to inhabit plants, soil and water, and are considered in most cases to be enteric bacteria (Brenner, 1992). They are both in the order Enterobacteriales, family Enterobacteriaceae. They have been isolated from all life stages of tephritid species that have been surveyed (Hagen, 1966; Jang and Nishijima, 1990; Lauzon, 1991; Potter, 2001; Marchini *et al.*, 2002; Lauzon and Potter, 2009), host-plant structures (Drew *et al.*, 1983; Drew and Lloyd, 1987; Lauzon, 1991) and in natural food sources (Hendrichs *et al.*, 1993) such as bird faeces (Prokopy *et al.*, 1993; Lauzon *et al.*, 2000). Recent taxonomic structure places strains of *Enterobacter agglomerans* into the genus *Pantoea*. To avoid confusion, and given the numerous historical references of *E. agglomerans* and tephritids, *P. agglomerans* will be referred to as *E. agglomerans* in this chapter.

The adaptive significance of bacteria

The intimacy and/or adaptive nature of any physiological relationship that existed between bacteria and pest tephritids was not addressed in depth until Lauzon and Sjogren (1998, 2000) found that attraction to *E. agglomerans* was linked to the ability of bacterium to degrade uric acid, a major source of nitrogen for fruit flies in nature. Fruit flies lack the capability to catabolize uric acid, and strains of *E. agglomerans* that lacked the ability to utilize uric acid did not attract fruit flies. Therefore, it appears that fruit flies seek out and acquire bacteria horizontally to degrade uric acid in the adult fly midgut to a usable form.

Further use of bacteria to afford nutrients occurs when a female lays her eggs into host tissue. *Enterobacter agglomerans* and *Klebsiella pneumoniae* are inoculated into the oviposition site (Lauzon, 1991; Potter, 2001). This has been shown by culturing oviposition sites and through visual documentation of these bacteria as biofilms on the egg surface (Lauzon *et al.*, 2009). Visual proof was achieved through the transformation of *E. agglomerans* and *K. pneumoniae* to contain fluorescent proteins and the subsequent ingestion of these bacteria by Medflies, *Ceratitis capitata* (Wiedemann) (Diptera: Tephritidae). On examination, female Medflies contained eggs that had a biofilm of the fluorescent bacteria that existed at the apical end of the eggs, the end that first entered host-fruit tissue during oviposition. How ingested bacteria migrate to the egg is unknown; however, these findings support the direct inoculation of symbiont bacteria into host fruit when eggs are laid. In keeping with the notion that fruit flies benefit from this relationship, Marchini *et al.* (1997) found antibacterial peptides on the laid egg chorion of Medflies that were antagonistic to *Escherichia coli*, a bacterium closely related to *E. agglomerans* and *K. pneumoniae*. Most antibiotics that affect *E. coli* also affect bacterial relatives. This finding suggests efficient selection toward the preservation of *E. agglomerans* and *K. pneumoniae*.

Bacterial growth and rot that follow oviposition provide eggs with available nitrogen (Drew, 1988; MacCollom *et al.*, 2007, unpublished results), which comes either through consumption of microbial waste products, microbes themselves, or through the nitrogen-fixing ability of both *Enterobacter* and *Klebsiella* spp. (Brenner, 1992; Murphy *et al.*, 1994; Behar *et al.*, 2005). As eggs mature into voracious larvae, the larvae likely consume bacteria and fruit rot for nutritional purposes (Miyazaki *et al.*, 1968; Drew *et al.*, 1983), but also, bacteria assist larvae with managing the physical confinements of the internal host-fruit environment, such as softening fruit tissue. It is not known if bacteria affect the development of a maturing fly in the pupal stage; however, visual inspection reveals bacteria in the egg micropyle. This would suggest that bacteria have an opportunity to become intracellular or exist in close proximity to differentiating cells. Few bacteria, however, appear to inhabit internal pupal structures (Lauzon *et al.*, 2009) and these are likely to seed the emergent adult fly with important alimentary canal bacteria.

On emergence, the adult fruit-fly gut is not fully developed and lacks the numerous bacteria typically observed in the gut of 18-hour-old adults (Lauzon and Potter, 2009). In nature, newly eclosed fruit flies generally remain relatively inactive during their first 24 hours of life. Perhaps this is to allow their gut to develop prior to ingestion of food and other noxious/toxic compounds, and for bacteria carried over from the pupal stage to grow and increase in number. Once a peritrophic membrane is formed along with a companion biofilm of beneficial bacteria, the fly appears ready to begin to digest food and other compounds properly. The cycle of horizontal and vertical transmission begins again.

This mixed pattern of symbiont acquisition probably reflects both a need to acquire the bacteria and the complex means in which the bacteria are acquired and transferred through tephritid life stages, and may be associated with differences in nutritional requirements among life stages. While the highest density of symbionts are associated with adult flies, the lower numbers of bacteria generally found in larval stages (Lauzon, unpublished results) are likely due to a lower requirement for nitrogen at this stage and the relatively homogeneous environment within the fruit (Lauzon *et al.*, 2009). In addition, the turbulent larval gut, as a result of frequent food passage and moulting, may diminish the ability of symbionts to establish and grow. At this stage, transmission efficiency is diminished and this reduction in bacterial numbers may explain why adults are attracted to sources that contain these bacteria. This replenishment of certain bacteria is probably based on need. Adults require high levels of protein for sexual maturation and egg production. Attraction to food, particularly faecal matter, results in the acquisition of nutrients and symbionts. This behaviour enhances symbiont transmission efficiency. In addition, high numbers of symbionts likely assist adults with other catabolic activities, such as detoxification of the noxious and/or toxic compounds they encounter while feeding on a variety of substrates in nature (Lauzon *et al.*, 2009).

In recent work, Lauzon *et al.* (unpublished results) suggest that a third bacterial species may be an important symbiont in pest tephritids. The bacterium, a member of the genus *Pseudomonas*, differs from *Enterobacter* and *Klebsiella* spp. in that it is a non-enteric, aerobic rod. Typically inhabitants of plants, *Pseudomonas* spp. are keen cyclers of nitrogen in the environment. We hypothesize that *Pseudomonas* spp. help regulate pH in the fly gut by utilizing excess ammonia that may occur from the degradation of urea by *Klebsiella* spp. Studies are currently under way in our laboratory to test this hypothesis.

Risk of infection

Foraging for food frequently entails risk of predation (Lima and Dill, 1990). For organisms such as fruit flies that forage for microbes, risk takes a slightly different path. Thus, some mention of fruit flies and their relationship with pathogens is required to complete the symbiotic story. Fruit flies consume and are exposed to insect pathogens such as *Serratia marcescens* (Steinhaus, 1959; Grimont and Grimont, 1978; Lauzon *et al*., 2003), *P. aeruginosa* (Jacques *et al*., 1969), Cricket Paralysis Virus (Manousis and Moore, 1987), reoviruses and reo-like viruses (Plus *et al*., 1981a,b; Lauzon *et al*., 2007, unpublished results), *Wolbachia* (Kittayapong *et al*., 2000; Riegler and Stauffer, 2002; Selivon *et al*., 2002; Zabalou *et al*., 2004; Rocha *et al*., 2005) and microsporidia (Fujii and Tamashiro, 1972). Relatively little information exists in the area of tephritid pathology because, in part, most microorganisms cannot be used as biocontrol agents due to their effects on non-target species. Products produced by these organisms, however, show some potential for use. For example, Lauzon *et al*. (2003) showed the lethality of products produced by *S. marcescens*, a bacterium isolated originally from dead and diseased apple maggot flies. More information needs to be acquired on pathogens because mass-rearing programmes must include plans to control and/or eliminate these microorganisms within the facility. During studies that showed that a diet for Medflies that included beneficial symbionts improved the gut of irradiated flies used in the sterile insect technique (SIT) (Lauzon and Potter, 2009) and their mating performance (Niyazi *et al*., 2004), Lauzon *et al*. (unpublished results) also found that bacteria that resided typically in the facility diet were eliminated or did not become established when the beneficial symbionts were present. This dynamic decreases the need for antimicrobial use in mass rearing and may reflect a protective mechanism exerted by beneficial symbionts for fruit flies; yet another reason for acquisition.

The other side of the association

It is not known what beneficial tephritid symbionts gain from their close and deliberate association with fruit flies. The likely explanation resides with the fact that in nature, bacteria exist typically in a state of starvation (Groat *et al*., 1986; Moriarty and Bell, 1993; Huisman and Kolter, 1994; Tappe *et al*., 1999). Competition for resources, for example, on the phylloplane is keen (Burrage, 1971; Lindow, 1991) due to the diverse and abundant microbial populations that reside there. Internal extracellular bacteria, such as *E. agglomerans* and *K. pneumoniae*, associated with fruit flies receive nearly constant nutrient availability and, due to the limited number of bacterial species that inhabit the tephritid alimentary canal, competition is reduced. The bacteria are equipped genetically to manage nutrient limitation (Barrow *et al*., 1996) and thus are capable of surviving the fruit-fly pupal stage.

Mixed patterns of symbiont transmission behaviour in insects are not uncommon (Agnew *et al*., 2003) and are likely to reflect a host's adaptation to various ecological niches. Multiple means and replenishments of symbionts in pest tephritids suggests that these bacteria serve important roles in fruit-fly nutrition and help overcome biological constraints encountered by these insects in nature. Knowledge and comprehension of symbiont inheritance and/or the relationship tephritids possess with bacteria should lead to a more complete understanding of pest tephritid dispersal and biology and management.

F is for Fighting

If one were to survey fruit for ovipositions by some focal tephritid, one would often find that the spatial distributions of such ovipositions are not random. There are at least two reasons why this might be so: (i) fruit quality varies in some non-random manner and flies discriminate among fruit (see section on Finickiness); and/or (ii) fly competition (Fights) for fruit forces the flies to exploit fruit in some non-random manner. Do females fight for host fruit and, if so, what are the outcomes of such interactions? The answers to these questions are not so simple because females employ various mechanisms to acquire and defend resources. Thus, this section is broken into three parts: (i) a primer on contest theory; (ii) a section on direct fighting by females; and (iii) a section on chemically mediated competition for hosts in female fruit flies.

A primer on contest theory

Imagine that a female fruit fly alights on a fruit, assesses it to be of sufficient quality to nourish her young and prepares to lay a clutch of eggs in that fruit. Now, imagine that a conspecific female subsequently alights on that same fruit. What is likely to happen? The field of behavioural ecology and, more specifically, game theory (Maynard-Smith, 1982), has considered this issue as one in which two contestants vie for a single resource (the fruit). The key to game theory is that net pay-off for any given behavioural strategy depends not just on the expression of that strategy by the focal individual but also on the form and frequency of strategies expressed by competitors. Cheating, in particular, has frequency-dependent pay-offs. For example, the males of many species of fruit flies assemble at leks, where females evaluate those males and then choose a mate. An alternative male mating tactic (a kind of cheating known as the sneaker strategy) is to settle just outside the lek and attempt to induce females forcibly into mating before they enter the lek. Although we have no evidence for the sneaker strategies in lek-forming tephritids, they are common in other taxa (Taborsky, 1994). Such a tactic will have negative frequency-dependent pay-offs because the more common the alternative tactic: (i) the fewer leks there will be to exploit; and (ii) the greater the competition among cheaters for females. The degree of frequency dependence will be influenced by ecological circumstance, when, for example, habitat structure will determine the number of possible lek sites. Below, we focus on female–female interaction and the possible outcomes of such interactions according to game theory.

Fighting flies

War of attrition

Here, two (or more) individuals attempt to wait each other out wherein the one that out-waits the other accrues the resource. These kinds of contests could be energy-based if flies signalled their presence to one another continually and such signals were energetically costly. Wars of attrition are most likely to evolve when competition is exploitation-based wherein resources are divisible; this does not seem to describe single resource-based contests (e.g. individual host fruits). We are not aware of any publications demonstrating wars of attrition between females of any species of fruit flies.

Fierce contests

Here, females fight each other until one of the combatants suffers such severe injuries that it must withdraw from the contest; death of one combatant frequently ensues. Enquist and Leimar (1990) show that such contests are likely to evolve when expected future reproductive success (V_0) is very low relative to the value of the current resource (V). Under such conditions, it pays for almost any individual, regardless of physical prowess, to engage in deadly battles; after all, there is little or nothing to lose. For most species of fruit flies, other opportunities for oviposition generally exist, thus $V_0 >> 0$ and fierce fighting is not favoured. In fact, we are not aware of mortal combats among adult female tephritids. By contrast, mortal combats are common in tephritid larvae. Why? For most larvae, conceding a contest and exiting a fruit means that death is near certain or, in the lingo of game theory, $V_0/V \approx 0$.

Uncorrelated contests

Here, individuals use some cue that is *not* correlated with fighting ability to determine the winner. For example, combatants use a simple rule such as 'resident always wins and intruder always loses', the so-called bourgeois rule: fight when owner, retreat when intruder (Maynard-Smith and Parker, 1976). In fact, residents frequently do win contests with intruders (e.g. Medflies – Papaj and Messing, 1998) but before one accepts these studies as critical support for the bourgeois strategy, the details associated with resident victory must be considered. After all, residents might be owners because they are good fighters! In the case of the aforementioned Medfly study, outcomes were strongly dependent on resident age and activity at the time of resident arrival, i.e. a simple, arbitrary 'resident wins' rule does not apply. Similarly, in oriental fruit flies, *Dacus dorsalis*

(Hednel) (Diptera: Tephritidae) and Anastrepha flies, protracted fights occur and size and experience, respectively, are the primary determinants of victory (Shelly, 1999; Diaz-Fleischer and Aluja, 2003). All of these facts argue against a simple arbitrary rule for settling contests.

Ritualized contests with continuous assessment

Here, individuals engage in ritualized, non-injurious displays wherein they assess the fighting of their opponent (e.g. pushing contests in beetles – Eberhard, 1979). Many of these ritualized, repetitive displays provide honest signals of ability, thus allowing combatants to obtain multiple samples and increase accuracy of assessment without fierce fighting. For example, in both oriental and Queensland fruit flies, *D. tryoni* (Froggatt) (Diptera: Tephritidae), nearly every agonistic encounter at fruits proceeds wherein the winner is determined without any physical contact (Shelly, 1999).

Another interpretation of fly response to competitors, which has its basis more in the psychological than behavioural ecology literature, is coined 'social facilitation', wherein individuals demonstrate a strengthening in response to a stimulus in the presence of conspecifics. For example, *R. basiola* (Osten Sacken) (Diptera: Tephritidae) females were more likely to accept recently attacked fruit in the presence of a conspecific than in its absence (Robertson *et al.*, 1995). Prokopy was particularly interested in this effect and devoted many studies towards unravelling this complex phenomenon. He and colleagues were able to demonstrate increased tendency to alight and oviposit in fruit in groups of apple maggot flies, Medflies, Queensland fruit flies and Mexican fruit flies, *Anastrepha ludens* (Loew) (Diptera: Tephritidae) (Prokopy and Reynolds, 1998; Prokopy *et al.*, 1999, 2000; Diaz-Fleischer and Aluja, 2003; but see Dukas *et al.*, 2001). It is not clear why social facilitation should be displayed consistently by tephritids (or any insects – see Prokopy and Roitberg, 2001) and as Robertson *et al.* (1995) and Diaz-Fleischer and Aluja (2003) point out, there are simple, non-psychological explanations. The former suggest that the presence of a conspecific provides information regarding the quality of the resource and the latter suggest that competition is present and requires a change in tactics. In fact, as described above, many of these interactions involve contest-like displays. The jury is out on this one.

Chemically mediated exploitation of hosts

In a large number of fruit flies, oviposition is usually followed by the individual dragging her ovipositor over the fruit surface. In the early 1970s, working with *R. pomonella*, Prokopy (1972) showed that concomitant with dragging was release of a species-specific host-marking pheromone. Since that time, oviposition pheromones have been reported for more than 20 species of fruit flies, in all three tephritid subfamilies – Dacinae, Trypetinae and Tephritidnae – though they are virtually absent from the Bactrocerines, which account for a large number of important tropical tephritid pests (but see Girolami *et al.* (1981) for *Bactrocera oleae* (Gmelin)). Originally, such marks were called oviposition-deterring pheromones, or ODPs, because females that encountered these marked fruit often rejected them. A decade or so later, Roitberg and Mangel (1988) cautioned against the use of the term, oviposition-deterring pheromone, and suggested the use of the more general term, host-marking pheromone, or HMP.

What is wrong with the term ODP and why should we refrain from using it? A little behavioural ecology can once again provide insights. Suppose we begin with a population of flies that does not employ marks of any kind. Now, suppose that a mutant fly arises that produces an ODP. What is likely to happen? Would that mutant's genes spread at the expense of the 'normals'? Assuming that production and release of ODPs has a cost, what are the benefits that would offset such costs? The primary benefit of an ODP would be the presumed reduced fitness of females that were prevented from ovipositing by the ODP. In other words, in this scenario, ODPs are spiteful pheromones and it turns out that typically spite is not adaptive in most systems (Rothstein, 1979). In fact, the analyses of Roitberg and Mangel (1988) and Roitberg and Prokopy (1987) show that the benefits of spiteful oviposition pheromones, especially when rare (i.e. as a rare mutation), are so slight that they cannot offset their costs of production and dissemination.

So, why do some tephritids employ oviposition-marking pheromones? The answer is twofold. First and foremost, individuals benefit from marking pheromones by reducing possible competition among their own offspring. In other words, females mark fruit for themselves and not necessarily for others. Second, marking pheromones can also function as 'public information' wherein they provide an honest signal (Nufio and Papaj, 2001) that says in effect, 'One or more of my offspring is present and by definition, it/they must be older than any offspring that you might deposit.' That latter section of this message is important because, as noted above, tephritid larvae may engage in intense fighting with the advantage going to older, larger larvae.

Other aspects of HMPs in tephritids support the notion of a dual private–public information scenario. First, these pheromones are relatively cheap with limited half-life, as one might expect if they function primarily for a depositor who will garner the greatest benefit soon after deposition (i.e. due to area-restricted foraging following oviposition) (Roitberg and Mangel, 1988). Second, we might expect females to invest more when the pay-off for marking increases. This is, in fact, the case (e.g. Nufio and Papaj, 2004a). Third, if marking pheromones do function as public information, then they should be open to exploitation; they are, as demonstrated by parasitoids that employ such marks as host-finding kairomones (e.g. Hoffmeister *et al.*, 2000).

Finally, if marking pheromones function as sources of information, then we should expect that female fruit flies would alter their response to such information, depending on ecological (external) and physiological (internal) conditions or states. For example, according to optimal diet theory (Speirs *et al.*, 1991), if host fruit are rare, females should ignore marks. This is the case (e.g. Mangel and Roitberg, 1989). Similarly, females with large egg loads should be insensitive to marks in the same manner that a hungry forager will ignore distasteful substances on their food; they are (van Randen and Roitberg, 1996). Moreover, game theory tells us that response to marking pheromones should differ in the presence and absence of conspecifics due to different fitness implications from sibling versus intraspecific competitive interactions; in fact, an *R. basiola* female is more likely to accept a marked fruit when she finds herself on a fruit with an ovipositing conspecific (Robertson *et al*., 1995).

In the concluding section, we deal specifically with how one might exploit fruit-fly 'fighting' behaviour in a pest management context. For example, one might apply marking pheromones to fruit trees to reduce oviposition rates, as has been done with cherry fruit flies (*R. cerasi* L. (Diptera: Tephritidae)) (Aluja and Boller, 1992). As we noted above, however, marking pheromone efficacy at reducing oviposition likely will not be a linear function of dose or coverage. Fly response to marking pheromone, synthetic or natural, will be impacted by ecological circumstance and fly perception of competitor density, i.e. when high-quality oviposition sites are rare and competitors common, flies will ignore marks (Roitberg and Prokopy, 1983).

F is for Fleeing

If the Bard had been an entomologist, he would have written 'To flee or not to flee: that is the question'. The evolution of flight has enabled insects to become highly successful organisms, able to flee from threats or challenging situations and to find resources scattered far and wide. Although, how wings and flight evolved in insects is a matter of some debate (see, for example, Brodsky, 1994; Kingsolver and Koehl, 1994; Daly *et al*., 1998), the adaptive value of wings and flight is unquestionable. Nevertheless, for any chosen group of insects, behavioural and functional aspects of flight are often poorly understood and tephritid flies are no exception. Here, we will examine some of the factors or state variables, internal and external, that may influence flight in fruit flies and review patterns of flight that have been recorded for selected species. An excellent historical review of studies of tephritid movements has been provided by Diaz-Fleischer and Aluja (2001) and will not be repeated here. Instead, we will attempt to address selected questions that include: Why does a fruit fly flee and, on fleeing, where and how far does a fruit fly fly?

Why does a fruit fly flee?

Asking a 'why' question means to ask about the adaptive significance or ultimate function of a given trait. In the case of fruit flies, the adaptive significance of fleeing a given site is to improve resource acquisition and/or avoid natural enemies, i.e. abandon the current site in search of a better one. Below, we describe the factors that facilitate emigration.

Once within a host patch, factors that include resource abundance and quality (including presence or absence of HMP), physical patch structure and the physiological state of the fly will serve to determine whether a fly continues to exhibit localized, appetitive behaviours or engages in longer distance dispersal flight and subsequently flees a host. In tephritids, many studies have examined factors that attract tephritids to hosts from some distance, including food odours, plant (non-host and/or host) odours and visual stimuli, male sex pheromones and volatiles from oviposition sites (as outlined in Prokopy, 1983) and those factors that influence foraging and patch residence time (e.g. Aluja, 1993), but fewer have sought to determine what factors will elicit dispersive flight *away* from a host. Fewer still have examined the difficult question of where a fruit fly goes when it leaves a host plant.

Clearly, throughout the life of a fly, feeding and flying (i.e. foraging) are closely linked activities. Studies have shown that fruit flies may spend considerable time off the host, particularly early in the season, presumably feeding on substrates to support reproductive development (Fletcher and Kapatos, 1981; Girolami *et al.*, 1983; Opp and Prokopy, 1987). It is this propensity of fruit flies to fly in search of food sources that has resulted in the use of proteinaceous baits in combination with insecticides for fruit-fly control. Availability of resources such as food, in addition to fly physiological state, have also been shown to affect the time spent by a fly on intra-tree movements and foraging (Malavasi and Prokopy, 1992; Prokopy *et al.*, 1994; Diaz-Fleischer and Aluja, 2001 – see section on biorational management).

While nitrogenous foods are required for reproduction in most adult tephritids (Cangussu and Zucoloto, 1992; Tsiropoulos, 1992; Warburg and Yuval, 1996; Blay and Yuval, 1997; Simmons and Bradley, 1997; Drew and Yuval, 2000; Shelly and Kennelly, 2002; Yuval *et al.*, 2002), carbohydrates are generally the primary fuels used for flight in Diptera (Remund *et al.*, 1977; Cooter, 1982; Rankin and Burchsted, 1992; Drew and Yuval, 2000; Zermeño, 2005). Foraging activities that take tephritids away from host plants are poorly understood, but are presumed to be related to the need for specific nutrients necessary to support survival, reproduction and flight.

The tendency of fruit flies to flee a host will also be influenced by extrinsic factors such as predation pressure, competition for resources and abiotic factors. Predators of fruit flies may include wasps (Vespidae), dragon- and damselflies (Odonata), mantids and spiders (Hendrichs and Hendrichs, 1990, 1998; Hendrichs *et al.*, 1991, 1994). While predator avoidance has been documented in the field, the distance tephritids move to escape predators is unclear. Competition, especially for suitable oviposition sites, also has been documented clearly to influence female emigration from host trees in search of new oviposition sites (Prokopy and Roitberg, 1984; Hendrichs and Hendrichs, 1990).

Abiotic factors, particularly temperature, may regulate the behaviours of tephritids as well (Yuval and Hendrichs, 2001). High temperatures have been shown to inhibit behaviours, including movements, in tephritids such as *C. capitata* (Hendrichs and Hendrichs, 1990; Cayol, 1999) and *B. oleae* (Kapatos and Fletcher, 1983; Xingeng Wang and Marshall W. Johnson, personal communication). In addition, as temperatures rise, flies may move toward cooler parts of host trees (Hendrichs *et al.*, 1990; Baker and van der Valk, 1992) or may change activity patterns to cooler times of day (Hendrichs and Hendrichs, 1990; Cayol, 1999).

On fleeing, where and how far does a fruit fly fly?

Flight is an energetically costly activity (Beenakkers *et al.*, 1984; Candy *et al.*, 1997; Dudley, 2000). The energetic demands of sustained migratory or dispersive flight may reduce the somatic energy stores that insects have available for other activities, including survival and reproduction (Slansky, 1982; Mason *et al*., 1989; Rankin and Burchsted, 1992; Zera and Denno, 1997). Thus, insects in flight may face energetic trade-offs. But, do such energetic trade-offs occur in tephritids in response to long-distance flight? Zermeño (2005) studied survival of the walnut husk fly, *R. completa* (Cresson) (Diptera:Tephritidae) and *C. capitata* adults following long periods of sustained flight from tethering on a laboratory flight mill. In *C. capitata*, significant negative effects of flight on subsequent lifespan were seen only in the longest distance fliers (i.e. those flying 1000 m or greater), while in *R. completa*, no significant negative effects of flight on survival were found, regardless of flight distance. While neither of these species is considered to be migratory or even exceptionally strong fliers, not surprisingly, the potential exists for extended flights to take a toll on other life activities. The extent to which this occurs in tephritids and influences movements or other activities is unknown because, to our knowledge, no other studies have looked for energetic trade-offs associated with flight in either laboratory or field studies of tephritids.

Not surprisingly, adult diet has been found to affect the flight, with respect to the propensity to fly and distance flown, of the tephritids in which it has been studied; when deprived of carbohydrate in the adult diet, *B. (Dacus) oleae* (Remund *et al*., 1977), *C. capitata* and *R. completa* (Zermeño, 2005) flew shorter distances in the laboratory than when carbohydrates were provided. However, similar influences of diet on flight distances in the field have not been so clear. Wilkerson (2007) released marked *R. completa* that had been fed different adult diets and found no influence of diet on recapture distances using baited sticky traps. While recapture distances are, at best, inferential measures of potential dispersal distances, this study seems to indicate that any dietary influence on flight is weak in this species.

Although most movements of temperate fruit flies are generally thought to be non-dispersive (Boyce, 1934; Fletcher, 1989; but see Zermeño, 2005), tropical tephritids generally are strong fliers that are capable of long-distance dispersal flights (see Díaz-Fleischer and Aluja, 2001, for review). Using flies tethered to laboratory flight mills, researchers have reported flights of tropical tephritids of 2–6 km in length (Chambers and O'Connell, 1969; Sharp, 1976; Sharp and Chambers, 1976; Remund *et al*., 1977). Surprisingly, *R. completa* (Zermeño, 2005) and *R. indifferens* Curran (Diptera: Tephritidae) (Senger *et al*., 2007), two temperate, oligophagous species, were also found capable of flights over 2 km when tethered on a flight mill. Even for tropical tephritids, these flight distances are beyond the typical dispersal distances for fruit flies, aside from *B. (Dacus) oleae* that have been carried by prevailing winds to islands in the Mediterranean. As pointed out by Díaz-Fleischer and Aluja (2001) and Wilkerson (2007), the discrepancies between flight distances recorded from tethered flight versus field studies calls into question the validity of flight mill studies for assessing flight capacity in terms of distance.

Flight mills, on the other hand, can be powerful tools to study the effects of state variables on relative flight capabilities and propensity of fruit flies. For example, X. Wang and M.W. Johnson (personal communication) have investigated the influence of high temperatures on the flight performance of adult

B. oleae using flight mills. Both flight distance and duration of flight were reduced significantly after the exposure of flies to pre-flight temperatures of 35–37°C when compared to flies held at 23°C. Without access to water during these temperature treatments, flight distance and duration were further reduced. In studies of *R. completa* and *C. capitata*, Zermeño (2005) found that adult diets low or lacking in carbohydrates inhibited significantly the flight distance and duration of flies tethered to flight mills. These effects of food and environmental factors would have been difficult, if not impossible, to study without a bioassay tool such as a flight mill.

Although not a state variable per se, fly gender has also been found to influence flight propensity and flight distance strongly in most tephritids studied. In many cases, females have been found to fly farther than males, presumably because females search for oviposition sites and proteinaceous food sources (Hendrichs *et al.*, 1991; Prokopy *et al.*, 1996). In general, this sexual dimorphism in flight propensity and/or distance has held true, whether studies are conducted in the field (e.g. *B. (Dacus) oleae* (Fletcher and Economopoulos, 1976), *R. pomonella* (Johnson, 1983), *C. capitata* (Hendrichs *et al.*, 1991; Prokopy *et al.*, 1996)), or in the laboratory using flies tethered on a flight mill (e.g. *A. suspense* (Loew) (Diptera: Tephritidae) (Sharp, 1976), *D. dorsalis* and *D. cucurbitae* Coquillett (Diptera: Tephritidae) (Sharp *et al.*, 1975), *C. capitata* (Zermeño, 2005)). Nevertheless, this female fruit fly flight superiority is not universal. For example, male *R. completa* tethered on a flight mill in the laboratory generally flew the same distances, or in some cases even farther, than females (Zermeño, 2005). Of still further interest, when the same species of fly was tested in the field, sometimes males flew farther than females, but other times no significant differences in flight (dispersal) distances were found (Wilkerson, 2007). Other factors related to gender, such as reproductive maturity and location of hosts (i.e. oviposition sites), may also affect these and other flies.

Thus, while the evolution of flight has provided fruit flies and other insects with an outstanding ability to flee, considerable gaps exist in our understanding of the fleeing fruit fly. We have some evidence that indicates fruit flies flee to escape predators and poor circumstances, to find mates and food and to locate favourable egg-laying sites, but we have yet to develop a comprehensive understanding of the behavioural and functional ecology of fleeing.

F is for Finickiness

Finickiness is common

While most often considered in the context of larval feeding or adult oviposition, frugivorous tephritid flies display a finickiness, or propensity to discriminate, in virtually all behavioural contexts. Even with respect to the use of host fruit, behaviour other than larval feeding and adult oviposition may show signs of selectivity. For example, *R. pomonella* males are choosy as to which fruit, apple or hawthorn they will defend in territorial contests, a choosiness that is dictated by previous experience (Prokopy *et al.*, 1989). We will restrict our discussion of selectivity in host use to patterns in oviposition (i.e. reproduction) by female fruit flies and, in so doing, will demonstrate the power of the functional approach to this multifaceted problem.

How is selectivity in host use expressed in fruit flies? With respect to oviposition, selectivity can be defined at multiple levels: among tree-fruit species, among trees of a given species within a habitat, among fruit varieties or genotypes (Rull and Prokopy, 2004), among fruit within a tree (Henneman and Papaj, 1999) and even among parts of a given fruit (Papaj, 1994). At the level of host species, some fly species are generalists, attacking the fruit of many dozens, or even hundreds, of host species, whereas others are extreme specialists, attacking not just a single species of fruit but some subset of fruit of that species (for example, the largest and/or ripest fruit and/or, as mentioned above, fruit that is free of fly damage). However, even the most extreme generalists, such as the Medfly, *C. capitata*, do not use all possible fruit species. As a general rule, a female's preference ranking for fruit species reflects the relative suitability of different species for juvenile survival and fecundity.

Within a given host species, fruit flies may show differences in the degree to which different trees of a given species are attacked in a given population. In some sites in southern Arizona, for example, fruit of some trees escape infestation routinely in a given season, while all of the fruit of others are attacked (Nufio *et al.*, 2000; D. Papaj, personal observation). In species that attack economically important hosts, there is evidence of oviposition preferences with respect to fruit biotype (Prokopy and Papaj, 1988; Rull and Prokopy, 2004). Rarely have the functional consequences of such preference for the fitness of offspring been evaluated.

Over fruit within a biotype, and even over fruit within a given tree, tephritid fly females can be selective. One criterion is the ripeness stage of the fruit, with riper fruit commonly preferred for oviposition (Bierbaum and Bush, 1990; Papaj and Alonso-Pimentel, 1997). Another criterion is fruit size, with flies tending to prefer to oviposit into fruit models that are similar in size to or larger than the actual host fruit (Prokopy and Papaj, 2000). Such preference is functional, because larger fruit tend to support more larvae to pupation and to result in larger offspring of higher fecundity.

Finally, females may be selective as to where oviposition occurs on a fruit. For example, members of the suavis group within the genus *Rhagoletis* tend to oviposit in or near an existing oviposition site (Papaj, 1994; Papaj and Alonso-Pimentel, 1997). Oviposition site preference within a fruit is an area of fruit-fly behaviour in need of more research.

Finickiness is based on multiple stimuli

How is host selectivity at these various levels mediated? One of the greatest research contributions of Ronald Prokopy was by defining the role of various stimuli, notably visual and chemical stimuli, in the catenary process of host selection in tephritid fruit flies. The combination of naturalistic observation and experimental manipulation in field and laboratory studies in the work of Prokopy and colleagues is a model for the characterization and identification of behavioural stimuli (Prokopy, 1968). Physical and chemical stimuli are important at various stages of host selection. In apple maggot fly, odour was found to play an important role in arresting a female fly in a tree bearing apples, but a small role in locating fruit within the tree (Aluja and Prokopy, 1992, 1993). In contrast, visual stimuli play a role at each stage. Tree silhouettes enable flies to locate trees from a distance (Green *et al.*, 1994) and visual fruit models are effective both in attraction to acceptable fruit (Aluja and Prokopy, 1993) and avoidance of unacceptable fruit (Henneman and Papaj, 1999) at closer range. Olfactory and visual host cues interact in sometimes complex ways in tephritid flies (Aluja and Prokopy, 1993; Green *et al.*, 1994; Henneman and Papaj, 1999; Piñero *et al.*, 2006). Host acceptance after alightment likewise is affected both by physical stimuli (Prokopy, 1966) and contact chemical stimuli (Bierbaum and Bush, 1990). Broadly speaking, the use of multimodal stimuli in host selection, versus stimuli in a single sensory modality, is functional. For example, olfactory stimuli act to arrest a female apple maggot fly in a host tree bearing ripe fruit, while visual stimuli enable females to locate fruit within the tree (Aluja and Prokopy, 1993; Green *et al.*, 1994).

Finickiness can be expressed in subtle ways

A more subtle manifestation of host discrimination in fruit flies is clutch size. As an adaptive trait, clutch size should relate to the number of larvae that can be supported in a given fruit. Such a pattern is observed across *Rhagoletis* species. *Rhagoletis pomonella* deposits eggs singly into hawthorn (*Crataegus* sp.) fruit, consistent with data showing that, on average, such fruit support only 1–2 larvae to pupation (Averill and Prokopy, 1987). By contrast, *Rhagoletis* species depositing eggs in walnut husks lay large clutches; clutch size in the walnut-infesting *R. juglandis* Cresson (Diptera: Tephritidae) was estimated in one field study to average 13 eggs. Such large clutches are consistent with data indicating that walnut fruit can support dozens of larvae to pupation (Nufio and Papaj, 2004b).

There is limited evidence that individual females within a given fly species can adjust clutch size in different kinds of fruit. For example, in the Medfly, *C. capitata*, clutch size is diminished in the presence of HMP, which is an indicator of eggs laid previously (Papaj *et al.*, 1990). In *A. ludens*, larger clutches are deposited in larger fruit models (Berrigan *et al.*, 1988). Not all fly species appear able to adjust clutch size at the level of the individual female. For example, *R. pomonella* deposits a single egg into apple, as it does into its native hawthorn fruit, even though many more larvae can be supported to pupation in an apple than in a hawthorn berry. The cues involved in clutch size adjustment are not known.

Finickiness in fruit use is variable at many levels

Species, races and individuals potentially can vary at each stage of host selection. What are the underlying causes of variation in host-selection behaviour? Such variation could reflect genetically based differences, environmentally based differences, both kinds of differences summed together, or an interaction between genetic and environmental sources of variation. Species differences in host breadth and host preference reflect genetically based differences, at least in part. This has been demonstrated in cases in which species hybrids have been obtained that show fruit preferences which vary from parental species. For example, a fly population attacking honeysuckle (*Lonicera* spp.: Caprifoliaceae) appears to represent a case of homoploid speciation in which a new fly species has been generated from hybridization of the blueberry maggot, *R. mendax* (Curran) (Diptera: Tephritidae), and the snowberry maggot, *R. zephyria* (Snow) (Diptera: Tephritidae), which use blueberries and huckleberries (*Vaccinium* spp. (Ericaceae) and *Gaylussacia* spp. (Ericaceae)) and snowberries (*Symphoricarpos* spp. (Caprifoliaceae)), respectively (Schwarz *et al.*, 2003, 2007). The *Lonicera* fly, which has the same chromosome number as its parental species, has a distinctly different profile of fruit acceptance for oviposition than either of the parental species. Both parental species accept honeysuckle readily in no-choice presentations; however, in choice assays, the *Lonicera* fly prefers honeysuckle to the hosts of the parental species, whereas parental species prefer their respective native hosts (Schwarz *et al.*, 2007). Work to date does not strictly rule out an effect of larval-rearing origin (test subjects were collected in the field as larvae in their respective host-fruit species); however, authors note that earlier efforts to define such effects on oviposition in *Rhagoletis* flies (Linn *et al.*, 2003; Dambroski *et al.*, 2005) have been fruitless. Although it would be useful to exclude conclusively a possible environmental basis for species differences, at present, they are thought to reflect genetic differences.

Strong evidence for genetic variation in host-use behaviour within species has been documented with respect to *R. pomonella* host races. In the apple maggot fly, apple race individuals show relatively higher host selection responses to apple versus hawthorn than do hawthorn race individuals (Prokopy *et al.*, 1988). These differences are associated with variations in responses to fruit volatiles. Apple race individuals show stronger responses to apple volatiles than to hawthorn volatiles, as compared to hawthorn race individuals (Linn *et al.*, 2003, 2005; Forbes and Feder, 2006). The apple race appears to have evolved a stronger attraction to apple fruit odour, but also some avoidance of ancestral hawthorn fruit odour. There is likewise a preference in the dogwood race for volatiles from its host of origin (Linn *et al.*, 2005). Forbes and Feder (2006) determined that both visual and olfactory cues contributed to apple and hawthorn race differences. Odour discrimination differences among apple, hawthorn and dogwood races are genetically based and may involve only a modest number of allelic differences at a small number of loci (Dambroski *et al.*, 2005).

In addition to genetic factors, a great deal of evidence suggests that host-fruit preference in tephritid flies is influenced by environmental factors. Prokopy, for example, conducted extensive work on the role of fruit experience in oviposition preference by fruit flies (Prokopy and Papaj, 2000). In a series of field and laboratory assays, acceptance of host fruit by *R. pomonella* females was shown to be influenced by previous oviposition experience; females that had oviposited previously in hawthorn were relatively more likely to oviposit in hawthorn fruit than apple fruit than naïve females or females given experience with apples (Prokopy *et al.*, 1982; Papaj and Prokopy, 1988). Learning appears to affect both the host-finding and host-acceptance components of host selection. Learning of this type can facilitate host-race formation (Papaj and Prokopy, 1988).

In addition to the effects of experience, host selectivity can be affected by physiological state. In particular, the degree of fruit preference is affected by how many mature oocytes are borne in a female fly's ovaries. Females that carry more eggs are predicted to be less discriminating about the kind of fruit into which they lay eggs. Consistent with this expectation, Medfly females were less discriminating about kumquats relative to grapefruit as egg load (= the number of mature oocytes) increased. In *R. zephyria*, egg load was manipulated by protein availability; the degree to which host pheromone-marked fruit (a low-quality fruit) were accepted for oviposition increased with egg load (van Randen and Roitberg, 1996).

Finickiness is not only behavioural

Experience and physiological state can interact to influence oviposition behaviour in fruit flies, when a female's ovarian development is influenced by fruit stimuli (Papaj, 2000). In *R. juglandis*, for example, maturation of oocytes in the first cycle of egg production is faster and more extensive when females are held in the presence of host walnut fruit (Alonso-Pimentel *et al.*, 1998; Lachmann and Papaj, 2001). Host foliage has no effect in and of itself and does not interact with the host-fruit effect (Alonso-Pimentel *et al.*, 1998). The effect of host-fruit stimuli on oogenesis in *R. juglandis* has parallels with the effects on oviposition behaviour. Just as fruit-mimicking spheres will elicit oviposition, so too, fruit models consisting of a sphere similar in size to native walnuts will promote oogenesis. A non-fruit-like cube of the same colour and surface area has no effect. Just as colour similar to host fruit promotes attraction by tephritid fly females to fruit models (Henneman and Papaj, 1999), so too spheres of colours similar to walnuts (yellow or green) enhance egg maturation more than colours like red or blue.

Parallels between oviposition and oogenesis in *R. juglandis* extend to variation among fruit in characteristics pertinent to larval fitness. Tephritid fly females often prefer to oviposit in larger than smaller fruit or fruit models (Prokopy and Papaj, 2000); likewise, egg maturation is more pronounced in *R. juglandis* females presented from emergence with a sphere of the size of a large walnut than when presented with a sphere of the size of a small walnut (Papaj, 2005). Tephritid fly females often avoid oviposition in larvae-infested fruit (Hennemann and Papaj, 1999); likewise, egg maturation in females held with a larvae-infested fruit was no greater than that of females held without fruit altogether (Papaj, 2005). Both fruit size and larval infestation patterns in ovarian dynamics were duplicated with visually mimicking host-fruit models.

Ovarian responses to fruit quality are thought to be functional in the same way that oviposition responses to fruit quality are functional. Large fruit tend to yield larger adults. Larger males may be more fit in terms of territorial contests, while large females mature more eggs faster. Large adults may also have better dispersal capacities. Likewise, eggs probably have little success when deposited in larval-infested fruit, due both to risk of depletion of fruit resources and to risk of cannibalism.

The functional interpretation of selectivity at the level of ovaries is thus straightforward. Less clear is why egg maturation should be condition-dependent in the first place. Cuing of egg maturation in relation to the presence or quality of hosts suggests a trade-off that differs from trade-offs associated with oviposition. For example, if an adult female emerges beneath a tree that is not bearing fruit, it may cost the female either in terms of dispersing efficiently to a tree that does bear fruit or surviving until a tree bearing fruit is found. In the olive fly, *B. oleae*, host sensory cuing appears to be a consequence of trade-offs associated with adult survival. In *B. oleae*, the absence of fruit promotes oosorption (wherein eggs are broken down and nutrients redistributed to somatic tissues) (Fletcher *et al.*, 1978). Oosorption presumably helps the flies to survive a period of fruit scarcity, which occurs seasonally in their native Mediterranean habitats.

In a fly species that specializes on just one host species, as in *R. juglandis* or *D. oleae*, there is no possibility for host sensory cuing to influence selectivity at the level of host species. One needs to examine more polyphagous species. In two polyphagous *Anastrepha* species, *A. obliqua* and *A. ludens*, manipulation of the presence or absence of artificial host models and host-fruit odour provided little firm evidence of enhanced egg maturation (Aluja *et al.*, 2001). Aluja *et al.* (2001) suggested that generalist species might depend less on host-derived sensory cuing and more on uncued adjustment of their 'standing crop' of eggs. For example, *A. obliqua*'s hosts are superabundant but ephemeral and unpredictable. *A. obliqua* maintains very high numbers of mature oocytes, as though physiologically 'prepared' to attack a large number of fruit when they suddenly become available (Aluja *et al.*, 2001).

In closing, there is good evidence that genetic and environmental factors each influence host selectivity in oviposition by tephritid fruit flies. Such finickiness may also reflect an interaction between genetic and environmental sources of variation in host-selection behaviour. Evidence for gene-by-environment interactions in host use by tephritid flies is scant. A notable exception is found in Prokopy *et al.* (1986), who report that the degree to which oviposition experience affects fruit acceptance in *R. pomonella* depends on the host race of origin for a test female. The interaction was in the predicted direction: females of a given race (apple or hawthorn) were relatively more likely to be affected by experience with their fruit of origin. More research is needed to determine the extent to which such interactions account for variation in host use at the level of fly species, host races and populations. These interactions mean that behaviour-based pest management tactics may need to be 'customized' according to variation among species, races and populations. Below, we evaluate some behaviour-based management tactics and we remind the reader to keep host discrimination in fruit flies in mind throughout.

The Four Fs and Biorational Management of Tree-Fruit Pests

In the preceding sections of this chapter, we dealt with individual components of the life histories of a variety of fruit flies from a functional perspective. We can now ask whether a deep understanding of fruit-fly behavioural ecology can be used to improve the methods by which we manage fruit flies as pests of fruit tree agroecosystems. This is a question to which Ron Prokopy gave considerable thought. The answer is a conditional 'yes', the condition being that the behavioural plasticity of fruit flies that we explicated earlier be made an explicit part of the programme. Below we provide a detailed example.

Prokopy was a great proponent of the concept of exploiting the natural behaviour of pest insects to reduce or eliminate their damage to fruit; however, he was quite aware of the difficulties (and opportunities) that the aforementioned variability of fruit-fly behaviour presented/afforded. An example of this concept is explored in Prokopy and Roitberg (2007). Here, the question was how to minimize the number of ovipositions by wild, immigrant apple maggot flies that ventured into commercial orchards. There are two ways to achieve this goal: (i) decrease the survival rates of female flies; and (ii) decrease fly per capita oviposition rate. Prokopy and Roitberg chose (i) by exploiting fruit-fly food and host search wherein flies would be trapped *before* they could cause economic damage (Prokopy *et al.*, 2000) to fruit. Here, the goal is to minimize the time it takes for flies to locate a trap. Where does the deep understanding of fruit-fly behavioural ecology come into play? In several places: first, as described in the earlier section on feeding, they used the relationship between proteins, carbohydrates and fly Darwinian fitness, along with the apple maggot's evolutionary history, with food and fruit resources as an explicit component of a dynamic life-history model. This model allowed them to calculate the optimal placement of artificial food- and fruit-laced traps (Prokopy and Roitberg, 2007). By optimal, we mean the trap density and distribution that maximizes net profit from fruit sales. These net profits are based on marginal costs from exploiting the fly's food- and fruit-finding behaviour and the marginal returns from increased fruit value via reduced infestation rates. As an example of this economic concept, as the number of odour-based traps increases, so does the death rate of immigrant flies, but in a decelerating manner. The slope of this curve can only be derived by deducing how a wild immigrant fly's search behaviour changes as a function of its protein reserves.

Second, Prokopy and Roitberg considered fly varietal preference, something we referred to earlier as a kind of finickiness. They considered how fly oviposition (acceptance) rates would vary in orchards, with different varieties leading to variable encounter rates by flies with traps and once again impacting the marginal return curves. This means that the optimal number and placement of traps will vary among orchards according to apple variety and distributions of varieties within orchard blocks. Thus, a deep understanding of finickiness is required to develop varietal-specific management policies.

Third, the Prokopy and Roitberg (2007) model was tactical in that it assumed that fly response to fruit quality and density would be independent of the presence of conspecifics (see Robertson *et al.*, 1995). As we noted earlier, flies do, in fact, alter their foraging behaviour when either contacting conspecifics directly or through encounters with marking pheromone. The exact form of fly response (e.g. change in acceptance rate of different varieties of fruit) to varietal ratios depends on how maternal fitness via larval survival has shaped the evolution of adult behaviour. It is important to remember that competition in natural settings could be much more intense than in orchards because, in the former, populations are often stable and dense and fruit are small (e.g. small, non-commercial apples or hawthorns).

The likelihood that a given fly will emigrate from a particular orchard can also be calculated based on the concepts discussed in the section titled 'F is for Fleeing'. Of course, one grower may not be so concerned about a fly leaving his or her orchard and going to his neighbour's trees, but the reduction in fly density due to death or emigration again impacts marginal returns on investment. Ultimately, growers should choose profit-maximizing pest management strategies. With regard to within-orchard movement and optimal placement of traps among trees, Roitberg and Mangel (1997) show how within-habitat movement rates can be predicted for *Rhagoletis* females as a function of habitat structure and life-history parameters, in particular survivorship and fitness from ovipositing in high- and low-quality fruit.

Finally, as Prokopy and Roitberg (2007) point out, the key to employing their model in orchards is to identify the structure of resources in the fly's natural habitat, use those data to calculate the expected foraging strategies of flies and then assemble odour-laced traps in the appropriate pattern.

Of course, the discussion above highlights just one of several examples of the application of the functional ecology of fruit flies to their management and these applications are not restricted to the temperate zone flies. Recently, Aluja has produced a series of papers that mirror the aforementioned approach for managing anastrephan and toxotrypanan flies on avocados (Aluja *et al.*, 2004), grapefruits (Birke *et al.*, 2006) and papayas (Aluja *et al.*, 1997). In addition, throughout this book, one can find examples of applications of functional ecology to fruit-tree management, in particular, Chapters 2, 5 and 7.

Conclusions

Ron Prokopy's second true love, the fruit fly, is a complex organism whose physiological and behavioural response to cues is highly contextual. It is clear, however, that despite great variance across species in feeding habits, site tenacity, pugnacity and choosiness, their commonalities can be well understood through the power of the functional approach. As an avid practitioner of natural selection theory, Ron would have lauded this line of attack.

Acknowledgements

The support, expertise and encouragement of Ronald J. Prokopy, to whom this book is dedicated, are particularly acknowledged. Bernard D. Roitberg's work is supported by an NSERC (Canada) operating grant. Carol R. Lauzon (CL) gratefully acknowledges the various contributions by several colleagues and researchers who have furthered the understanding of fruit fly–bacteria interactions. Funding for the work of CL and her students and colleagues was provided by the United States Department of Agriculture, the International Atomic Energy Agency, the California Department of Food and Agriculture, the Citrus Research Board, the Binational Science Foundation, and California State University. Susan Opp's work is supported by funding from USDA, the Agricultural Research Initiative of California State University, the Citrus Research Board and the Walnut Marketing Board, and by agreements with the East Bay Regional Parks District.

References

Agnew, P., Becnel, J.J., Ebert, D. and Michalakis, Y. (2003) Symbiosis of *Microsporidia* and insects. In: Bourtzis, K. and. Miller, T.A. (eds) *Insect Symbiosis*. CRC Press, Boca Raton, Florida, pp. 145–163.

Allen, T.C. and Riker, A.J. (1932) A rot of apple fruit caused by *Phytomonas melopthora* N. sp., following invasion by the apple maggot. *Phytopathology* 22, 557–571.

Allen, T.C., Pinckard, J.A. and Riker, A.J. (1934) Frequent association of *Phytomonas melopthora* with various stages in the life cycle of the apple maggot, *Rhagoletis pomonella. Phytopathology* 24, 228–238.

Alonso-Pimentel, H., Korer, J.B., Nufio, C. and Papaj, D.R. (1998) Role of color and shape stimuli in host-enhanced oogenesis in the walnut fly, *Rhagoletis juglandis*. *Physiological Entomology* 23, 97–104.

Aluja, M. (1993) The study of movement in tephritid flies: review of concepts and recent advances. In: Aluja, M. and Liedo, P. (eds) *Fruit Flies: Biology and Management.* Springer-Verlag, New York, pp. 105–113.

Aluja, M. and Boller, E. (1992). Host-marking pheromone of *Rhagoletis cerasi*: field deployment of synthetic pheromone as a novel cherry fruit fly management strategy. *Entomologia Experimentalis et Applicata* 65, 141–147.

Aluja, M. and Prokopy, R.J. (1992) Host search behavior by *Rhagoletis pomonella* flies: intertree movement patterns in response to windborne fruit volatiles under field conditions. *Physiological Entomology* 17, 1–8.

Aluja, M. and Prokopy, R.J. (1993) Host odor and visual stimulus interaction during intratree host-finding behavior of *Rhagoletis pomonella* flies. *Journal of Chemical Ecology* 19, 2671–2696.

Aluja, M., Jiménez, A., Camino, M., Piñero, J., Aldana, L., Castrejón, V. and Valdés, M.E. (1997) Habitat manipulation to reduce papaya fruit fly (Diptera: Tephritidae) damage: orchard design, use of trap crops and border trapping. *Journal of Economic Entomology* 90, 1567–1576.

Aluja, M., Díaz-Fleischer, F., Papaj, D.R., Lagunes, G. and Sivinski, J. (2001) Effects of age, diet, female density, and the host resource on egg load in *Anastrepha ludens* and *A. obliqua* (Diptera: Tephritidae). *Journal of Insect Physiology* 47, 975–988.

Aluja, M., Díaz-Fleischer, F. and Arredondo, J. (2004) Non-host status of commercial *Persea americana* cultivar 'Hass' to *Anastrepha ludens*, *Anastrepha obliqua*, *Anastrepha serpentina*, and *Anastrepha striata* (Diptera: Tephritidae) in Mexico. *Journal of Economic Entomology* 97, 293–309.

Averill, A.L. and Prokopy, R.J. (1987) Intraspecific competition in the tephritid fruit fly, *Rhagoletis pomonella. Ecology* 68, 878–886.

Baker, P.S. and van der Valk, H. (1992) Distribution and behavior of sterile Mediterranean fruit flies in a host tree. *Journal of Applied Entomology* 114, 67–76.

Barrow, P.A., Lovell, M.A. and Barber, L.Z. (1996) Growth suppression in early-stationary-phase nutrient broth cultures of *Salmonella typhimurium* and *Escherichia coli* is genus specific and not regulated by σ^S. *Journal of Bacteriology* 178, 3072–3076.

Beenakkers, A.M.T., Van der Horst, D.J. and Van Marrewijk, W.J.A. (1984) Insect flight muscle metabolism. *Insect Biochemistry* 14, 243–260.

Behar, A., Yuval, B. and Jurkevitch, E. (2005) Enterobacteria-mediated nitrogen fixation in natural populations of the fruit fly, *Ceratitis capitata. Molecular Ecology* 14, 2637–2643.

Berrigan, D.A., Carey, J.R., Guillen, J. and Celedonio, H. (1988) Age and host effects on clutch size in the Mexican fruit fly, *Anastrepha ludens*. *Entomologia Experimentalis et Applicata* 47, 73–80.

Bierbaum, T.J. and Bush, G.L. (1990) Host fruit chemical stimuli eliciting distinct ovipositional responses from sibling species of *Rhagoletis* fruit flies. *Entomologia Experimentalis et Applicata* 56, 165–177.

Birke, A., Aluja, M., Greany, P., Bigurra, E., Pérez-Staples, D. and McDonald, R. (2006) Long aculeus and behavior of *Anastrepha ludens* renders gibberellic acid ineffective as an agent to reduce 'Ruby Red' grapefruit susceptibility to the attack of this pestiferous fruit fly in commercial citrus orchards. *Journal of Economic Entomology* 99, 1184–1193.

Blay, S. and Yuval, B. (1997) Nutritional correlates of reproductive success of male Mediterranean fruit flies (Diptera: Tephritidae). *Animal Behavior* 54, 59–66.

Boyce, A.M. (1934) Bionomics of the walnut husk fly, *Rhagoletis completa*. *Hilgardia* 8, 363–579.

Brenner, D.J. (1992) Facultatively anaerobic Gram negative rods. In: Holt, J.G. and Krieg, N.R. (eds) *Bergey's Manual of Determinative Bacteriology*, 10th edition, volume 1. Williams and Wilkins Co., Baltimore, Maryland, pp. 408–516.

Brodsky, A.K. (1994) *The Evolution of Insect Flight*. Oxford University Press, New York.

Burrage, S.W. (1971) The micro-climate at the leaf surface. In: Preece, T.F and Dickinson, C.H (eds) *Ecology of Leaf Surface Microorganisms*. Academic Press, London, pp. 89–101.

Buss, D. (2006) True love. In: Eggers, D. (ed.) *The Best American Non-required Reading*. Houghton Mifflin, New York, p. 10.

Candy, D.J., Becker, A. and Wegener, G. (1997) Coordination and integration of metabolism in insect flight. *Comparative Biochemistry and Physiology* 117B, 497–512.

Cangussu, J.A. and Zucoloto, F.S. (1992) Nutritional value and selection of different diets by adult *Ceratitis capitata* flies (Diptera: Tephritidae). *Journal of Insect Physiology* 38, 485–491.

Cayol, J.P. (1999) Changes in sexual behavior and life history traits of tephritid species caused by mass-rearing processes. In: Aluja, M. and Norrbom, A.L. (eds) *Fruit Flies (Tephritidae): Phylogeny and Evolution of Behavior.* CRC Press, Boca Raton, Florida, pp. 843–860.

Chambers, D.L. and O'Connell, T.B. (1969) A flight mill for studies with the Mexican fruit fly. *Annals of the Entomological Society America* 62, 917–920.

Cooter, R.J. (1982) Studies on the flight of black-flies (Diptera: Simuliidae). I. Flight performance of *Simulium ornatuum* Meigen. *Bulletin of Entomological Research* 72, 303–317.

Daly, H.V., Doyen, J.T. and Purcell, A.H. (1998) *Introduction to Insect Biology and Diversity*. Oxford University Press, New York.

Dambroski, H.R., Linn, C., Berlocher, S.H., Forbes, A.A., Roelofs, W. and Feder, J.L. (2005) The genetic basis for fruit odor discrimination in *Rhagoletis* flies and its significance for sympatric host shifts. *Evolution* 59, 1953–1964.

Díaz-Fleischer, F. and Aluja, M.A. (2001) Behavior of tephritid flies: a historical perspective. In: Aluja, M. and Norrbom, A.L. (eds) *Fruit Flies (Tephritidae): Phylogeny and Evolution of Behavior.* CRC Press, Boca Raton, Florida, pp. 39–69.

Díaz-Fleischer, F. and Aluja, M. (2003) Influence of conspecific presence, experience, and host quality on oviposition behavior and clutch size determination in *Anastrepha ludens* (Diptera: Tephritidae). *Journal of Insect Behavior* 16, 537–554.

Drew, R.A.I. (1988) Amino acid increases in fruit infested by fruit flies of the family Tephritidae. *Zoological Journal of the Linnean Society* 93, 107–112.

Drew, R.A.I. and Faye, H.A. (1988) Elucidation of the roles of ammonia and bacteria in the attraction of *Dacus tryoni* (Froggatt) (Queensland fruit fly) to proteinaceous suspensions. *Journal of Plant Protection in the Tropics* 5, 127–130.

Drew, R.A.I. and Lloyd, A.C. (1987) Relationship of fruit flies (Diptera: Tephritidae) and their bacteria to host plants. *Annals of the Entomological Society of America* 80, 629–636.

Drew, R.A.I. and Yuval, B. (2000) The evolution of fruit fly feeding behavior. In: Aluja, M.

and Norrbom, A.L. (eds) *Fruit Flies (Tephritidae): Phylogeny and Evolution of Behavior.* CRC Press, Boca Raton, Florida, pp. 731–749.

Drew, R.A.I., Courtice, A.C. and Teakle, D.S. (1983) Bacteria as a natural source of food for adult fruit flies (Diptera: Tephritidae). *Oecologia* 60, 279–284.

Dudley, R. (2000) *The Biomechanics of Insect Flight: Form, Function, Evolution*. Princeton University Press, Princeton, New Jersey.

Dukas, R., Prokopy, R.J., Papaj, D.R. and Duan, J.J. (2001) Egg laying behavior of Mediterranean fruit flies (Diptera: Tephritidae): is social facilitation important? *The Florida Entomologist* 84, 665–671.

Eberhard, W.G. (1979) *Sexual Selection and Reproductive Competition in Insects*. Academic Press, New York.

Enquist, M. and Leimar, O. (1990) The evolution of fatal fighting. *Animal Behavior* 39, 1–9.

Epsky, N.D., Dueben, B.D., Heath, R.R., Lauzon, C.R. and Prokopy, R.J. (1997) Attraction of *Anastrepha suspensa* (Diptera: Tephritidae) to volatiles from avian fecal material. *The Florida Entomologist* 80, 270–277.

Epsky, N.D., Heath, R.R., Dueben, B.D., Lauzon, C.R., Proveaux, A.T. and MacCollom, G.B. (1998) Attraction of 3-methyl-1-butanol and ammonia identified from *Enterobacter agglomerans* to *Anastrepha suspensa. Journal of Chemical Ecology* 24, 1867–1880.

Fletcher, B.S. (1989) Movements of tephritid fruit flies. In: Robinson, A.S. and Hooper, G. (eds) *Fruit Flies: Their Biology, Natural Enemies and Control*, Volume 3B. Elsevier Science, Amsterdam, pp. 195–219.

Fletcher, B.S. and Economopoulos, A.P. (1976) Dispersal of normal and irradiated laboratory strains and wild strains of the olive fly, *Dacus oleae,* in an olive grove. *Entomologia Experimentalis et Applicata* 20, 183–194.

Fletcher, B.S. and Kapatos, E. (1981) Dispersal of the olive fly, *Dacus oleae*, during the summer period on Corfu. *Entomologia Experimentalis et Applicata* 29, 1–8.

Fletcher, B.S., Pappas, S. and Kapatos, E.T. (1978) Changes in the ovaries of olive flies (*Dacus oleae* [Gmelin]) during the summer, and their relationship to temperature, humidity and fruit availability. *Ecological Entomology* 3, 99–107.

Forbes, A.A. and Feder, J.L. (2006) Divergent preferences of *Rhagoletis pomonella* host races for olfactory and visual fruit cues. *Entomologia Experimentalis et Applicata* 119, 121–127.

Fujii, J.K. and Tamashiro, M. (1972) *Nosema tephritidae* sp. N., a microsporidium pathogen of the oriental fruit fly, *Dacus dorsalis* Hednel. *Proceedings of the Hawaiian Entomological Society* 21, 191–203.

Girolami, V., Vianello, A., Strapazzon, E., Ragazzi, E. and Veronese, G. (1981) Ovipositional deterrents in *Dacus oleae*. *Entomologia Experimentalis et Applicata* 29, 177–188.

Girolami, V., Strapazzon, A. and DeGerloni, P.F. (1983) Insect/plant relationships in olive flies: general aspects and new findings. In: Cavalloro, R. (ed.) *Fruit Flies of Economic Importance*. A.A. Balkema, Rotterdam, The Netherlands, pp. 258–267.

Gow, P.L. (1954) Proteinaceous bait for the Oriental fruit fly (Diptera: Tephritidae): biology and control. *Journal of Economic Entomology* 47, 153–160.

Green, T.A., Prokopy, R.J. and Hosmer, D.W. (1994) Distance of response to host tree models by female apple maggot flies: interaction of visual and olfactory stimuli. *Journal of Chemical Ecology* 20, 2393–2413.

Grimont, P. and Grimont, F. (1978) The genus *Serratia. Revista de Microbiologia* 32, 221–248.

Groat, R.G., Schultz, J.E., Zychlinsky, E., Bockman, A. and Matin, A. (1986) Starvation proteins in *Escherichia coli*: kinetics of synthesis and role in starvation survival. *Journal of Bacteriology* 168, 486–493.

Hagen, K.S. (1966) Dependence of the olive fruit fly, *Dacus oleae*, larvae on symbiosis with *Pseudomonas savastanoi* for the utilization of olive. *Nature* 109, 423–425.

Hendrichs, J. and Hendrichs, M.A. (1990) Mediterranean fruit fly (Diptera: Tephritidae) in nature; location and diel pattern of feeding and other activities on fruiting and non-fruiting hosts and non-hosts. *Annals of the Entomological Society of America* 83, 632–641.

Hendrichs, J. and Hendrichs, M.A. (1998) Perfumed to be killed: interception of Mediterranean fruit fly (Diptera: Tephritidae) sexual signaling by predatory foraging wasps (Hymenoptera: Vespidae). *Annals of the Entomological Society of America* 91, 228–234.

Hendrichs, J., Lauzon, C., Cooley, S.S. and Prokopy, R.J. (1990) What kind of fruit do the apple maggot flies need for survival and reproduction? *Massachusetts Fruit Notes* 55, 9–11.

Hendrichs, J., Katsoyannos, B.I., Papaj, D.R. and Prokopy, R.J. (1991) Sex differences in movement between natural feeding and mating sites and trade-offs between food consumption, mating success and predator evasion in Mediterranean fruit flies (Diptera: Tephritidae). *Oecologia* 86, 223–231.

Hendrichs, J., Lauzon, C.R., Cooley, S.S. and Prokopy, R.J. (1993) Contribution of natural food sources to adult longevity and fecundity of *Rhagoletis pomonella* (Diptera: Tephritidae). *Annals of the Entomological Society of America* 86, 250–264.

Hendrichs, J., Katsoyannos, B.I., Wornoayporn, V. and Hendrichs, M.A. (1994) Odor-mediated foraging by yellowjacket wasps (Hymenoptera: Vespidae): predation on leks of pheromone-calling Mediterranean fruit fly males (Diptera: Tephritidae). *Oecologia* 99, 88–94.

Henneman, M.L. and Papaj, D.R. (1999) Role of host fruit color in the behavior of the walnut fly *Rhagoletis juglandis*. *Entomologia Experimentalis et Applicata* 93, 249–258.

Hoffmeister, T., Roitberg, B.D. and Lalonde, R. (2000) Catching Ariadne by her thread: how a parasitoid exploits the herbivore's marking trails to locate its host. *Entomologia Experimentalis et Applicata* 95, 77–85.

Howard, D.J., Bush, G.L. and Breznak, J.A. (1985) The evolutionary significance of bacteria associated with *Rhagoletis*. *Evolution* 39, 405–417.

Huisman, G.W. and Kolter, R. (1994) Sensing starvation: a homoserine lactone-dependent signaling pathway in *Escherichia coli*. *Science* 265, 537–539.

Jacques, R.P., Neilson, W.T.A. and Huston, F. (1969) A bloating disease of adults of the apple maggot. *Journal of Economic Entomology* 62, 850–851.

Jang, E.B. and Nishijima, K.A. (1990) Identification and attractancy of bacteria associated with *Dacus dorsalis* (Diptera: Tephritidae). *Environmental Entomology* 62, 850–851.

Johnson, P.C. (1983) Response of adult apple maggot (Diptera: Tephritidae) to Pherocon AM traps and red spheres in a non-orchard habitat. *Journal of Economic Entomology* 76, 1279–1284.

Kapatos, E.T. and Fletcher, B.S. (1983) Seasonal changes in the efficiency of McPhail traps and a model for estimating olive fly densities from trap catches using temperature data. *Entomologia Experimentalis et Applicata* 32, 20–26.

Kingsolver, J.G. and Koehl, M.A.R. (1994) Selective factors in the evolution of insect wings. *Annual Review of Entomology* 39, 425–451.

Kittayapong, P., Milne, J.R., Tigvattananont, S. and Baimai, V. (2000) Distribution of the reproduction-modifying bacteria, *Wolbachia*, in natural populations of tephritid flies in Thailand. *Science Asia* 26, 93–103.

Lachmann, A.D. and Papaj, D.R. (2001) Effect of host stimuli on ovariole development in the walnut fly, *Rhagoletis juglandis* (Diptera, Tephritidae). *Physiological Entomology* 26, 38–48.

Lauzon, C.R. (1991) Microbial ecology of a fruit fly pest, *Rhagoletis pomonella* (Walsh) (Diptera: Tephritidae). PhD dissertation, The University of Vermont, Burlington, Vermont.

Lauzon, C.R. and Potter, S.E. (2009) Visual description of radiation-induced tissue and cellular damage in the midgut of *Ceratitis capitata* Wiedemann (Diptera: Tephritidae) and *Anastrepha ludens* (Diptera: Tephritidae) used for sterile insect technique. *Annals of the Entomological Society of America* (in press).

Lauzon, C.R. and Sjogren, R.E. (1998) Attraction of *Rhagoletis pomonella* (Diptera: Tephritidae) to odors of bacteria: apparent confinement to specialized members of Enterobacteriaceae. *Environmental Entomology* 27, 853–857.

Lauzon, C.R. and Sjogren, R.E. (2000) Enzymatic capabilities of bacteria associated with apple maggot flies: a postulated role in attraction. *Journal of Chemical Ecology* 26, 953–967.

Lauzon, C.R., Sjogren, R.E. and Prokopy, R.J. (2000) Enzymatic capabilities of bacteria associated with apple maggot flies: a postulated role in attraction. *Journal of Chemical Ecology* 26, 953–967.

Lauzon, C.R., Bussert, T.C., Sjogren, R.E. and Pokopy, R.J. (2003) *Serratia marcescens* as a bacterial pathogen of *Rhagoletis pomonella* flies (Diptera : Tephritidae). *European Journal of Entomology* 100, 87–92.

Lauzon, C.R., McCombs, S.D. and Potter, S.E. (2009) Vertical passage of *Enterobacter agglomerans* and *Klebsiella pneumoniae* in *Ceratitis capitata* Weidemann, the Mediterranean fruit fly. *Annals of the Entomological Society of America* (in press).

Lima, S.L. and Dill, L.M. (1990) Behavioral decisions made under the risk of predation: a review and prospectus. *Canadian Journal of Zoology* 68, 619–640.

Lindow, S.E. (1991) Determinants of epiphytic fitness in bacteria. In: Andrews, J.H. and Hirano, S.S. (eds) *Microbial Ecology of Leaves*. Springer-Verlag, New York, pp. 295–314.

Linn, C., Feder, J.L., Nojima, S., Dambroski, H.R., Berlocher, S.H. and Roelofs, W. (2003) Fruit odor discrimination and sympatric host race formation in *Rhagoletis*. *Proceedings of the National Academy of Sciences of the United States of America* 100, 11490–11493.

Linn, C., Nojima, S. and Roelofs, W. (2005) Antagonist effects of non-host fruit volatiles on discrimination of host fruit by Rhagoletis flies infesting apple (*Malus pumila*), hawthorn (*Crataegus* spp.), and flowering dogwood (*Cornus florida*). *Entomologia Experimentalis et Applicata* 114, 97–105.

Lloyd, A.C., Drew, R.A.I., Teakle, D.S. and Howard, A.C. (1986) Bacteria associated with some *Dacus* species (Diptera: Tephritidae) and their host fruit in Queensland. *Australian Journal of Biological Sciences* 39, 361–368.

Luthy, P., Struder, D., Jaquet, F. and Yamvrias, C. (1983) Morphology and *in vitro* cultivation of the bacterial symbiote of *Dacus oleae*. *Mitteilungen der Schweizerische Entomologische Gesellschaft* 56, 67–72.

MacCollom, G.B., Lauzon, C.R., Weires, R.W. Jr and Rutkowski, A.A. (1992) Attraction of adult apple maggot (Diptera: Tephritidae) to microbial isolates. *Journal of Economic Entomology* 85, 83–87.

Malavasi, A. and Prokopy, R.J. (1992) Effect of food deprivation on the foraging behavior of *Rhagoletis pomonella* females for food and host hawthorn fruit. *Journal of Entomological Science* 27, 185–193.

Mangel, M. and Roitberg, B.D. (1989) Dynamic information and oviposition decisions in a fruit fly. *Ecological Entomology* 14, 181–189.

Manousis, T. and Moore, N.F. (1987) Cricket paralysis virus, a potential control agent for olive fruit fly, *Dacus oleae* Gmel. *Applied Environmental Microbiology* 53, 142–148.

Marchini, D., Marri, L., Rosetto, L., Manetti, A. and Marri, L. (1997) Presence of antibacterial peptides on the laid egg chorion of the medfly, *Ceratitis capitata*. *Biochemical and Biophysical Research Communications* 240, 657–663.

Marchini, D., Rosetto, L., Dallai, R. and Marri, L. (2002) Bacteria associated with the oesophageal bulb of the medfly *Ceratitis capitata* (Diptera: Tephritidae). *Current Microbiology* 44, 120–124.

Mason, L.J., Johnson, S.J. and Woodring, J.P. (1989) Influence of carbohydrate deprivation and tethered flight on stored lipid, fecundity, and survivorship of the soybean looper (Lepidoptera; Noctuidae). *Environmental Entomology* 18, 1090–1094.

Maynard-Smith, J. (1982) *Evolution and the Theory of Games*. Cambridge University Press, Cambridge, UK.

Maynard-Smith, J. and Parker, G.A. (1976) The logic of asymmetric contests. *Animal Behavior* 24, 159–175.

Miyazaki, S., Boush, G.M. and Baerwald, R.J. (1968) Amino acid synthesis by *Pseudomonas melopthora* bacterial symbiote of *Rhagoletis pomonella* (Diptera). *Journal of Insect Physiology* 14, 513–518.

Moriarty, D.J.W. and Bell, R.T. (1993) Bacterial growth and starvation in aquatic environments. In: Kjelleberg, S. (ed.) *Starvation in Bacteria*. Plenum Press, New York, pp. 25–53.

Murphy, K.M., Teakle, D.S. and MacRae, I.C. (1994) Kinetics of colonization of adult Queensland fruit flies (*Bactrocera tryoni*) by dinitrogen fixing alimentary canal tract bacteria. *Applied Environmental Microbiology* 60, 2508–2517.

Niyazi, N., Lauzon, C.R. and Shelly, T.E. (2004) The effect of a probiotic adult diet on fitness components of sterile male Mediterranean fruit flies (Diptera: Tephritidae) under laboratory and field conditions. *Journal of Economic Entomology* 97, 1581–1586.

Nufio, C.R. and Papaj, D.R. (2001) Host marking behavior in phytophagous insects and parasitoids. *Entomologia Experimentalis et Applicata* 99, 273–293.

Nufio, C.R. and Papaj, D.R. (2004a) Host-marking behaviour as a quantitative signal of competition in the walnut fly *Rhagoletis juglandis. Ecological Entomology* 29, 336–344.

Nufio, C.R. and Papaj, D.R. (2004b) Superparasitism of larval hosts by the walnut fly, *Rhagoletis juglandis*, and its implications for female and offspring performance. *Oecologia* 141, 460–467.

Nufio, C.R., Papaj, D.R. and Alonso-Pimentel, H. (2000) A field analysis of host fruit utilization by the walnut fly, *Rhagoletis juglandis. Environmental Entomology* 93, 994–1001.

Opp, S.B. and Prokopy, R.J. (1987) Seasonal changes in resighting of marked wild *Rhagoletis pomonella* (Diptera: Tephritidae) flies in nature. *The Florida Entomologist* 70, 450–457.

Papaj, D.R. (1994) Oviposition site guarding by male walnut flies and its possible consequences for mating success. *Behavioral Ecology and Sociobiology* 34, 187–195.

Papaj, D.R. (2000) Ovarian dynamics and host use. *Annual Review of Entomology* 45, 423–448.

Papaj, D.R. (2005) Ovarian dynamics in relation to host quality in the walnut-infesting fly, *Rhagoletis juglandis. Functional Ecology* 19, 396–404.

Papaj, D.R. and Alonso-Pimentel, H. (1997) Why walnut flies superparasitize: time savings as a possible explanation. *Oecologia* 109, 166–174.

Papaj, D.R. and Messing, R.H. (1998) Asymmetries in physiological state as a possible cause of resident advantage in contests. *Behaviour* 135, 1013–1030.

Papaj, D.R. and Prokopy, R.J. (1988) The effect of prior adult experience on components of habitat preference in the apple maggot fly (*Rhagoletis pomonella*). *Oecologia* 76, 538–543.

Papaj, D.R., Roitberg, B.D., Opp, S.B., Aluja, M., Prokopy, R.J. and Wong, T.T.Y. (1990) Effect of marking pheromone on clutch size in the Mediterranean fruit fly. *Physiological Entomology* 15, 463–468.

Piñero, J.C., Jácome, I., Vargas, R. and Prokopy, R.J. (2006) Response of female melon fly, *Bactrocera cucurbitae*, to host-associated visual and olfactory stimuli. *Entomologia Experimentalis et Applicata* 121, 261–269.

Plus, N., Gissman, L., Veyrunes, J.C., Pfister, H. and Gateff, E. (1981a) Reoviruses of *Drosophila* and *Ceratitis* populations and of *Drosophila* cell lines: a possible new genus of the Reoviridae family. *Annales de l'Institut Pasteur Virology* 132, 261–270.

Plus, N.L, Veyrunes, J.C. and Cavallaro, R. (1981b) Endogenous viruses of *Ceratitis capitata* Wied. J.R.C. Ispra strain, and of *C. capitata* permanent cell lines. *Annales de l'Institut Pasteur Virologie* 132, 91–100.

Potter, S.E. (2001) Bacterial associations in the walnut husk fly, *Rhagoletis completa* Cresson. MSc thesis, California State University, East Bay, Hayward, California.

Prokopy, R.J. (1966) Artificial oviposition devices for apple maggot. *Journal of Economic Entomology* 59, 384–387.

Prokopy, R.J. (1968) Visual responses of apple maggot flies, *Rhagoletis pomonella* (Diptera: Tephritidae): orchard studies. *Entomologia Experimentalis et Applicata* 11, 403–422.

Prokopy, R.J. (1972) Evidence for a marking pheromone deterring repeated oviposition in apple maggot flies. *Environmental Entomology* 1, 326–332.

Prokopy, R.J. (1983) Tephritid relationships with plants. In: Cavalloro, R. (ed.) *Fruit Flies of Economic Importance.* A.A. Balkema, Rotterdam, The Netherlands, pp. 230–239.

Prokopy, R.J. and Papaj, D.R. (1988) Learning of apple fruit biotypes by apple maggot flies. *Journal of Insect Behavior* 1, 67–74.

Prokopy, R.J. and Papaj, D.R. (2000) Behavior of flies of the genera *Rhagoletis*, *Zonosemata*, and *Carpomya* (Trypetinae: Carpomyina). In: Aluja, M. and Norrbom, A.L. (eds) *Fruit Flies (Tephritidae): Phylogeny and*

Evolution of Behavior. CRC Press, Boca Raton, Florida, pp. 219–252.
Prokopy, R.J. and Reynolds, A. (1998) Ovipositional enhancement through socially facilitated behavior in *Rhagoletis pomonella*. *Entomologia Experimentalis et Applicata* 86, 281–286.
Prokopy, R.J. and Roitberg, B.D. (1984) Foraging behavior of true fruit flies. *American Scientist* 72, 41–49.
Prokopy, R.J. and Roitberg, B.D. (2001) Joining and avoidance behavior in non-social insects. *Annual Review of Entomology* 46, 631–665.
Prokopy, R.J. and Roitberg, B.D. (2007) Behavioral control of insect pests. In: Kogan, M. and Jepson, P. (eds) *Ecological Approaches to Pest Management*. Cambridge University Press, Cambridge, UK, pp. 87–121.
Prokopy, R.J., Averill, A.L., Cooley, S.S., Roitberg, C.A. and Kallet, C. (1982) Associative learning in egg-laying site selection by apple maggot flies. *Science* 218, 76–77.
Prokopy, R.J., Papaj, D.R., Cooley, S.S. and Kallet, C. (1986) On the nature of learning in oviposition site acceptance by apple maggot flies. *Animal Behavior* 34, 98–107.
Prokopy, R.J., Diehl, S.R. and Cooley, S.S. (1988) Behavioral evidence for host races in *Rhagoletis pomonella* flies. *Oecologia* 76, 138–147.
Prokopy, R.J., Cooley, S.S. and Opp, S.B. (1989) Prior experience influences the fruit residence of male apple maggot flies, *Rhagoletis pomonella*. *Journal of Insect Behavior* 2, 39–48.
Prokopy, R.J., Cooley, S.S., Galarza, L., Bergweiler, C. and Lauzon, C.R. (1993) Bird droppings in natural food complete with bait sprays against *Rhagoletis pomonella* flies (Diptera: Tephritidae). *Canadian Entomologist* 125, 413–422.
Prokopy, R.J., Roitberg, B.D. and Vargas, R.I. (1994) Effects of egg load on finding and acceptance of host fruit in *Ceratitis capitata* flies. *Physiological Entomology* 19, 124–132.
Prokopy, R.J., Duan, J.J. and Vargas, R.I. (1996) Potential for host range expansion in *Ceratitis capitata* flies: impact of proximity of adult food to egg-laying sites. *Ecological Entomology* 21, 295–299.
Prokopy, R.J., Romig, M.C. and Drew, A.I. (1999) Facilitation in ovipositional behavior of *Bactrocera tryoni* flies. *Journal of Insect Behavior* 12, 815–832.
Prokopy, R.J., Miller, N.W., Duan, J.J. and Vargas, R.T. (2000) Local enhancement of arrivals of *Ceratitis capitata* females on fruit mimics. *Entomologia Experimentalis et Applicata* 97, 211–217.
Rankin, M.A. and Burchsted, J.C.A. (1992) The cost of migration in insects. *Annual Review of Entomology* 37, 533–559.
Remund, U., Boller, E.F., Economopoulos, A.P. and Tsitsipis, J.A. (1977) Flight performance of *Dacus oleae* reared on olives and artificial diet. *Zeitschrift fur Angewandte Entomologie* 82, 330–339.
Riegler, M. and Stauffer, C. (2002) *Wolbachia* infections as superinfections of cytoplasmic incompatibility populations of the European cherry fruit fly, *Rhagoletis cerasi* (Diptera: Tephritidae). *Molecular Ecology* 11, 2425–2434.
Robacker, D.C. and Bartelt, R.J. (1997) Chemicals attractive to Mexican fruit fly from *Klebsiella pneumoniae* and *Citrobacter freundii* cultures sampled by solid-phase microextraction. *Journal of Chemical Ecology* 23, 2897–2915.
Robacker, D.C. and Flath, R.A. (1995) Attractants from *Staphylococcus aureus* cultures for the Mexican fruit fly, *Anastrepha ludens*. *Journal of Chemical Ecology* 28, 1549–1563.
Robacker, D.C. and Lauzon, C.R. (2002) Purine metabolizing capability of *Enterobacter agglomerans* affects volatiles production and attractiveness to Mexican fruit fly. *Journal of Chemical Ecology* 28, 1549–1563.
Robacker, D.C., Lauzon, C.R. and He, X. (2004) Volatiles production and attractiveness to the Mexican fruit fly of *Enterobacter agglomerans* isolated from apple maggot and Mexican fruit flies. *Journal of Chemical Ecology* 30, 1329–1347.
Robertson, I., Roitberg, B.D., Williamson, I. and Senger, S. (1995) Contextual chemical ecology. *American Entomologist* 41, 237–240.
Rocha, L.S., Mascarenhas, R.O., Perondini, A.L.P. and Selivon, D. (2005) Occurrence of *Wolbachia* in Brazilian samples of

Ceratitis capitata (Wiedemann) (Diptera: Tephritidae). *Neotropical Entomology* 34, 1013–1015.

Roitberg, B.D. and Mangel, M. (1988) On the evolutionary ecology of marking pheromones. *Evolutionary Ecology* 2, 289–315.

Roitberg, B. and Mangel, M. (1997) Individuals on the landscape: behavior can mitigate differences among habitats. *Oikos* 80, 234–240.

Roitberg, B.D. and Prokopy, R.J. (1983) Host deprivation influence on response of *Rhagoletis pomonella* to its oviposition deterring pheromone. *Physiological Entomology* 8, 69–72.

Roitberg, B.D. and Prokopy, R.J. (1987) Insects that mark host plants. *Bioscience* 37, 400–406.

Rull, J. and Prokopy, R.J. (2004) Host-finding and ovipositional-boring responses of apple maggot (Diptera: Tephritidae) to different apple genotypes. *Environmental Entomology* 33, 1695–1702.

Rothstein, S.I. (1979) Gene frequencies and selection for inhibitory traits, with special emphasis on the adaptiveness of territoriality. *American Naturalist* 113, 317–331.

Schwarz, D., McPheron, B.A., Hartl, G.B., Boller, E.F. and Hoffmeister, T.S. (2003) A second case of genetic host races in Rhagoletis? A population genetic comparison of sympatric host populations in the European cherry fruit fly, *Rhagoletis cerasi*. *Entomologia Experimentalis et Applicata* 108, 11–17.

Schwarz, D., Shoemaker, K.D., Botteri, N.L. and McPheron, B.A. (2007) A novel preference for an invasive plant as a mechanism for animal hybrid speciation. *Evolution* 61, 245–256.

Selivon, D., Perondini, A.L.P., Ribeiro, A.F., Marino, C.I., Lima, M.M.A. and Coserato, V.F. (2002) *Wolbachia* endosymbionts in a species of *Anastrepha fraterculus* complex. *Invertebrate Reproduction and Development* 42, 121–127.

Senger, S., Roitberg, B.D. and Thistlewood, H. (2007) Relative flight responses of *Rhagoletis indifferens* (Diptera: Tephritidae) as influenced by crowding, sex, and resources. *Entomologia Experimentalis et Applicata* 123, 91–100.

Sharp, J.L. (1976) Comparison of flight ability of wild-type and laboratory-reared Caribbean fruit flies on a flight mill. *Journal of the Georgia Entomological Society* 11, 255–258.

Sharp, J.L. and Chambers, D.L. (1976) Gamma irradiation effect on the flight performance of *Dacus dorsalis* and *Ceratitis capitata*. *Proceedings of the Hawaiian Entomological Society* 22, 335–344.

Sharp, J.L., Chambers, D.L. and Haramoto, F.H. (1975) Flight mill and stroboscopic studies of Oriental fruit flies and melon flies, including observations of Mediterranean fruit flies. *Proceedings of the Hawaiian Entomological Society* 22, 137–144.

Shelly, T.E. (1999) Defense of oviposition sites by female Oriental fruit flies (Diptera: Tephritidae). *The Florida Entomologist* 82, 339–346.

Shelly, T.E. and Kennelly, S. (2002) Influence of male diet on male mating success and longevity and female remating in the Mediterranean fruit fly (Diptera: Tephritidae) under laboratory conditions. *The Florida Entomologist* 85, 572–579.

Simmons, F.H. and Bradley, B.J. (1997) An analysis of resource allocation in response to dietary yeast in *Drosophila melanogaster*. *Journal of Insect Physiology* 43, 779–788.

Slansky, F. Jr (1982) Insect nutrition: an adaptationist's perspective. *The Florida Entomologist* 65, 45–71.

Speirs, D., Sherratt, T. and Hubbard, S.F. (1991) Parasitoid diets: does superparasitism pay? *Trends in Ecology and Evolution* 6, 22–25.

Steinhaus, E.A. (1959) *Serratia marcescens* Bizio as an insect pathogen. *Hilgardia* 28, 351.

Taborsky, M. (1994) Sneakers, satellites, and helpers: parasitic and cooperative behaviour in fish reproduction. *Advances in the Study of Behavior* 23, 1–100.

Tappe, W., Laverman, A., Bohland, M., Braster, M., Ritterhaus, S., Groeneweg, J. and van Verseveld, H.W. (1999) Maintenance energy demand and starvation recovery dynamics of *Nitrosomonas europaea* and *Nitrobacter winogradskyi* cultivated in a retentostat with complete biomass retention. *Applied and Environmental Microbiology* 65, 2471–2477.

Tsiropoulos, G. (1992) Feeding and dietary requirements of the tephritid fruit flies. In: Anderson, T.E. and Leppla, N.C. (eds) *Advances in Insect Rearing for Research and Pest Management.* Westview Press, Boulder, Colorado, pp. 93–118.

Van Randen, E. and Roitberg, B.D. (1996) Egg load affects superparasitism in the snowberry fly. *Entomologia Experimentalis et Applicata* 79, 241–245.

Warburg, M.S. and Yuval, B. (1996) Effects of diet and activity on lipid levels of adult Mediterranean fruit flies. *Physiological Entomology* 21, 151–158.

Wilkerson, C. (2007) Effects of diet and mark–recapture methods on field dispersal of walnut husk fly. MSc thesis, California State University, East Bay, California.

Yuval, B. and Hendrichs, J. (2001) Behavior of flies in the genus Ceratitis (Dacine: Ceratitidini). In: Aluja, M. and Norrbom, A.L. (eds) *Fruit Flies (Tephritidae): Phylogeny and Evolution of Behavior*. CRC Press, Boca Raton, Florida, pp. 429–457.

Yuval, B., Kaspi, R., Field, S.A., Blay, S. and Taylor, P. (2002) Effects of post-teneral nutrition on reproductive success of male Mediterranean fruit flies (Diptera: Tephritidae). *The Florida Entomologist* 85, 165–170.

Zabalou, S., Riegler, M., Theodorakopoulou, M., Stauffer, C., Savakis, C. and Bourtzis, K. (2004) *Wolbachia*-induced cytoplasmic incompatibility as a means for insect pest control. *Proceedings of the National Academy of Sciences of the United States of America* 101, 15042–15045.

Zera, A.J. and Denno, R.F. (1997) Physiology and ecology of dispersal polymorphism in insects. *Annual Review of Entomology* 42, 207–231.

Zermeño, J. (2005) Laboratory flight and survivorship of two tephritid species. MSc thesis, California State University, East Bay, California.

4 How Do Key Tree-Fruit Pests Detect and Colonize Their Hosts: Mechanisms and Applications for IPM

Silvia Dorn[1] and Jaime C. Piñero[2]

[1]ETH Zurich, Institute of Plant Sciences/Applied Entomology, CH 8092 Zurich, Switzerland; [2]College of Tropical Agriculture and Human Resources, University of Hawaii at Manoa, Honolulu, Hawaii, USA

General Introduction

Insect herbivores of fruit trees are faced with the challenging task of locating suitable host plants among the array of stimuli that are associated with host and non-host plants and with the abiotic environment. Fruit trees as a perennial system are expected to provide reliable stimuli for a herbivore approaching at a specific time in the growing season. Though typically designed as monocultures, fruit crops may contain a large number of different plant species growing between and outside the tree rows, hence altering the microclimate as well as the olfactory and visual characteristics of host plants for an approaching herbivore. Stimuli from host plants are very likely to be more important than those from non-hosts in fruit-tree monocultures, given that herbivores are able to locate the tree resources (e.g. fruit) efficiently, often reaching pest status. Insect herbivores colonize trees by crawling or by flight and their mode of transport is likely to impact the types of monitoring and/or control systems that will be effective.

Fruits are high-value crops and are often consumed with little or no further processing, thus consumers expect a high-quality product in terms of visual appearance and organoleptic properties. Accordingly, the economic threshold for carpophagous pests is considerably, in some cases even outrageously, low. In the past, control has relied on frequent applications of broad-spectrum insecticides, which, however, contradicts the actual demand for healthy fruit grown in a healthy environment. To meet such demands, pioneers such as R.J. Prokopy have laid a scientific basis for the behavioural monitoring and control of key tree-fruit pests (Prokopy, 1994, 2003; Prokopy *et al.*, 2005).

Knowing the types of stimuli used by herbivores to locate resources such as mates, food, oviposition sites and shelter is essential for the development of effective monitoring and control tools (Prokopy and Owens, 1978; Prokopy *et al.*, 1991). Hence, a comprehensive understanding of the orientation and movement of the insect species (Dorn *et al.*, 1999; Prokopy *et al.*, 2005), as well as the factors that influence host selection, are crucial for the development of such behaviourally based control systems.

In their selection of host plants, insect herbivores are guided by a rich diversity of abiotic and biotic stimuli. Abiotic cues include weather-related parameters, such as temperature, that inform herbivores about suitable microhabitats (Kührt *et al.*, 2006). Biotic cues include physical information on the host plant, such as visual contrast against the background, shape, colour and texture (Prokopy and Owens, 1978, 1983), as well as olfactory and gustatory semiochemical cues (Masson and Mustaparta, 1990; Bernays and Chapman, 1994; Bruce, 2005). Some cues, such as achromatic visual stimuli (Prokopy and Owens, 1983) and olfaction (Visser, 1986; Hern and Dorn, 1999), are of particular importance for long-range host location. Others are only relevant after landing on a host plant, such as tactile physical cues and gustatory cues, and several stimuli, such as colour vision and olfaction, are of potential significance both before and after landing. The different sensory inputs are received at the periphery (e.g. by the antennae) and processed and integrated in the central nervous system of the insect herbivore, which subsequently responds by accepting or rejecting host resources, depending on their suitability (Schoonhoven *et al.*, 1998).

R.J. Prokopy, to whom this book is dedicated, studied the ecology and behaviour of key herbivore fruit-tree pests, in particular the apple maggot fly, *Rhagoletis pomonella* (Walsh) (Diptera: Tephritidae), and the plum curculio, *Conotrachelus nenuphar* (Herbst) (Coleoptera: Curculionidae), and successfully applied the data collected for monitoring and control purposes. As stated accurately by Papaj (2004), R.J. Prokopy invariably took 'a holistic perspective in which every aspect of the insect's biology from oviposition, mating and feeding, chemical cues and visual cues, larval and adult stages, to the effects of experience and physiological state was subject to scrutiny...'. The basic knowledge gained from this research, carried out with colleagues and students, is exemplary in breadth as well as depth. Two additional characteristics render his approach even more outstanding. At a regional level, this knowledge was applied in integrated pest management (IPM) systems for the direct benefit of growers. At an international level, R.J. Prokopy recognized early the related approaches developed by colleagues and embedded their contributions into symposia planned and chaired by him. This refers particularly to the work on the apple blossom weevil, *Anthonomus pomorum* (L.) (Coleoptera: Curculionidae), and also the interesting projects evolving with the fruit-damaging sap beetles (*Carpophilus* spp.) (Coleoptera: Nitidulidae), which did not escape his attention (e.g. Piñero and Prokopy, 2003). The four fruit-tree herbivore species mentioned: the apple maggot fly, the plum curculio, the apple blossom weevil and the Australian sap beetles, represent excellent examples where ecological and behavioural research has led to innovative recommendations for monitoring and control. While findings on fruit flies (Diptera: Tephritidae) are summarized by Aluja and Mangan (2008), by Leskey *et al.* (Chapter 5, this volume) and by Aluja and Rull (Chapter 7, this volume), we present here the three coleopteran species listed in Table 4.1. This chapter provides the first comprehensive picture on the plum curculio's host-location

Table 4.1. Three coleopteran herbivore species associated with fruit trees exemplifying the highly successful stepwise procedures from ecological and behavioural research to application in monitoring and/or control in IPM systems.

Herbivore	Family	Host fruit tree	Continent
Conotrachelus nenuphar (Herbst) Plum curculio	Curculionidae	Plum, peach, cherry, apricot, apple	North America
Anthonomus pomorum (L.) Apple blossom weevil	Curculionidae	Apple, pear	Europe
Carpophilus davidsoni Dobson Australian sap beetle	Nitidulidae	Apricot, cherry, peach, nectarine	Australia

behaviour and its application in IPM, and the first review on the apple blossom weevil and the Australian sap beetle, thereby providing a global perspective on fruit-damaging coleopteran pests.

We describe how the three coleopteran species disperse and colonize fruit trees and how detailed analyses of the behaviour of these herbivores led to the development of novel approaches for monitoring and control. These insects exploit different habitats in their annual cycle and move towards agricultural areas in the spring. The three coleopteran species selected only recently have been considered pests of fruit trees and each of these three herbivores is endemic to fruit-growing regions either in North America, Europe or Australia (Table 4.1). Our review will synthesize and interpret results from field surveys, field experiments and laboratory studies on the host-location processes of these pests, as well as the application of these findings in IPM systems in different continents.

Key Coleopteran Tree-Fruit Pests

The behaviours exhibited by herbivores during dispersal and tree colonization, as well as relevant stimuli (introduced above) used during the host-selection process before and after landing, are described. Particular emphasis is placed on demonstrating how knowledge gained from basic research is applied to monitoring and control.

In agreement with recent literature, we will apply the terms 'dispersal', 'immigration' and 'colonization' as follows:

- *Dispersal* is the movement of individuals to new breeding habitats (Hanski and Gilpin, 1997; Clobert *et al.*, 2001). It is a ubiquitous life-history trait of mobile organisms, with profound consequences on the spatial and temporal dynamics of populations (including immigration of a species into a distinct habitat).
- *Immigration* refers to the movement of insects into a distinct area, such as into an orchard (Nestel *et al.*, 2004).
- *Colonization* refers to the settlement of a species in the new habitat (Ims and Yoccoz, 1997).

The plum curculio, *Conotrachelus nenuphar*, a case study with a focus on visual and olfactory stimuli

Native to North America, the plum curculio, *C. nenuphar* (Herbst) (Curculionidae), has been the subject of intensive research for more than a century, given the serious threat it poses to the production of all stone and pome fruit in eastern and central North America (Racette *et al*., 1992; Vincent *et al*., 1999). Nearly 200 published research articles report on different aspects of its biology, ecology, behaviour and management. In this section, we provide a synopsis of the most representative experimental results on the visual and chemical ecology of plum curculio published in recent years, with applications for the monitoring and control of this pest in eastern North America. Particular emphasis will be on the behaviours exhibited by the northern strain plum curculios during the process of immigration into apple orchard blocks in response to semiochemicals in association with traps. For information about the biology and life cycle and other details on weevil–tree interaction, as well as a more in-depth overview of monitoring and control tactics currently available, see Leskey *et al*. (Chapter 5, this volume).

Dispersal

Most relevant for later monitoring and pest management decisions is the characterization of the process of immigration of the plum curculio into fruit orchards in early spring. Traps baited with attractive lures have been used in Massachusetts to characterize the seasonal course of plum curculio immigration into an unsprayed section of a commercial apple orchard over a 6-year period (Piñero and Prokopy, 2006). Traps evaluated were Tangletrap coated clear plexiglas panels (i.e. flight traps) and black pyramidal traps (Tedders and Wood, 1994). They were deployed in pairs along the periphery of the orchard block in close proximity to adjacent forested areas that represent the main overwintering sites. By intercepting immigrant weevils approaching host trees by means of flying or crawling, these two trap types were expected to provide more insight into the main mode of movement used by plum curculios during immigration, as well as the impact of daily ambient temperatures. According to Dixon *et al*. (1999), the mode of movement is temperature-dependent. Three key findings from this work with direct implications for management are: (i) the trap types evaluated are complementary in function: flight traps are very efficient at capturing immigrant plum curculios on warm days, whereas pyramid traps, which are less temperature-dependent (Piñero *et al*., 2001), detect immigration during cool days and during scotophase; (ii) most immigration by adult plum curculios has already taken place by petal fall, when resources for oviposition become available and when the first insecticide application commonly targeting all trees within a block takes place in many areas of eastern North America (Vincent *et al*., 1997); and (iii) the effect of air temperature on dispersal behaviour of this weevil is much stronger during the period preceding petal fall, which may encompass as few as 16 and as many as 29 days after first weevil captures in traps depending on weather, than during the period following petal fall. Using this information, the key stages of plum curculio immigration can be predicted using cumulative degree days (Piñero and Prokopy, 2006).

A full-block spray by the time of petal fall is needed, given the ability of overwintered plum curculios to penetrate into the interior of the blocks (Lafleur and Hill, 1987; Piñero, 2005). After petal fall, plum curculios have shown a tendency to remain in perimeter-row trees (Quaintance and Jenne, 1912; Rings, 1952; Chouinard *et al.*, 1992). This behavioural observation has allowed researchers to implement perimeter-row sprays for the successful control of plum curculio after petal fall under some conditions (Vincent *et al.*, 1997; Prokopy *et al.*, 2003, 2004).

Pre-landing: stimuli and behaviours

How adult plum curculios locate essential resources such as feeding and oviposition sites was not well understood until fairly recently. Intensive work conducted primarily by R.J. Prokopy, colleagues and students (e.g. Butkewich and Prokopy, 1993, 1997; Prokopy and Wright, 1998; Leskey and Prokopy, 2002) revealed that adult plum curculios responded positively to olfactory and visual cues, some of which are described below (for a thorough description see Leskey *et al.*, Chapter 5, this volume). The level of responsiveness of adult plum curculios to host-associated stimuli, however, is much higher for overwintered individuals that are imigrating from forested areas into orchard blocks than for weevils that are already present on host trees (Leskey and Wright, 2004; Piñero, 2005).

Visual cues may complement olfactory cues during the process of host-tree location by overwintered plum curculios (Butkewich and Prokopy, 1997). Trunk mimics (Prokopy and Wright, 1998), as well as branch mimics (Leskey and Prokopy, 2002), both of which offer strong contrast against background, have been shown to offer visual cues to approaching plum curculios.

The attraction of adult plum curculios to semiochemicals has been documented under both laboratory and field conditions (Leskey and Prokopy, 2000, 2001; Leskey *et al.*, 2001; Prokopy *et al.*, 2001). Remarkably, of several plant volatiles evaluated, benzaldehyde was the only one that enhanced synergistically the response of plum curculios to the synthetic, male-produced aggregation pheromone, grandisoic acid (Piñero and Prokopy, 2003). Benzaldehyde is a compound prevalent in plant species belonging to the plant family Rosaceae (Macht, 1922; Pettersson, 1970; Buchbauer *et al.*, 1993; Leskey *et al.*, 2001; Natale *et al.*, 2003; Vallat and Dorn, 2005), the most commonly attacked by plum curculio (Leskey and Wright, 2007).

Post-landing stimuli and behaviours

Once within the tree canopy, detection of individual fruit for oviposition apparently occurs at a distance of no greater than a few centimetres and may be based largely on gustatory cues (Butkewich and Prokopy, 1993). The olfactory response of adult plum curculios to synthetic lures decreases substantially after colonizing host trees, probably due to odour competition (Leskey and Wright, 2004). In order to achieve better management during fruit set and development, there is a current need to increase the attractive power of lures for plum curculio (see Leskey *et al.*, Chapter 5, this volume).

Application for IPM: exploiting a combination of visual and olfactory cues

As described above, both visual (e.g. tree-structure mimicking) and olfactory (e.g. host-plant volatile, aggregation pheromone) stimuli have been shown to be valuable components of plum curculio monitoring devices. When baited, panel and pyramid traps complement each other by capturing both males and females immigrating by either flying or crawling. This allows for accurate monitoring of plum curculio immigration, regardless of spring weather conditions.

The synergistic binary combination of benzaldehyde plus grandisoic acid has also been used to monitor the seasonal course of injury to fruit by plum curculio using the 'trap tree' approach (Prokopy *et al.*, 2003, 2004; Leskey *et al.*, Chapter 5, this volume). This efficient monitoring strategy has allowed growers to accurately determine the need for and timing of post-petal fall insecticide applications against plum curculio (Prokopy *et al.*, 2003, 2004; A. Agnello *et al.*, unpublished results). Trap trees also have the potential for use as a control strategy (Leskey *et al.*, 2008, Chapter 5, this volume). Hence, the exploitation of the sensory orientation of the plum curculio has proved to be very successful for the development of semiochemically based monitoring of adult immigration (Piñero and Prokopy, 2006) and egg-laying activity (Prokopy *et al.*, 2003, 2004; A. Agnello *et al.*, unpublished results), as well as for direct management of this pest (Leskey *et al.*, 2008), in eastern North America.

The apple blossom weevil, a case study with a focus on visual stimuli and behavioural responses to abiotic conditions

The plum curculio is not the only important curculionid pest of fruit trees. A related species attacking the very early growth stages of fruit trees is the apple blossom weevil, *A. pomorum* (L.). It is distributed widely throughout Europe and its primary hosts are apple and pear that are damaged at their bud stage (reviewed by Duan *et al.*, 1998).

Known as a key pest of apple of similar importance as the codling moth *Cydia pomonella* (L.), the apple blossom weevil caused complete fruit loss in several seasons in the 19th century (Ormerod, 1890). Since that time, it has been suppressed largely by broad-spectrum insecticides used to control codling moth and aphids (Wildbolz, 1992). Following the introduction of more selective pest management tools towards the end of the 20th century, this weevil has attained pest status again (Blommers, 1994; Hausmann *et al.*, 2004a,b). Infestations by the apple blossom weevil account for up to 50% bud losses (Hausmann *et al.*, 2004b) and such injury results in major economic losses, particularly in years with low fruit set. Initially, natural enemies were scarce. Their populations expanded with a time delay of several years following the increase of the herbivore's population densities. For example, populations of the prevailing ichneumonid parasitoid, *Scambus pomorum* (Ratz), rose by one order in magnitude within a decade. In 1996 and 1997, 0.8–1.5% infested blossoms contained *S. pomorum* in IPM orchards in northern Switzerland (n = 40,000 infested blossoms; Cross *et al.*, 1999), while in 2005 and 2006, a total of 4.8–16.7% infested blossoms contained *S. pomorum* (n = 10,000; K. Mody and S. Dorn, unpublished data).

The apple blossom weevil is univoltine. Weevils overwinter in the adult stage (Ormerod, 1890), then colonize pome fruit trees in early spring, where they subsequently mate. Females feed on buds to acquire nutrients for ovarian maturation. They then oviposit single eggs into flower buds and the hatched larvae feed inside the bud (Duan *et al.*, 1996). Eventually, this causes sterile and capped flowers (Ormerod, 1890).

Most knowledge regarding the apple blossom weevil was anecdotal, dating back to 1890–1930. Little was known about how this curculionid detects and colonizes host fruit trees prior to the studies carried out by the applied entomology group at ETH Zurich. Accordingly, monitoring was difficult and accurate timing of insecticide application was often impossible (S. Dorn, personal observation). There was an urgent need for research on behavioural ecology in this species.

Dispersal

'The red (= warm) winds bring the apple blossom weevil into the orchard' was often the causal explanation for the sudden appearance of populations of apple blossom weevils in orchards in the 19th century (de Boisduval, 1867, cited in Bain *et al.*, 1995). Although this insect is, of course, not windborne, later scientific experiments documented the significance of abiotic parameters on its spatial and temporal dynamics (Duan *et al.*, 1996, 1998; Toepfer *et al.*, 1999). Similar to the plum curculio, the apple blossom weevil in the adult stage disperses from its overwintering sites and colonizes fruit trees. Multiple choice bioassays indicated that its preferred overwintering sites are dry leaf litter found at forest borders, rough bark like that found on traditionally grown tall host trees, but not smooth bark that is characteristic of modern dwarf trees (Toepfer *et al.*, 2000). Thus, modern fruit-tree plantations offer few within-habitat overwintering sites but, in Europe, many of them are adjacent to a forest, where dry leaf litter is abundant. In fact, trees in rows closest to the forest bear higher blossom infestation by *A. pomorum*, supporting the hypothesis that most weevils immigrate from the forest border (Brown *et al.*, 1993). But how does the weevil disperse in space and time between agricultural and other habitats? How is dispersal characterized in spring when reproduction sites and food are the resources needed, and how does this occur during autumn when the need for shelter is the key issue? Such research is relevant for providing baseline data for the development of monitoring tools. In fact, the adult herbivore's needs for food resources, oviposition sites, mates or refuges ultimately underlie each effective monitoring system that has been developed (Prokopy *et al.*, 1999; Hausmann *et al.*, 2004a).

To characterize dispersal in spring, mark–release–recapture studies were carried out with individually marked weevils in two orchards in 2 subsequent years. Following release at the forest border, the spatial coordinates of the weevil were assessed regularly using limb jarring on sunny afternoons, i.e. under conditions yielding reliable results. Circular statistics (Toepfer *et al.*, 1999) were applied for, for data analysis. The weevils consistently moved from the forest border in the direction of the centre of the orchard. This oriented dispersal pattern was independent of variation in climatic conditions from year to year, and of the structural orchard design. Analysis of the environmental parameters indicated that ambient temperature was the driving factor for this dispersal. Maximal daily temperature determined the daily speed of dispersal (Toepfer *et al.*, 1999) and key phases of weevil immigration could be predicted using cumulative degrees days (Toepfer *et al.*, 2002). Hence, the abiotic environment is the driving factor for the dispersal of this species in spring. The documented dispersal behaviour between agricultural and other habitats indicates a high adaptation of the apple blossom weevil to the fruit orchard ecosystem that has existed in Europe for more than 1000 years (Toepfer *et al.*, 1999).

To evaluate weevil dispersal to overwintering sites in summer/autumn, individually marked weevils were released in two parts of an orchard. The first part contained tall trees with rough bark, the second part dwarf trees with smooth bark. The weevils were released in the centre of the two sites in 2 successive years. In the habitat with the rough bark trees, two-thirds of the weevils remained in close vicinity to the release site and only one-third dispersed, travelling short distances within the orchard (Toepfer *et al.*, 2000). In contrast, weevils from the site with the smooth bark trees travelled longer distances, but only a small proportion displayed a directional dispersal towards the forest and some weevils moved towards the area with rough bark trees (Toepfer *et al.*, 2000). This is likely an adaptive behaviour, as minimal dispersal reduces exposure to predators. In conclusion, dispersal from the reproduction site to the overwintering site late in the season is not reversing spring dispersal. A large proportion of the weevils does not undertake directional movement towards the forest border, but attempts to remain within the orchard. Experimental evidence indicates that apple orchards with traditional tall trees provide better hibernation sites than modern dwarf trees due to abundant leaf litter and rough bark, resulting in reduced mortality of overwintered adults (Toepfer *et al.*, 2000). Hence, these findings on dispersal in summer/autumn indicate a relatively low adaptation of the apple blossom weevil to dwarf trees introduced only few decades ago.

Pre-landing stimuli and behaviours

During orchard colonization, the apple blossom weevil uses both flight and crawling for locomotion (Hausmann *et al.*, 2004a). Flight events seem to be limited due to low flight proclivity (Toepfer *et al.*, 1999) and elevated temperature requirements (Duan *et al.*, 1998). Crawling occurs already at temperatures near the freezing point (Duan *et al.*, 1996) and trap captures by means of commercially available screen traps document the significance of this behaviour in the field (Hausmann *et al.*, 2004a). A comparative field study carried out over several weeks showed that there was a significant and positive correlation between the dynamics of captures in screen traps and in flight traps. By extrapolating the size of the passive flight trap to cover the complete vertical projection of each tree, a similar total number of weevils colonizing a tree by flight was estimated using screen traps as was assessed for weevils colonizing the tree by crawling (Hausmann *et al.*, 2004a). In contrast, captures by limb jarring were highly dependent on ambient temperature. They were correlated positively with maximum daily temperature and thus only reliable when used during warm days (Hausmann *et al.*, 2004a).

Vision and olfaction are the likely sensory modalities used by the apple blossom weevil approaching fruit trees (Kalinova *et al.*, 2000; Hausmann *et al.*, 2004a). Plant-derived chemical cues probably contribute to the weevil's orientation, although experimental evidence on the behavioural effect of single or combined compounds is lacking. Visual orientation relies predominantly on achromatic cues at large distances, but when the herbivore is in close proximity to a plant, it seems that the spectral quality of the light emitted by the plant becomes the predominant cue eliciting detection and alighting (Prokopy and Owens, 1983). The visual stimuli the apple blossom weevil encounters during the colonization of apple trees at the beginning of the growing period comprise the grey or brown bark of the tree, the green undercover vegetation and the sky. Experimentally, most weevils orient toward a high contrast (dark grey) vertical silhouette placed in front of green transparencies, when given a choice with a low contrast (light green) silhouette (Hausmann *et al.*, 2004c). These findings are of potential value for monitoring and control when behaviourally effective semiochemicals are identified. However, at this time, nothing is known about either bioactive plant volatile constituents or pheromones. Hence, behaviours driven by abiotic stimuli are currently of particular interest.

Post-landing stimuli and behaviours

The early spring season during which apple blossom weevils immigrate into orchards is typically characterized by some sunny warm afternoons, but temperatures may drop to freezing, particularly at night. Many insects have evolved morphological, physiological or behavioural means of coping with unsuitable abiotic environments. Some species actively exploit the thermal heterogeneity of their environment by moving from unfavourable to more favourable microhabitats. They engage in behavioural thermoregulation, thus regulating their body temperature by distinct behavioural steps (Kührt *et al.*, 2005). If such a behaviour was found in the apple blossom weevil, its thermal orientation could be used in a monitoring system.

To test the hypothesis of thermoregulation in *A. pomorum*, weevils were placed in a temperature gradient arena with a range of approximately 0–15°C. Both sexes of overwintered weevils were found to perform active behavioural thermoregulation by orienting towards the warm sectors. This preference for warm microhabitats was observed for several days after termination of diapause; later, females no longer showed thermal preference (Hausmann *et al.*, 2005). Hence, we expect weevils to choose warm microhabitats as a shelter, particularly at night, during colonization of the orchard. Field trials designed to develop a shelter trap verified this behaviour (Hausmann *et al.*, 2004b).

Application for IPM: exploiting thermal orientation

Exploiting thermal orientation led to the development of a monitoring system for the apple blossom weevil. The goal in trap design was that it should meet the weevil's need for a thermally favourable shelter. A shelter trap was developed in a stepwise manner, with the final type consisting of a transparent bubble wrap that allowed the weevils to enter. This simple design was intended to mimic rough bark. The wrap generates a shelter for the weevil to hide. Underneath the wrap, the tree bark was covered tightly with masking tape. A removable brown plastic band was fastened around the trap to protect the weevils from light (Hausmann *et al.*, 2004b). Diel refuge dynamics of the weevil culminated in the early morning between 3 am and 9 am. The number of weevils caught was correlated negatively to the ambient air temperature (Hausmann *et al.*, 2004b). The temperature within the shelter trap was slightly, but significantly, higher than that directly on the trunk (on average +0.25°C) and this small difference was sufficient to elicit oriented movement of the weevil to the shelter (Hausmann *et al.*, 2004b).

This finding indicates that weevils in the field exploit microclimate differences, as documented previously in the laboratory by Hausmann *et al.* (2005). They occupy thermally favourable refuges. The seasonal dynamics of shelter trap captures were not correlated with any climatic parameters. Captures were independent of weather conditions, in contrast to limb jarring, which required dry and sunny conditions to yield useful data (Toepfer *et al.*, 1999; Hausmann *et al.*, 2004a). The shelter trap offers, for the first time, the opportunity to monitor orchard colonization by the apple blossom weevil independent of weather conditions.

The next step was to assess the quantitative relationship between the number of weevils caught and injury on the fruit tree. As outlined above, fertilized females deposit single eggs into blossom heads, leading to severe injury, i.e. to capped blossoms. To test whether the shelter trap was suitable for forecasting injury, field trials were carried out in different orchards to evaluate whether few weevils caught indicated few blossoms affected, and vice versa. We determined that orchards with low numbers of weevils in traps also had a low number of infested buds on the apple trees, while orchards with a high number of trapped weevils showed a high number of infested buds later in the season. Similarly, there was a positive linear relationship between weevil counts and relative injury, which was expressed as a percentage of infested buds relative to all buds on the tree (Hausmann *et al.*, 2004b). Shelter trap captures are thus relevant for monitoring purposes. The method is suitable for predicting potential damage, as well as for indicating the best time for intervention with insecticides (Hausmann *et al.*, 2004b) and will contribute to more sustainable pest management in the framework of an IPM system in fruit orchards.

This case study on the apple blossom weevil is an example of the basic approach to solving insect pest problems favoured by R.J. Prokopy (e.g. Prokopy and Owens, 1978; Prokopy, 1994, 2003).

The Australian sap beetle, a case study with a focus on intra- and interspecific olfactory cues

Nitidulid beetles, primarily *Carpophilus* species, are worldwide pests of fruit and grain, both before and after harvest (James *et al.*, 1994). Economic losses due to infested ripening fruit in orchards have often been recorded as a consequence of *Carpophilus* spp. reservoirs in fallen stone fruit (Hossain and Williams, 2003), citrus (James *et al.*, 1996) or dates (Bartelt *et al.*, 1994).

Carpophilus spp. has attained pest status in stone-fruit orchards in southern Australia over the past few years (James *et al.*, 1997; James and Vogele, 2000). Previously, the beetle populations were suppressed by frequent applications of broad-spectrum insecticides applied to control Lepidoptera such as the oriental fruit moth, *Grapholita* (Cydia) *molesta* (Busck), and sucking pests such as the two-spotted spider mite, *Tetranychus urticae* (Koch) (James and Vogele, 2000).

Selective pest management with mating disruption for the lepidopteran species and biological control for the mite pest removed population suppression, resulting in increases in *Carpophilus* populations.

Three species of *Carpophilus*, *C. davidsoni* (Dobson), *C. hemipterus* (L.) and *C. mutilatus* (Erichson), have been identified in fruit orchards in Australia (James and Vogele, 2000). Beetles are attracted to ripening stone fruit, particularly apricot, peach and nectarine. They enter the fruit by chewing through the skin, usually around the stem end or in sutures, or through sites of mechanical damage (James *et al.*, 1994). Adults that penetrate the fruit cause direct damage and contribute to indirect damage as they passively transfer the spores of brown rot (*Monilia* spp.) and open an entry port for this pathogen into the fruit. Fruit losses of 5–20% and more have been noted (Hossain *et al.*, 2006). Different research groups reported consistently that *C. davidsoni* was the dominant species in all stone-fruit growing regions of Australia. They used a three-component pheromone blend attracting *C. davidsoni*, *C. hemipterus* and *C. mutilatus* (James and Vogele, 2000; Hossain *et al.*, 2006). *C. davidsoni* accounts for at least 70% (James *et al.*, 1997) or even 90–98% of the nitidulid beetles captured (James and Vogele, 2000; Hossain *et al.*, 2005). It was first described by Dobson (1952), who predicted that *C. davidsoni* would grow in economic importance, given the pest status of related species and its host range, which included sweetcorn, fig and peach. Apparently, that prediction has been fulfilled within less than half a century (Bartelt and James, 1994).

Dispersal

Little is known about the Australian sap beetle's natural hosts, overwintering, dispersal and colonization of new habitats. The original description of *C. davidsoni* does not mention any natural hosts, but states that the species has been collected from crop plants (Dobson, 1952), none of which is native to Australia. While the overwintering site is yet unknown, evidence suggests that *C. davidsoni* hibernate as non-reproductive adults (James *et al.*, 1994, 1997; Hossain and Williams, 2003). *Carpophilus* beetles are capable of moving over considerable distances, as captures in attractant-baited traps have been observed at distances of 350 m (James *et al.*, 1996) and of 500 m (James *et al.*, 2001) from the source.

Pre-landing stimuli and behaviours

Olfactory orientation has been well documented in the Australian sap beetle (Bartelt and James, 1994; Bartelt and Hossain, 2006), while other sensory modalities in host-location behaviour have not yet been investigated. Analysis of the male-released pheromone in *C. davidsoni* yielded a total of eight hydrocarbons, of which the following four prevailed quantitatively: (2*E*,4*E*,6*E*)-5-ethyl-3-methyl-2,4,6-nonatriene, (3*E*,5*E*,7*E*)-6-ethyl-4-methyl-3,5,7-decatriene, (2*E*,*E*,6*E*,8*E*)-3,5,7-trimethyl-2,4,6,8-undecatetraene and (2*E*,4*E*,6*E*,8*E*)-7-ethyl-3,5-dimethyl-2,4,6,8-undecatetraene (Bartelt and James, 1994). The relative abundance of these compounds is 100:7:9:31.

However, ratios of these hydrocarbons are altered when large groups of about 50 individuals assemble (Bartelt and James, 1994). It is assumed that the skewed ratio of compounds is detectable to approaching conspecifics, indicating suboptimal (i.e. crowded) feeding/breeding sites (Bartelt and James, 1994). To date, no study is published that compares the attraction of males and females to these odours.

Indications that the behavioural effect of the described aggregation pheromone can be synergized by food-related cues (Bartelt and James, 1994) and that fermenting food odours are attractive (Mansfield and Hossain, 2004) led to the study and elucidation of host-plant-derived attractants (Bartelt and Hossain, 2006). They are termed 'co-attractants', as the growers respond positively to the terms 'pheromones' or 'attractants' but not to 'semiochemicals' (Hossain *et al.*, 2006, 2007).

While it was found that few *C. davidsoni* beetles were attracted to volatiles emitted from unripe peach fruit, attraction was high to odours from overripe peach fruit and fermented peach juice (Bartelt and Hossain, 2006). Subsequent chemical and behavioural investigations focused on fermented peach fruit odours. These blends contained short-chain aldehydes and alcohols, as well as esters. A series of bioassays was conducted to understand the contribution of these constituents to the overall effect. The synthetic mixture simulating the aldehydes and alcohols was synergized effectively by the ester mixture, which was almost ineffective when tested alone. Reduction of the number of esters to the most abundant one, ethyl acetate, proved to be sufficient to elicit a behavioural response when combined with the aldehydes and alcohols in aqueous solution. This bioactive mixture contained acetaldehyde, ethanol, 2-methyl-1-propanol, 3-methyl-1-butanol, 2-methyl-1-butanol and ethyl acetate, at a ratio of 0.051:35:0.094:0.027:0.060:0.020 (Bartelt and Hossain, 2006). Field experiments in peach orchards validated the efficacy of this host-plant-derived attractant for *C. davidsoni*, the species that accounted for over 95% of all trap captures (Bartelt and Hossain, 2006).

These studies on intra- and interspecific attractants provide important baseline data for the development of an environmentally friendly IPM system for the monitoring and control of *C. davidsoni* and related species of the same genus in Australian stone-fruit trees.

Post-landing stimuli and behaviours

Investigations of the Australian sap beetle have been confined largely to the beetle's response to biotic cues before landing, rather than on behaviour after landing.

Application for IPM: exploiting a combination of intra- and interspecific cues

A unique approach has been taken to monitor and control *C. davidsoni*. Monitoring relies on interspecific plant-derived cues, while control is based on both intraspecific herbivore- and plant-derived olfactory stimuli. This fine-tuned application of behaviourally effective volatiles, using odours from conspecifics for monitoring and a combination of odours from conspecifics and host plant for control, is remarkable as it is not known from other IPM systems in fruit trees. Fruit moths, for example, are controlled using one single sex pheromone constituent only that is identical to the compound commonly used for monitoring (Witzgall *et al.*, 2008), and this odour attracts only males. Odour-baited traps developed for the Australian sap beetle attract both females and males. To which degree visual cues provided by the traps contribute to the overall effect is yet unknown. Insecticides are included in the traps to kill the beetles captured (Hossain *et al.*, 2006).

The monitoring device contains fermented fruit juice in a container covered tightly with a mosquito net and placed into a commercially available funnel trap (Magnet™ trap; Agrisense, Pontypridd, Glamorgan, UK) (Hossain *et al.*, 2006). Traps are positioned along a transect within the orchard. They are suspended on a branch in the tree canopy at about 1.5 m above the ground. The fruit juice is replaced at regular intervals (Hossain *et al.*, 2006). Weekly captures did not exceed a few dozen to a few hundred *Carpophilus* beetles during the period, when weekly captures in nearby traps used for control amounted to more than 200,000 beetles (Hossain *et al.*, 2006), suggesting that the monitoring tool baited only with pheromone was less attractive than the control tool.

This control tool, termed attract-and-kill station, contains aggregation pheromone in combination with plant-derived olfactory cues, either natural or nature-identical synthetic analogues (Hossain *et al.*, 2006, 2007). The three-component pheromone that attracts *C. davidsoni* as well as *C. hemipterus* and *C. mutilatus* is dispensed from rubber septa that are placed above the plant odour and are replaced at regular intervals. The plant-derived olfactory cues were subject to continuous improvement: first, to yield more consistent results; second, to compete more effectively with surrounding environmental stimuli, particularly with the attractive odours of ripening peach (Bartelt and Hossain, 2006); and third, to render the system more user friendly and to minimize working time to establish and service the attract-and-kill station (Hossain *et al.*, 2007). To fulfil all these requirements, most recent studies moved from a system containing fermented peach juice and ripening peach (set-up illustrated in Hossain *et al.*, 2006) to the use of a water solution of synthetic compounds mimicking fermented peach juice (Hossain *et al.*, 2007).

Trap positioning is important and can be optimized by adopting a systematic approach in which the behaviour and the response of the target insect are observed closely (Cook *et al.*, 2007). To control *Carpophilus* beetles effectively, the odour-baited traps containing an insecticide (i.e. the attract-and-kill stations) are always positioned at the upwind end of the plot to be treated. To assess the effect of trap height on protecting fruit from beetle attack, traps were positioned at ground level or at 2.5 m above ground level (approximately at upper canopy height). The height of the traps did not appear to influence beetle response (James *et al.*, 1996). To assess the effect of the distance of the traps from the trees on damage prevention, traps were hung on sticks either within the plot at a distance of 1 m from the tree canopy, or outside the plot at a distance of 20–30 m from the perimeter of the plot. Trap positioning outside the plot reduced fruit damage significantly compared to the untreated control, while positioning adjacent to trees increased fruit infestation (James *et al.*, 1996). Accordingly, subsequent trials relied on trap positioning outside the plot to be treated. They were placed a distance of 10–15 m from the perimeter of the plots (Hossain *et al.*, 2006, 2007).

Excellent *Carpophilus* control was obtained with a combination of optimal trap positioning and the best natural or synthetic plant odours used together with the aggregation pheromone in the attract-and-kill stations. In a field trial with natural fruit odours consisting of fermented peach juice and ripening peach, fruit damage was reduced to 0.1% in the tree canopy and to 0.6% on the ground, compared to 5.2% and 19.9% in the control, which received half of the usually applied organophosphate sprays (Hossain *et al.*, 2006). The fact that behavioural control in the odour bait-treated plots reduces fruit damage to almost zero under such strong infestation pressure indicates the potential for practical application of the attract-and-kill station (Hossain *et al.*, 2006). In a subsequent field trial, the natural fruit odours were replaced by the synthetic odour mixture described above. Plots with the attract-and-kill stations were compared to plots subject to organophosphate treatments that reflect common growers' practice. Results indicate a fivefold lower level of fruit damage in the experimental plots with the odour-baited traps (0.24%), compared to control plots (1.38%) (Hossain *et al.*, 2007). Hence, the concept is proven that behavioural control can protect fruit orchards successfully from *Carpophilus* damage. Control efficacy of the attract-and-kill station even outcompeted that of the standard insecticide treatment. It is expected that implementation of the behavioural control tool into IPM systems will be facilitated by the planned further reduction of labour costs needed to service the traps (Hossain *et al.*, 2007).

General Discussion

The successful development of monitoring and control tools in all three herbivore pests of fruit trees discussed above relies on knowledge of how the herbivores immigrate into the orchard and colonize the host trees, as well as how they meet their needs for resources.

Key aspects of immigration and trap types

Both the plum curculio and the apple blossom weevil leave their overwintering sites in large numbers when a certain temperature constant is reached and immigrate into the orchard. Degree-day models can thus be used as an indication of when monitoring traps should be installed (Toepfer *et al.*, 2002; Piñero and Prokopy, 2006). These weevils colonize the trees by crawling and by flight (Blanchet, 1987; Prokopy *et al.*, 1999; Hausmann *et al.*, 2004a). A clear temperature dependence was found for plum curculio captured by odour-baited flight (panel) traps that were more effective during warm days, compared to odour-baited pyramid traps that were less dependent on air temperature (Piñero *et al.*, 2001). Hence, a combination of both trap types baited with benzaldehyde plus grandisoic acid is currently recommended for monitoring purposes (for a novel application of semiochemicals for direct control of plum curculio, see Leskey *et al.*, Chapter 5, this volume). In the apple blossom weevil, meaningful application of the traditionally used limb jarring is limited to warm and sunny afternoons (Toepfer *et al.*, 1999; Hausmann *et al.*, 2004a). In contrast, the newly developed shelter trap captures crawling weevils seeking shelter in the tree bark.

Irrespective of weather conditions, there was a strong and positive correlation with captures in the flight traps (Hausmann *et al.*, 2004a). The more sensitive and easy to handle shelter trap (Hausmann *et al.*, 2004a,b) evoked the interest of extension entomologists in many European countries for implementation as a monitoring tool. In conclusion, the most successful trapping systems captured the coleopteran pests under a wide range of ambient temperature conditions and they were developed based on recent advances made in the understanding of immigration and tree colonization by the target insect species.

The herbivore's need for resources and its exploitation for IPM

The need of any herbivore for essential resources is exploited in different ways in the three exemplary cases presented. The monitoring tool developed for the curculionid *A. pomorum* relies simply on this weevil's need for shelter during night in very early spring, and the shelter trap offered fulfils its need (Hausmann *et al.*, 2004a,b). As emphasized above, neither pheromones nor behaviourally active host-plant volatile constituents have been identified in this species thus far. Development of shelter traps might be an option for other tree herbivores that colonize their host plant to a significant degree by crawling and for which olfactory stimuli have not been identified.

Monitoring of the curculionid *C. nenuphar* and control of the nitidulid *C. davidsoni* rely on a synergistic interaction of herbivore-derived and plant-derived semiochemicals. Thus, pheromones and host-plant kairomones, either natural blends or synthetic constituent analogues, are released simultaneously from traps to attract both genders of these coleopteran fruit-tree herbivores (Piñero and Prokopy, 2003, 2006; Bartelt and Hossain, 2006). These stimuli exploit these species' need for mates, food and oviposition sites. Pest control by mass trapping can also rely on such cues, as documented by current IPM practice of other coleopteran pests in other non-tree-fruit agroecosystems (e.g. *Rhynchophorus palmarum* L. (Alpizar *et al.*, 2002; Oehlschlager *et al.*, 2002a), *Metamasius hemipterus* L. (Alpizar *et al.*, 2002; Oehlschlager *et al.*, 2002b)). Pest control systems like these ought to be developed for other fruit-tree pests based on the type of knowledge gained with the three coleopteran species described above. A novel approach to plum curculio management using odour-baited trap trees (Prokopy *et al.*, 2003, 2004), which results in a substantial reduction in insecticide use, has been developed recently by Leskey *et al.* (2008) (see Leskey *et al.*, Chapter 5, this volume). Visual cues should be added to these systems, for example in the form of trunk mimics (Tedders and Wood, 1994) that yield a strong contrast against the background (Leskey, 2006).

Interactions among cues and their significance for reliable IPM strategies

Synergism was demonstrated between herbivore-associated pheromones and plant-derived kairomones that mediate host location in the plum curculio and the Australian sap beetle (Table 4.2). We hypothesize that these herbivores of

Table 4.2. Interactions among and within sensory modalities during host-location behaviour in four selected fruit-tree pests belonging to the orders Coleoptera, Diptera and Lepidoptera.

Insect species	Key host plants	Main sensory modality	Type of synergism or interaction reported	Reference(s)
Conotrachelus nenuphar (plum curculio)	Plum, peach, cherry, apricot, apple	Olfaction	Synergism between the plant-derived compound benzaldehyde and the synthetic male-produced pheromone (grandisoic acid)	Piñero and Prokopy (2003)
Carpophilus davidsoni (Australian sap beetle)	Peach, nectarine	Olfaction	Synergism No. 1 within plant-derived compounds: short-chain alcohols and aldehydes synergized by esters; synergism No. 2 between plant-derived compounds and the 4-compound male-produced pheromone	Bartelt and Hossain (2006)
Rhagoletis pomonella (apple maggot fly)	Hawthorn, apple	Olfaction/vision	Interaction between olfactory (5 synthetic apple esters) and visual (red spheres) stimuli	Aluja and Prokopy (1993), Green *et al.* (1994), Zhang *et al.* (1999)
Grapholita (Cydia) molesta (oriental fruit moth)	Peach, nectarine, apple	Olfaction	Synergism between three general green-leaf volatiles and two specific aromatic compounds	Piñero and Dorn (2007), Piñero *et al.* (2008)

fruit trees use synergistic interactions within the single sensory modality of olfaction as a fail-safe mechanism. Simultaneous perception of odours from two different sources elicits a significantly stronger behavioural response in these herbivores than either of the two constituents alone. In fact, many behaviourally effective volatiles emitted by plants are ubiquitous (Tasin *et al.*, 2006). They are released in distinct ratios from the plant, but such ratios are modified with the progressing season (Vallat and Dorn, 2005), as well as with changing weather conditions (Vallat *et al.*, 2005). Hence, a combination of host-plant volatiles with pheromones from conspecific insects conveys highly reliable information for an adult herbivore foraging for a suitable habitat and for a mate.

Interactions between insect sex pheromones and plant semiochemicals have been investigated in numerous crop pests (reviewed by Landolt and Phillips, 1997; Reddy and Guerrero, 2004). Chemicals from host plants often enhance synergistically the response of an insect to sex pheromones. For example, host-plant compounds identified in volatile blends collected in apple orchards (Vallat and Dorn, 2005) synergize the attraction of codling moth males to synthetic sex pheromone (Witzgall *et al.*, 2008). However, no plant volatile constituents are added to the pheromone dispensers that are well-established tools for codling moth control. This tool relies solely on one pheromone constituent, codlemone, and affects only males (Witzgall *et al.*, 2008).

Interactions between insect aggregation pheromones and plant semiochemicals have been discovered in the plum curculio and the Australian sap beetle, but not in other herbivore species of fruit trees so far. Aggregation pheromones attract both sexes and simultaneous perception of these insect-derived cues, together with host-plant volatiles, may indicate an ideal combination of essential resources (for feeding, mating and oviposition) to male and female herbivores in nature. The exploitation of such combinations in IPM holds great promise for particularly effective traps. In forest pest beetles, aggregation pheromones represent the primary pull stimulus used in push–pull strategies (Cook *et al.*, 2007) and these cues are often enhanced or synergized by host-plant odours (Lindgren and Borden, 1993; Reddy and Guerrero, 2004). Combinations of aggregation pheromones with natural blends of host-plant volatiles are also used in the IPM of coleopteran pests in sugarcane (Oehlschlager *et al.*, 2002b) and palm trees (Alpizar *et al.*, 2002), but the bioactive plant constituents have not been identified yet. Our postulate that the odour combination of constituents emanating from conspecific insects and from the host plant yield a more reliable signal for many herbivores, is supported by findings on pheromone combinations with non-host volatiles. The attractant effect of the aggregation pheromone in bark beetles is antagonized by diverse non-host trees and plants common in the beetle's habitat (Byers *et al.*, 2004). These authors conclude that the herbivore's olfactory avoidance of non-host plants, despite the presence of the pheromone, may have evolved to minimize mistakes during host selection. In a complex olfactory landscape, reliance on both the aggregation pheromone and the host-plant odour will benefit the herbivore. The door is wide open for investigations of such chemically mediated behaviours and their application in behavioural control of many fruit-tree herbivores.

Synergism was not only demonstrated between pheromones and plant semiochemicals, but also between two constituents of the host-plant odours in the Australian sap beetle (Bartelt and Hossain, 2006) (Table 4.2). Such sophisticated exploitation of olfactory cues from the biotic environment in the fruit-tree agroecosystem is not confined to the order Coleoptera. In fact, empirical evidence for synergism in the attraction of gravid females to synthetic host-plant-derived mixtures was documented recently in the oriental fruit moth, *Grapholita (Cydia) molesta* (Busck) (Lepidoptera: Tortricidae) (Table 4.2). The oriental fruit moth represents a serious threat to the production of all stone fruit and also, more recently, of pome fruit, particularly apple, in several regions (Pollini and Bariselli, 1993; Hughes and Dorn, 2002). Over 20 compounds have been identified in the headspace of peach shoot volatiles (Natale *et al.*, 2003), the main host plant of this herbivore. A particular mixture comprised of three general green-leaf volatiles and two specific aromatic compounds resulted in a bioactive mixture that was as attractive to gravid females as natural peach shoot volatiles (Piñero and Dorn, 2007). Exclusion of one compound resulted in a significant loss of attractiveness of the resulting mixtures. Remarkably, the synergistic behavioural responses of gravid females to the bioactive mixture in olfactometer bioassays were found to be mirrored by calcium activity in their antennal lobes in neurophysiological experiments involving optical imaging (Piñero *et al.*, 2008). A new type of glomerulus that is highly sensitive to odour mixtures has been identified, meaning that mixtures in this species are perceived as qualitatively different from single compounds or from 'incomplete' mixtures in the insect's antennal lobes.

The above behavioural and neurophysiological findings with the oriental fruit moth support strongly the postulated significance of a plant-odour combination consisting of general and specific odours (Harrewijn *et al.*, 1995) acting in concert to offer a high-quality signal to female moths. Thus, interactions between distinct odour constituents can render the information conveyed by these volatiles more reliable for the approaching herbivore. Such fine-tuned olfactory interactions hold the potential of being exploited in target insect-specific traps.

Potentially useful interactions occur not only within one sensory modality such as olfaction, but also can derive from different sensory modalities such as vision and olfaction. However, this possibility has not yet been investigated fully for the coleopteran species described in detail in this chapter. An interaction between visual and olfactory cues has been reported in different species, primarily in the dipteran family Tephritidae, including the fruit-tree herbivore *R. pomonella* (Aluja and Prokopy, 1993; Green *et al.*, 1994) (Table 4.2). For instance, Aluja and Prokopy (1993) documented that, after arrival on apple trees, flies seemed to locate individual fruit on the basis of vision, provided that a fruit offered strong visual cues (e.g. dark colour). However, the effect of odour became apparent when visual stimulus provided by a fruit was comparatively weak (e.g. light colour). Further information on visual and chemical stimuli interactions during the host-plant finding process in tephritid flies and applications for IPM is provided by Leskey *et al.* (Chapter 5, this volume) and by Aluja and Rull (Chapter 7, this volume).

Even an apparently additive effect of two different modalities may be exploited by the host-seeking insect, as demonstrated recently in the ichneumonid parasitoid *Pimpla turionellae* (L.), which is often found in fruit-tree orchards. Using both visual and mechanosensory cues that interact, female wasps improve their accuracy of host location in an additive way (Fischer *et al.*, 2001) under moderate temperature conditions. However, under low temperatures, this species relies almost exclusively on visual cues because mechanosensory cues are thermally sensitive (Kroder *et al.*, 2006, 2007). We postulate that multisensory orientation may represent another fail-safe mechanism leading to host recognition by insect herbivores. Exploitation of multiple sensory modalities can have important implications for pest monitoring and control, as IPM tools that do not rely unilaterally on one cue are likely to work more reliably under rapidly changing environmental conditions.

Conclusions

In IPM systems of coleopteran pests of fruit trees, day-degree models proved to be useful as a first approximation for the critical period when traps should be installed. Olfactory and/or visual location of the host tree is exploited with traps for monitoring and control. The interactions within or between cues were often found to be synergistic, thereby were likely to increase the chances of host and mate location in nature, and thus held high potential for IPM. Traps exploiting such interactions can lead to a more reliable attraction of the target insect under variable environmental conditions. Interactions among modalities of host-finding cues offer a wide field of opportunities for future research and tool development.

For a number of pest species of fruit trees, neither pheromones nor attractive plant volatiles have been identified yet. In case these insects colonize the trees mainly by crawling, their need for shelter can be exploited successfully. This type of trap relies on the herbivore's response to abiotic cues. Insects as ectotherms often show thermotactic behaviour that might be utilized more widely in future IPM systems.

Optimization of traps should rely on important external and internal factors related to the resource-foraging behaviour of the insect target. External factors include the current state of the environment, such as the spatial and temporal aspects of abundance, quality and distribution of resources, the presence of conspecifics and natural enemies, as well as abiotic conditions (Mangel and Roitberg, 1989; Prokopy and Roitberg, 2007). For example, traps may be placed optimally inside or outside the orchard plot near the herbivore's overwintering sites, as suggested for the plum curculio, or in the perimeter of the plot, as suggested for the Australian sap beetle. The internal state of the pest herbivore includes its genetic origin, its physiological state, such as hormone levels, level of starvation, mating status and egg load, as well as its informational state, including its memory (Prokopy *et al.*, 2005, and references therein). For example, the time window during which the apple blossom weevil responds to shelter traps is limited to the first days after termination of diapause (Hausmann *et al.*, 2005). Knowledge of these factors will contribute to the optimized use of traps in space and time.

Finally, spanning the thread from basic research to tool development, and even further to application in IPM systems, is not a simple task. This is illustrated by the very few reports on tree-fruit herbivores for which the process of basic research has led to applicable IPM tools. Additional successful biorational monitoring and control IPM systems should be developed based on ecological and behavioural knowledge gained for further pest species associated with fruit trees. To accomplish this enormous task, we very much need talented researchers such as R.J. Prokopy, an admirable entomologist who, throughout his inexhaustible academic career, had the outstanding ability of 'blurring the line between basic and applied research' (Papaj, 2004).

Acknowledgements

We thank Dr Kathrin Tschudi-Rein, Dr Mofakahr Hossain, Lester M. Klungness, the editors of this book and two anonymous reviewers for useful comments on earlier drafts of the manuscript. Silvia Dorn thanks DPI Victoria at Tatura, Australia, and its entomologists for their hospitality and fruitful scientific exchange during her recent sabbatical leave.

References

Alpizar, D., Fallas, M., Oehlschlager, A.C., Gonzalez, L.M., Chinchilla, C.M. and Bulgarelli, J. (2002) Pheromone mass trapping of the west Indian sugarcane weevil and the American palm weevil (Coleoptera: Curculionidae) in palmito palm. *Florida Entomologist* 85, 426–430.

Aluja, M. and Mangan, R.L. (2008) Fruit fly (Diptera: Tephritidae) host status determination: critical conceptual, methodological, and regulatory considerations *Annual Review of Entomology* 53, 473–502.

Aluja, M. and Prokopy, R.J. (1993) Host odor and visual stimulus interaction during intratree host finding behavior of *Rhagoletis pomonella* flies. *Journal of Chemical Ecology* 19, 2671–2696.

Bain, C., Bernard, J.L. and Fougeroux, A. (1995) *Protection des cultures et travail des hommes.* Le Carrousel, Paris.

Bartelt, R.J. and Hossain, M.S. (2006) Development of synthetic food-related attractant for *Carpophilus davidsoni* and its effectiveness in the stone fruit orchards in southern Australia. *Journal of Chemical Ecology* 32, 2145–2162.

Bartelt, R.J. and James, D.G. (1994) Aggregation pheromone of Australian sap beetle, *Carpophilus davidsoni* (Coleoptera: Nitidulidae). *Journal of Chemical Ecology* 20, 3207–3219.

Bartelt, R.J., Vetter, R.S., Carlson, D.G. and Baker, T.C. (1994) Responses to aggregation pheromones for 5 *Carpophilus* species (Coleoptera: Nitidulidae) in a California date garden. *Environmental Entomology* 23, 1534–1543.

Bernays, E.A. and Chapman, R.F. (1994) *Host-Plant Selection by Phytophagous Insects. Contemporary Topics in Entomology 2.* Chapman and Hall, New York.

Blanchet, R. (1987) Movement of plum curculio (*Conotrachelus nenuphar*) from a woodlot to an orchard in southwestern Quebec. BSc thesis, McGill University, Montreal, Canada.

Blommers, L.H.M. (1994) Integrated pest management in European apple orchards. *Annual Review of Entomology* 39, 213–241.

Brown, M.W., Szentkiralyi, F. and Kozar, F. (1993) Spatial and temporal variation of apple blossom weevil populations (Coleoptera: Curculionidae) with recommendations for sampling. *Journal of Applied Entomology – Zeitschrift für Angewandte Entomologie* 115, 8–13.

Bruce, T.J.A. (2005) Insect host location: a volatile situation. *Trends in Plant Science* 10, 269–274.

Buchbauer, G., Jirovetz, L., Wasicky, M. and Nikiforov, A. (1993) Headspace and essential oil analysis of apple flowers. *Journal of Agriculture and Food Chemistry* 41, 116–118.

Butkewich, S.L. and Prokopy, R.J. (1993) The effect of short-range host odor stimuli on host finding and feeding behavior of plum curculio adults (Coleoptera: Curculionidae). *Journal of Chemical Ecology* 19, 825–835.

Butkewich, S.L. and Prokopy, R.J. (1997) Attraction of adult plum curculios to host-tree odor and visual stimuli in the field. *Journal of Entomological Science* 32, 1–6.

Byers, J.A., Zhang, Q.H. and Birgersson, G. (2004) Avoidance of non-host plants by a bark beetle, *Pityogenes bidentatus*, in a forest of odors. *Naturwissenschaften* 91, 215–219.

Chouinard, G., Hill, S.B., Vincent, C. and Barthakur, N.N. (1992) Border-row sprays against the plum curculio (Coleoptera: Curculionidae) in apple orchards: a behavioral study. *Journal of Economic Entomology* 85, 1307–1317.

Clobert, J., Danchin, E., Dhondt, A.A. and Nichols, J.D. (2001) *Dispersal*. Oxford University Press, Inc, New York.

Cook, S.M., Khan, Z.R. and Pickett, J.A. (2007) The use of push–pull strategies in Integrated Pest Management. *Annual Review of Entomology* 52, 375–400.

Cross, J.V., Solomon, M.G., Babendreier, D., Blommers, L., Easterbrook, M.A., Jay, C.N., Jenser, G., Jolly, R.L., Kuhlmann, U., Lilley, R., Olivella, E., Toepfer, S. and Vidal, S. (1999) Biocontrol of pests of apples and pears in northern and central Europe: 2. Parasitoids. *Biocontrol Science and Technology* 9, 277–314.

Dixon, B.M., Prokopy, R.J. and Schultz, B.B. (1999) Influence of weather and time of day on plum curculio (Coleoptera: Curculionidae) tree canopy entry behaviors and evaluation of traps for predicting fruit injury. *Journal of Entomological Science* 34, 191–202.

Dobson, R.M. (1952) A new species of *Carpophilus stephens* (Coleoptera: Nitidulidae) from Australia. *Entomological Monograph Magazine* 88, 256–258.

Dorn, S., Schumacher, P., Abivardi, C. and Meyhöfer, R. (1999) Global and regional pest insects and their antagonists in orchards: spatial dynamics. *Agriculture Ecosystems and Environment* 73, 111–118.

Duan, J.J., Weber, D.C., Hirs, B. and Dorn, S. (1996) Spring behavioral patterns of the apple blossom weevil. *Entomologia Experimentalis et Applicata* 79, 9–17.

Duan, J.J., Weber, D.C. and Dorn, S. (1998) Flight behavior of pre- and postdiapause apple blossom weevils in relation to ambient temperature. *Entomologia Experimentalis et Applicata* 88, 97–99.

Fischer, S., Samietz, J., Wäckers, F.L. and Dorn, S. (2001) Interaction of vibrational and visual cues in parasitoid host location. *Journal of Comparative Physiology A – Neuroethology Sensory Neural and Behavioral Physiology* 187, 785–791.

Green, T.A., Prokopy, R.J. and Hosmer, D.W. (1994) Distance of response to host tree models by female apple maggot flies, *Rhagoletis pomonella* (Walsh) (Diptera: Tephritidae): interaction of visual and olfactory stimuli. *Journal of Chemical Ecology* 20, 2393–2413.

Hanski, I. and Gilpin, M.E. (1997) *Metapopulation Biology: Ecology, Genetics, and Evolution*. Academic Press, San Diego, California.

Harrewijn, P., Minks, A.K. and Mollema, C. (1995) Evolution of plant volatile production in insect–plant relationships. *Chemoecology* 2, 55–73.

Hausmann, C., Samietz, J. and Dorn, S. (2004a) Monitoring the dynamics of orchard colonisation by *Anthonomus pomorum* in spring. *Entomologia Experimentalis et Applicata* 110, 207–216.

Hausmann, C., Samietz, J. and Dorn, S. (2004b) Significance of shelter traps for spring monitoring of *Anthonomus pomorum* in apple orchards. *Entomologia Experimentalis et Applicata* 112, 29–36.

Hausmann, C., Samietz, J. and Dorn, S. (2004c) Visual orientation of overwintered *Anthonomus pomorum* (Coleoptera: Curculionidae). *Environmental Entomology* 33, 1410–1415.

Hausmann, C., Samietz, J. and Dorn, S. (2005) Thermal orientation of *Anthonomus pomorum* (Coleoptera: Curculionidae) in early spring. *Physiological Entomology* 30, 48–53.

Hern, A. and Dorn, S. (1999) Sexual dimorphism in the olfactory orientation of adult *Cydia pomonella* in response to alpha-farnesene. *Entomologia Experimentalis et Applicata* 92, 63–72.

Hossain, M.S. and Williams, D.G. (2003) Phenology of *Carpophilus* beetle populations (Coleoptera: Nitidulidae, *Carpophilus* spp.) in a fruit dump in northern Victoria. *Australian Journal of Experimental Agriculture* 43, 1275–1279.

Hossain, M.S., Williams, D.G. and Milner, A.D. (2005) *Carpophilus* spp. (Coleoptera: Nitidulidae) responses to aggregation pheromone plus decomposed fruit as co-attractant in stone fruit orchards in southern Australia. *General Applied Entomology* 34, 33–42.

Hossain, M.S., Williams, D.G., Mansfield, C., Bartelt, R.J., Callinan, L. and Il'ichev, A.L. (2006) An attract-and-kill system to control *Carpophilus* spp. in Australian stone fruit orchards. *Entomologia Experimentalis et Applicata* 118, 11–19.

Hossain, M.S., Williams, D.G., Hossain, M.A.B.M. and Norng, S. (2007) Comparison of trap designs for use with aggregation pheromone and synthetic co-attractant in a user-friendly attract-and-kill system to control *Carpophilus* spp. (Coleoptera: Nitidulidae). *Australian Journal of Entomology* 46, 244–250.

Hughes, J. and Dorn, S. (2002) Sexual differences in the flight performance of the oriental fruit moth, *Cydia molesta*. *Entomologia Experimentalis et Applicata* 103, 171–182.

Ims, R.A. and Yoccoz, N.G. (1997) Studying transfer processes in metapopulations: emigration, migration and colonization. In: Hanski, I.A. and Gilpin, M.E. (eds) *Metapopulation Biology: Ecology, Genetics, and Evolution*. Academic Press, San Diego, California, pp. 247–265.

James, D.G. and Vogele, B. (2000) Development and survivorship of *Carpophilus hemipterus* (L.), *Carpophilus mutilatus* Erichson and *Carpophilus humeralis* (F.) (Coleoptera: Nitidulidae) over a range of constant temperatures. *Australian Journal of Entomology* 39, 180–184.

James, D.G., Bartelt, R.J. and Faulder, R.J. (1994) Attraction of *Carpophilus* spp. (Coleoptera: Nitidulidae) to synthetic aggregation pheromones and host-related coattractants in Australian stone fruit orchards – beetle phenology and pheromone dose studies. *Journal of Chemical Ecology* 20, 2805–2819.

James, D.G., Bartelt, R.J. and Moore, C.J. (1996) Mass-trapping of *Carpophilus* spp. (Coleoptera: Nitidulidae) in stone fruit orchards using synthetic aggregation pheromones and a coattractant: development of a strategy for population suppression. *Journal of Chemical Ecology* 22, 1541–1556.

James, D.G., Faulder, R.J., Vogele, B., Bartelt, R.J. and Moore, C.J. (1997) Phenology of *Carpophilus* spp. (Coleoptera: Nitidulidae) in stone fruit orchards as determined by pheromone trapping: implications for prediction of crop damage. *Australian Journal of Entomology* 36, 165–173.

James, D.G., Vogelet, B., Faulder, R.J., Bartelt, R.J. and Moore, C.J. (2001) Pheromone-mediated mass trapping and population diversion as strategies for suppressing *Carpophilus* spp. (Coleoptera: Nitidulidae) in Australian stone fruit orchards. *Agricultural and Forest Entomology* 3, 41–47.

Kalinova, B., Stransky, K., Harmatha, J., Ctvrtecka, R. and, Zd'arek, J. (2000) Can chemical cues from blossom buds influence cultivar preference in the apple blossom weevil (*Anthonomus pomorum*)? *Entomologia Experimentalis et Applicata* 95, 47–52.

Kroder, S., Samietz, J. and Dorn, S. (2006) Effect of ambient temperature on mechanosensory host location in two parasitic wasps of different climatic origin. *Physiological Entomology* 31, 299–305.

Kroder, S., Samietz, J. and Dorn, S. (2007) Temperature affects interaction of visual and vibrational cues in parasitoid host location. *Journal of Comparative Physiology A – Neuroethology Sensory Neural and Behavioral Physiology* 193, 223–231.

Kührt, U., Samietz, J. and Dorn, S. (2005) Thermoregulation behaviour in codling moth larvae. *Physiological Entomology* 30, 54–61.

Kührt, U., Samietz, J. and Dorn, S. (2006) Thermal response in adult codling moth. *Physiological Entomology* 31, 80–88.

Lafleur, G. and Hill, S.B. (1987) Spring migration, within-orchard dispersal, and apple-tree preference of plum curculio (Coleoptera: Curculionidae) in southern Quebec. *Journal of Economic Entomology* 80, 1173–1187.

Landolt, P.J. and Phillips, T.W. (1997) Host plant influences on sex pheromone behavior of phytophagous insects. *Annual Review of Entomology* 42, 371–391.

Leskey, T.C. (2006) Visual cues and capture mechanisms associated with traps for plum curculio (Coleoptera: Curculionidae). *Journal of Entomological Science* 41, 97–106.

Leskey, T.C. and Prokopy, R.J. (2000) Sources of apple odor attractive to adult plum curculios (Coleoptera: Curculionidae). *Journal of Chemical Ecology* 26, 639–653.

Leskey, T.C. and Prokopy, R.J. (2001) Adult plum curculio (Coleoptera: Curculionidae) attraction to fruit and conspecific odors. *Annals of the Entomological Society of America* 94, 275–288.

Leskey, T.C. and Prokopy, R.J. (2002) Developing a branch-mimicking trap for adult plum curculios. *Entomologia Experimentalis et Applicata* 102, 253–259.

Leskey, T.C. and Wright, S.E. (2004) Influence of host tree proximity on adult plum curculio (Coleoptera: Curculionidae) responses to monitoring traps. *Environmental Entomology* 33, 389–396.

Leskey, T.C. and Wright, S.E. (2007) Host preference of the plum curculio. *Entomologia Experimentalis et Applicata* 123, 217–227.

Leskey, T.C., Prokopy, R.J., Wright, S.E., Phelan, P.L. and Haynes, L.W. (2001) Evaluation of individual components of plum odor as potential attractants for adult plum curculios. *Journal of Chemical Ecology* 27, 1–17.

Leskey, T.C., Piñero, J.C. and Prokopy, R.J. (2008) Odor-baited trap trees: a novel management tool for the plum curculio, *Conotrachelus nenuphar* (Herbst) (Coleoptera: Curculionidae). *Journal of Economic Entomology* 101, 1302–1309.

Lindgren, B.S. and Borden, J.H. (1993) Displacement and aggregation of mountain pine beetles, *Dendroctonus ponderosae* (Coleoptera: Scolytidae), in response to their anti-aggregation and aggregation pheromones. *Canadian Journal of Forest Research* 23, 286–290.

Macht, D. (1922) A pharmacological examination of benzaldehyde and mandelic acid. *Archives Internationales de Pharmacodynamie et de Thérapie* 27, 163–174.

Mangel, M. and Roitberg, B.D. (1989) Dynamic information and host acceptance by a tephritid fly. *Ecological Entomology* 14, 181–189.

Mansfield, C.M. and Hossain, M.S. (2004) The attractiveness of different fermenting food odours to *Carpophilus* spp. (Coleoptera: Nitidulidae). *General Applied Entomology* 33, 41–44.

Masson, C. and Mustaparta, H. (1990) Chemical information processing in the olfactory system of insects. *Physiological Reviews* 70, 199–245.

Natale, D., Mattiacci, L., Hern, A., Pasqualini, E. and Dorn, S. (2003) Response of female *Cydia molesta* (Lepidoptera: Tortricidae) to plant-derived volatiles. *Bulletin of Entomological Research* 93, 335–342.

Nestel, D., Carvalho, J. and Nemny-Lavy, E. (2004) The spatial dimension in the ecology of insect pests and its relevance to pest management. In: Horowitz, A.R. and Ishaava, I. (eds) *Insect Pest Management: Field and Protected Crops.* Springer-Verlag, Berlin, Heidelberg, Germany, pp. 45–64.

Oehlschlager, A.C., Chinchilla, C., Castillo, G. and Gonzalez, L. (2002a) Control of red ring disease by mass trapping of *Rhynchophorus palmarum* (Coleoptera: Curculionidae). *Florida Entomologist* 85, 507–513.

Oehlschlager, A.C., Gonzalez, L., Gomez, M., Rodriguez, C. and Andrade, R. (2002b) Pheromone-based trapping of West Indian sugarcane weevil in a sugarcane plantation. *Journal of Chemical Ecology* 28, 1653–1664.

Ormerod, E.A. (1890) *A Manual of Injurious Insects with Methods of Prevention and Remedy for their Attacks to Food Crops, Forest Trees, and Fruit.* Simpkin, Marshall, Hamilton, Kent and Co. Limited, London, pp. 300–303.

Papaj, D.R. (2004) In praise of a good colleague: Ronald John Prokopy. *Journal of Insect Behavior* 17, 569–577.

Pettersson, J. (1970) Studies on *Rhopalosiphum padi* (L.). I. Laboratory studies on olfactometric responses to the winter host *Prunus padus* L. *Lantbrukshogskolans* 36, 381–399.

Piñero, J.C. (2005) Dynamics of plum curculio, *Conotrachelus nenuphar* (Herbst.) (Coleoptera: Curculionidae), immigration into apple orchards. PhD thesis, University of Massachusetts, Amherst, Massachusetts.

Piñero, J.C. and Dorn, S. (2007) Synergism between aromatic compounds and green leaf volatiles derived from the host plant underlies female attraction in the oriental fruit moth. *Entomologia Experimentalis et Applicata* 125, 185–194.

Piñero, J.C. and Prokopy, R.J. (2003) Field evaluation of plant odor and pheromonal combinations for attracting plum curculios. *Journal of Chemical Ecology* 29, 2735–2748.

Piñero, J.C. and Prokopy, R.J. (2006) Temporal dynamics of plum curculio, *Conotrachelus nenuphar* (Herbst.) (Coleoptera: Curculionidae), immigration into an apple orchard in Massachusetts. *Environmental Entomology* 35, 413–422.

Piñero, J., Wright, S.E. and Prokopy, R.J. (2001) Response of plum curculio (Coleoptera: Curculionidae) to odor-baited traps near woods. *Journal of Economic Entomology* 94, 1386–1397.

Piñero, J.C., Galizia, C.G. and Dorn, S. (2008) Synergistic behavioral responses of female oriental fruit moths (Lepidoptera: Tortricidae) to synthetic host-plant derived mixtures are mirrored by odor-evoked calcium activity in their antennal lobes. *Journal of Insect Physiology* 54, 333–343.

Pollini, A. and Bariselli, M. (1993) *Cydia molesta*: pest on the increase and defence of pome fruits. *Informatore Agrario* 14, 19–21.

Prokopy, R.J. (1994) Integration in orchard pest and habitat management – a review. *Agriculture Ecosystems and Environment* 50, 1–10.

Prokopy, R.J. (2003) Two decades of bottom-up, ecologically based pest management in a small commercial apple orchard in Massachusetts. *Agriculture Ecosystems and Environment* 94, 299–309.

Prokopy, R.J. and Owens, E.D. (1978) Visual generalist with visual specialist phytophagous insects – host selection behavior and application to management. *Entomologia Experimentalis et Applicata* 24, 609–620.

Prokopy, R.J. and Owens, E.D. (1983) Visual detection of plants by herbivorous insects. *Annual Review of Entomology* 28, 337–364.

Prokopy, R.J. and Roitberg, B.D. (2007) Arthropod behavioural ecology and IPM. In: Kogan, M. and Jepson, P. (eds) *Perspectives in Ecological Theory and Integrated Pest Management.* Cambridge University Press, New York, pp. 87–121.

Prokopy, R.J. and Wright, S.E. (1998) Plum curculio responses to unbaited pyramid and cone traps. *Journal of Economic Entomology* 91, 225–234.

Prokopy, R.J., Drew, R.A.I., Sabine, B.N.E., Lloyd, A.C. and Hamacek, E. (1991) Effect of physiological and experiential state of *Bactrocera tryoni* flies on intra-tree foraging behavior for food (bacteria) and host fruit. *Oecologia* 87, 394–400.

Prokopy, R.J., Wirth, C.B. and Leskey, T.C. (1999) Movement of plum curculio adults toward host trees and traps: flight versus walking. *Entomologia Experimentalis et Applicata* 91, 385–392.

Prokopy, R.J., Phelan, P.L., Wright, S.E., Minalga, A.J., Barger, R. and Leskey, T.C. (2001) Compounds from host odor attractive to plum curculio adults (Coleoptera: Curculionidae). *Journal of Entomological Science* 36, 122–134.

Prokopy, R.J., Chandler, B.W., Dynok, S.A. and Piñero, J.C. (2003) Odor-baited trap trees: a new approach to monitoring plum curculio (Coleoptera: Curculionidae). *Journal of Economic Entomology* 96, 826–834.

Prokopy, R.J., Jácome, I., Gray, E., Trujillo, G., Ricci, M. and Piñero, J.C. (2004) Using odor-baited trap trees as sentinels to monitor plum curculio (Coleoptera: Curculionidae) in apple orchards. *Journal of Economic Entomology* 97, 511–517.

Prokopy, R.J., Jácome, I. and Bigurra, E. (2005) An index for assigning distances between odor-baited spheres on perimeter trees of orchards for control of apple maggot flies.

Entomologia Experimentalis et Applicata 115, 371–377.

Quaintance, A.L. and Jenne, E.L. (1912) The plum curculio. *United States Department of Agriculture, Bureau of Entomological Bulletin* 103, 1–250.

Racette, G., Chouinard, G., Vincent, C. and Hill, S.B. (1992) Ecology and management of plum curculio, *Conotrachelus nenuphar* (Coleoptera: Curculionidae), in apple orchards. *Phytoprotection* 73, 85–100.

Reddy, G.V.P. and Guerrero, A. (2004) Interactions of insect pheromones and plant semiochemicals. *Trends in Plant Science* 9, 253–261.

Rings, R.W. (1952) Experimental control of plum curculio on peaches. *Journal of Economic Entomology* 45, 436–444.

Schoonhoven, L.M., Jermy, T. and van Loon, J.J.A. (1998) *Insect–Plant Biology: From Physiology to Evolution*. Chapman and Hall, London.

Tasin, M., Backman, A.C., Bengtsson, M., Ioriatti, C. and Witzgall, P. (2006) Essential host plant cues in the grapevine moth. *Naturwissenschaften* 93, 141–144.

Tedders, W.L. and Wood, B.W. (1994) A new technique for monitoring pecan weevil emergence (Coleoptera: Curculionidae). *Journal of Entomological Science* 29, 18–30.

Toepfer, S., Gu, H. and Dorn, S. (1999) Spring colonisation of orchards by *Anthonomus pomorum* from adjacent forest borders. *Entomologia Experimentalis et Applicata* 93, 131–139.

Toepfer, S., Gu, H. and Dorn, S. (2000) Selection of hibernation sites by *Anthonomus pomorum*: preferences and ecological consequences. *Entomologia Experimentalis et Applicata* 95, 241–249.

Toepfer, S., Gu, H. and Dorn, S. (2002) Phenological analysis of spring colonisation of apple trees by *Anthonomus pomorum*. *Entomologia Experimentalis et Applicata* 103, 151–159.

Vallat, A. and Dorn, S. (2005) Changes in volatile emissions from apple trees and associated response of adult female codling moths over the fruit-growing season. *Journal of Agricultural and Food Chemistry* 53, 4083–4090.

Vallat, A., Gu, H. and Dorn, S. (2005) How rainfall, relative humidity and temperature influence volatile emissions from apple trees *in situ*. *Phytochemistry* 66, 1540–1550.

Vincent, C., Chouinard, G., Bostanian, N.J. and Morin, Y. (1997) Peripheral zone treatments for plum curculio management: validation in commercial apple orchards. *Entomologia Experimentalis et Applicata* 84, 1–8.

Vincent, C., Chouinard, G. and Hill, S.B. (1999) Progress in plum curculio management: a review. *Agriculture Ecosystems and Environment* 73, 167–175.

Visser, J.H. (1986) Host odor perception in phytophagous insects. *Annual Review of Entomology* 31, 121–144.

Wildbolz, T. (1992) Pest situations in orchards evolving from natural and man-made factors. *Acta Phytopathologica et Entomologica Hungarica* 27, 669–678.

Witzgall, P., Stelinski, L., Gut, L. and Thomson, D. (2008) Codling moth management and chemical ecology. *Annual Review of Entomology* 53, 503–522.

Zhang, A.J., Linn, C., Wright, S.E., Prokopy, R., Reissig, W. and Roelofs, W. (1999) Identification of a new blend of apple volatiles attractive to the apple maggot, *Rhagoletis pomonella*. *Journal of Chemical Ecology* 25, 1221–1232.

5 Monitoring and Management of the Apple Maggot Fly and the Plum Curculio: Honouring the Legacy of R.J. Prokopy

TRACY C. LESKEY,[1] GÉRALD CHOUINARD[2] AND CHARLES VINCENT[3]

[1]*USDA-ARS, Appalachian Fruit Research Station, Kearneysville, West Virginia, USA;* [2]*Institut de Recherche et de Développement en Agroenvironnement (IRDA), Saint-Hyacinthe, Quebec, Canada;* [3]*Horticultural Research and Development Centre, Agriculture and Agri-Food Canada, Saint-Jean-sur-Richelieu, Quebec, Canada*

Introduction

All phytophagous insects demonstrate some degree of selectivity in terms of the host plants they consume. Host-plant selection involves a behavioural sequence that allows the insects to distinguish between host and non-host plants. The process begins with host finding, detection and location of the host plant from a distance, and concludes with host acceptance, confirmation of host-plant species and quality after arrival (Bernays and Chapman, 1994).

Host-finding mechanisms generally include recognition of visual cues associated with and chemoreception of volatiles emitted by particular host plants. The identification of these attractive visual and olfactory stimuli has led to advances in sustainable pest management practices as they can provide the basis for the development of monitoring and management strategies using the principle of behavioural manipulation. Behavioural manipulation relies on a comprehensive understanding of specific behavioural responses from the targeted insect and subsequent manipulation of these responses to protect a valued resource. The attract-and-kill method is a behavioural manipulation method that requires a long-distance olfactory stimulus to attract a particular pest, in combination with some type of killing agent and/or collection device. The more specific the stimulus is to an insect species, the greater the likelihood that a particular behaviour can be manipulated successfully (Foster and Harris, 1997). A second approach, the push–pull strategy, uses repellents or deterrents to push insects away from a protected resource or crop, in combination with attractants to pull insects toward an attractive, but economically insignificant resource such as a trap crop to facilitate their removal (Cook *et al.*, 2007).

These strategies have become the cornerstone of some integrated pest management (IPM) systems. In these systems, the basic behaviours of key pests are studied in detail and specific attributes are recognized and manipulated in such a manner to create ecologically sound pest management strategies (Prokopy and Roitberg, 2007). Some of the best examples of this type of strategy in tree-fruit production systems were developed by the late Ron Prokopy for two of the key pests found in New England apple orchards (Prokopy *et al.*, 1990, 1996; Prokopy, 2003). He worked extensively on the host-finding behaviour of the apple maggot fly, *Rhagoletis pomonella* (Walsh), for most of his career and on plum curculio, *Conotrachelus nenuphar* (Herbst), in the last years of his career, with the aim to develop monitoring and/or management strategies for these two species. The key components of these strategies generally included: (i) synthetic versions of host-plant volatiles or other semiochemicals deployed as attractants; (ii) attractive visual cues integrated into a trapping design; (iii) a capture mechanism that is compatible with a particular trap design; and (iv) an effective deployment strategy (optimal location for and timing of trap installation).

In this chapter, we seek to highlight the pathway followed by the late Ron Prokopy, who tenaciously attempted to bridge the considerable gap between basic behavioural studies and crop protection applications, and whose research programme serves as a model system for others. Although Ron Prokopy aimed to serve New England apple growers and aid them in implementing a stepwise progression of increasingly sophisticated IPM strategies, as discussed by Cooley and Coli (Chapter 6, this volume), his research was of broader behavioural application, geographical scope and scientific interest and, likewise, we address the subject accordingly. We discuss early published results of basic behavioural studies by Ron Prokopy and continue to follow and highlight the advances he and others have made toward the development of effective behavioural-based monitoring and management strategies for the apple maggot fly and the plum curculio. We also suggest research questions that still remain but must be addressed in order to meet this ultimate goal. Finally, we conclude with a discussion of how fundamental studies of insect behaviour can be used as the basis for novel crop protection applications, an approach pioneered by Ron Prokopy, and how this approach can be applied to other key pests and cropping systems.

Key Pests of Apple Orchards in North-eastern North America

The apple maggot fly and the plum curculio are key pests of apple orchards in north-eastern North America. These species share a number of biological and ecological characteristics (Table 5.1); for example, both are native species to eastern North America that have expanded their host range to include cultivated host fruit species. However, there are also many important differences (Table 5.1) that have major implications for the development of behaviourally based monitoring and management systems. One of the most important is voltinism. The plum curculio is univoltine in northern locations, with an obligatory diapause,

Table 5.1. Shared (a) and differing (b) characteristics of the apple maggot fly and the plum curculio (compiled from Racette *et al.*, 1992; Vincent and Roy, 1992; and Vincent *et al.*, 1999).

(a) Shared characteristics

1. Species native to eastern North America
2. Eggs laid singly in fruitlets
3. Fruit-feeding larvae
4. Few natural enemies
5. Mate on or near host plants
6. No insecticide resistance reported
7. Insecticide treatments aimed at adults
8. Immigrating (non-residential) populations are primary problem in commercial orchards
9. No sex pheromone identified
10. Attractive host-plant volatiles identified
11. Major pests (>50% fruit damage at harvest) in absence of insecticide treatments in most areas

(b) Differing characteristics	Apple maggot fly (Diptera: Tephritidae)	Plum curculio (Coleoptera: Curculionidae)
1. Voltinism	Univoltine	Univoltine (northerly) and multivoltine (southerly) populations
2. Speciation	Host races	Geographic strains (northern univoltine and southern multivoltine)
3. Reproductive incompatibility	Pre-mating reproductive isolation due to assortive mating on hosts	Post-mating reproductive isolation reported (probably due to infection from different *Wolbachia* strains)
4. Seasonal phenology	Mid- to late-season pest	Early-season pest in northern locations and season-long threat in southern locations
5. Phytophagy	Oliphagous; a few hosts in the family Rosaceae	Polyphagous; large number of hosts in family Rosaceae and several in Ericaceae and Vitaceae
6. Economically important acquired hosts	Apple	Apple, peach, European and Japanese plum, apricot, blueberry, cherry, nectarine, pear, quince
7. Fruit-feeding adult	No	Yes
8. Appearance of damage on fruit epidermis	Not conspicuous, pinpoint sites more noticeable on mature fruit	Conspicuous, half moon-shaped damage of irregular size (*c.*1 cm^2)
9. Movement toward host trees	Principally by flight	Temperature-dependent movement. Mostly walking when <20°C and mostly flying at ≥20°C
10. Movement within host trees	Flight	Principally walking, some flight
11. Overwintering stage	Pupae	Adult
12. Adult diel activity	Diurnal	Primarily nocturnal early in the season, then diurnal and nocturnal

(*Continued*)

Table 5.1. continued

(b) Differing characteristics	Apple maggot fly (Diptera: Tephritidae)	Plum curculio (Coleoptera: Curculionidae)
13. Learning by adults	Yes	Not demonstrated
14. Adult visual ecology	Visual specialist	Visual generalist
15. Oviposition-deterring pheromone	Behavioural evidence	No
16. Aggregation pheromone	No	Yes, male-produced (grandisoic acid)
17. Host plant-based attractants	Butyl hexanoate, 5-component blend	Benzaldehyde Other blends
18. Commercially available chemical attractant	Ammonium-based baits, 5-component blend	Grandisoic acid, plum essence, benzaldehyde
19. Defensive strategies	Batesian mimic of salticids	Thanatosis
20. Sound production	No	Stridulation
21. Rearing for research purpose	Feasible, low output	Yes, viable multivoltine laboratory colonies
22. Trap for adults	Red spheres (several variants) and yellow traps set in trees	Pyramid traps deployed on ground
23. Behaviour exploited by trap	Flight	Walking and flight
24. Capture mechanism	Sticky coating	Non-return collection device (no sticky coating involved)
25. Threshold	1 fly per trap per ha	1% fruitlets attacked
26. Management of adults at the periphery of orchards	Possible with traps when populations are low	Possible by frequent examination of fruitlets
27. Larvae in fruit at harvest and in stored fruit	Yes	No in northern locations, yes in southern locations

but is multivoltine in southern portions of its geographic range, while the apple maggot fly is exclusively univoltine. Thus, the plum curculio is primarily an early-season pest in northern locations, attacking apple just after fruit have set, but poses a season-long threat in southern locations. The apple maggot fly becomes important from midseason until harvest. Biological differences related to phytophagy, visual ecology and behaviour also likely have a bearing on the development of behavioural-based monitoring and management strategies. The apple maggot fly is a visual specialist that uses a limited number of hosts, whereas the plum curculio has a broad host range and is likely a visual generalist (Table 5.1). Another notable difference between the two pests is that apple maggot fly is diurnal, whereas plum curculio adults are primarily nocturnal and shift towards diurnal activity as the season progresses.

Based on these characteristics, it appears that the development of behaviourally based strategies for the apple maggot fly may be accomplished more easily because the stimuli they respond to are likely to be more specific and because their diurnal activity does not limit the direct observation of behaviour, an approach much used by Ron Prokopy in his apple maggot fly research. In sum, owing to differences in the biological, behavioural and ecological characteristics of the apple maggot fly and the plum curculio, as well as agronomic considerations, IPM issues and research challenges posed by the apple maggot fly and the plum curculio are different.

The apple maggot fly

Biology and pest status

The apple maggot fly (Diptera: Tephritidae) is a native species that was found originally throughout eastern North America and Mexico and that bred principally in native hawthorn, *Crataegus* spp. Later, it expanded its host range to include European cultivars of apple, with reports of economic damage to New England apple crops as early as the 1860s (Walsh, 1867). The geographic range occupied by the apple maggot fly has expanded; as of 2008, this species is found throughout all regions in the USA where apples are grown (Ali Niazee, 1988), including California, Oregon, Washington, Idaho, Utah and Colorado, although it is not considered a serious pest in these western states. Principally, it is considered to be a key pest in apple production areas located in the north-eastern and midwestern USA and the eastern provinces of Canada. Other confirmed hosts of the apple maggot fly include pear, *Pyrus communis* (Prokopy and Bush, 1972), European plum, *Prunus domestica* (Herrick, 1920), tart cherry, *Prunus cerasus* (Shervis *et al.*, 1970), apricot, *Prunus armeniaca* (Lienk, 1970) and crabapple, *Malus* spp. In Washington state, confirmed rosaceous hosts include hawthorn, apple and crabapple, as well as Asian pear, *Pyrus serotina*, bitter cherry, *Prunus emarginata*, pear, plum and cranberry cotoneaster, *Cotoneaster apiculatus* (Yee and Goughnour, 2006).

This univoltine insect overwinters as a pupa in the soil and emerges from the puparium as an adult in early to mid summer. After emergence, females feed on carbohydrate- and protein-based resources such as aphid honeydew (Neilson and Wood, 1966; Roitberg *et al.*, 1982) and bird excrement (Prokopy *et al.*, 1993; Hendrichs *et al.*, 1993); they reach reproductive maturity in approximately 10–14 days. Males and female meet and mate on host fruit (Prokopy *et al.*, 1971, 1972). A mated female will deposit eggs singly beneath the skin of host fruit and then drag her ovipositor on the fruit surface in order to deposit an oviposition-deterring pheromone (Prokopy, 1972a). In apples, eggs hatch in less than 7 days and maggots feed internally in the fruit, leaving behind brown trails of decay. Mature maggots leave fallen fruit and pupate in the soil, where they will overwinter and remain until the following year.

Conventional management

In Massachusetts, growers typically apply an average of three insecticide applications to the entire apple orchard to control the apple maggot fly (Prokopy *et al.*,

1990) from mid-July through to late August. In Quebec, Canada, the pressure is much lower than in Massachusetts and, consequently, apple growers typically use one or two insecticidal applications. If left uncontrolled, flies can damage at least 90% of the fruit annually, based on data collected for over two decades from unmanaged apple trees in New England (Prokopy, 2003). In Quebec, maximal damage on fruit reached 61% (average 21.7%) based on a 13-year study (Vincent and Roy, 1992). Historically, flies have been managed using broad-spectrum insecticides and, in particular, the organophosphates, azinphosmethyl and phosmet. More recently, a number of 'reduced-risk' insecticides (mostly broad-spectrum neonicotinoids) have been labelled for use against apple maggot fly on apple; these include acetamiprid, indoxacarb, imidacloprid and spinosad in the USA and thiacloprid in the USA and Canada. Questions remain regarding the effects of these 'reduced-risk' insecticides on natural enemy populations and their effects on non-target pests. Therefore, many growers continue to use organophosphates for control.

Within conventionally managed commercial orchards, insecticide applications aimed at controlling apple maggot fly and removal of dropped fruit within the orchard can reduce the threat of resident populations building up within the orchard itself. Insecticide applications applied against the adult generally target immigrating rather than residential populations. Populations found in and around wild hosts can reach tremendous levels and these adults have the potential to travel great distances, up to 1.6 km in some cases, and arrive at commercial orchards (Maxwell and Parsons, 1968). Management of adult apple maggot fly using traps is possible as there is a fairly good mathematical relationship between captures and levels of risks. Over the years, Ron Prokopy made considerable efforts to understand the behaviour underlying this relationship, and few other insect pests have been researched so extensively in that respect.

Development of behaviourally based strategies

VISUAL ECOLOGY Ron Prokopy published extensively on the subject of the foraging behaviour of the apple maggot fly, as he, his graduate students and colleagues studied the behaviour of this insect under natural conditions and identified a number of cues used by the adult fly to locate host trees at a distance and host fruit within the canopy of those trees. His early studies, in particular, certainly are considered to be some of the seminal papers that helped define the field of visual ecology within the family Tephritidae, and even more broadly within the field of entomology.

Prokopy and Owens (1983) describe the process by which a phytophagous insect visually detects a host plant independent of olfactory stimuli. Initially, when an insect is located several metres or more from a host plant, it probably orients to the plant by perceiving the plant silhouette against the lit sky or background rather than by precise spectral qualities. Spectral qualities become more important at shorter distances and elicit detection and alightment, although similarities among plant species make it likely that these stimuli are not extremely specific. Dimensional characteristics of the host plant such as shape and size are discernible to the insect at distances of a metre or less once the insect has arrived on the host plant itself.

Toward host trees. The apple maggot fly uses visual stimuli as it approaches host trees. Prokopy (1968a) began by first demonstrating that flies were more attracted to yellow rectangles (30 × 40 cm) than to green, orange, blue, red, violet, black or white rectangles. Rectangles were suspended either from apple tree branches 1.5 m above the ground or from wooden crossbars at various distances from trees. A sticky coating applied to the surface of these rectangles served to capture alighting flies.

These results were followed by experiments aimed at learning if flies were responding principally to the true colour or hue or to the intensity of reflected light from the rectangles. Fly responses to 15 × 20 cm rectangles were, in this case, assayed in a flight tunnel and these results were complemented and confirmed with field-based studies in apple orchards. Prokopy (1972b) found that flies were responding primarily to the yellow hue compared with other hues and not to the intensity of the light reflected from each hue. He based this conclusion on spectral reflectance curves; these curves demonstrated that the intensity of light reflected by yellow was less than white and light grey stimuli, but greater than black or darker shades of grey, for example. Apple maggot flies do not respond to ultraviolet light; Prokopy (1972b) demonstrated this principle simply by using an aluminium foil-coated rectangle which reflected a large amount of ultraviolet light. He also stated that flies were orienting based not on repulsion by other hues or shades, but on positive attraction to yellow. He defined yellow as a 'supernormal' foliage-type stimulus attractive to apple maggot flies and eliciting food-seeking and/or host-finding behaviour (Prokopy, 1972b). Yellow emits peak energy in the same wavelength as foliage (between 500–580 nm), but at a greater intensity (Prokopy and Owens, 1983).

A subsequent study aimed at defining more specifically the visual stimuli eliciting the attraction of apple maggot flies to host trees was conducted in an open field using sticky-coated, two-dimensional tree models 122 × 244 cm in size. As with previous studies, apple maggot flies preferred the colour yellow compared with other colours, indicating that a tree was visually detectable based on the yellow foliage-mimicking hue. Studies associated with size revealed that a large dark silhouette against the background of a lit sky also provided a visual stimulus for foraging flies. This finding was based on the number of flies captured on red rectangles; the greater the size of the rectangle, the greater the captures per cm^2. The vertical orientation of yellow tree models was an important component of the tree-mimicking stimulus, as significantly greater numbers of flies were captured on vertical rather than on horizontal models. Apple maggot flies are low flying and as they approach a host tree, the taller the tree silhouette appears, the more distinguishable it becomes relative to the background (Moericke *et al.*, 1975). Moericke *et al.* (1975) concluded that visual stimuli guiding apple maggot flies to host trees were not very specific to the host trees themselves and that other stimuli must have been playing an important role in the host-finding process. Although apple maggot flies are attracted to sticky-coated, vertically hanging yellow rectangles, these traps are not particularly selective (Prokopy, 1975a).

Within host trees. Once an apple maggot fly reaches a host tree, rather than foraging as a visual generalist, it becomes a visual specialist within the canopy (Prokopy and Owens, 1978). Sexually mature apple maggot flies meet and mate on host

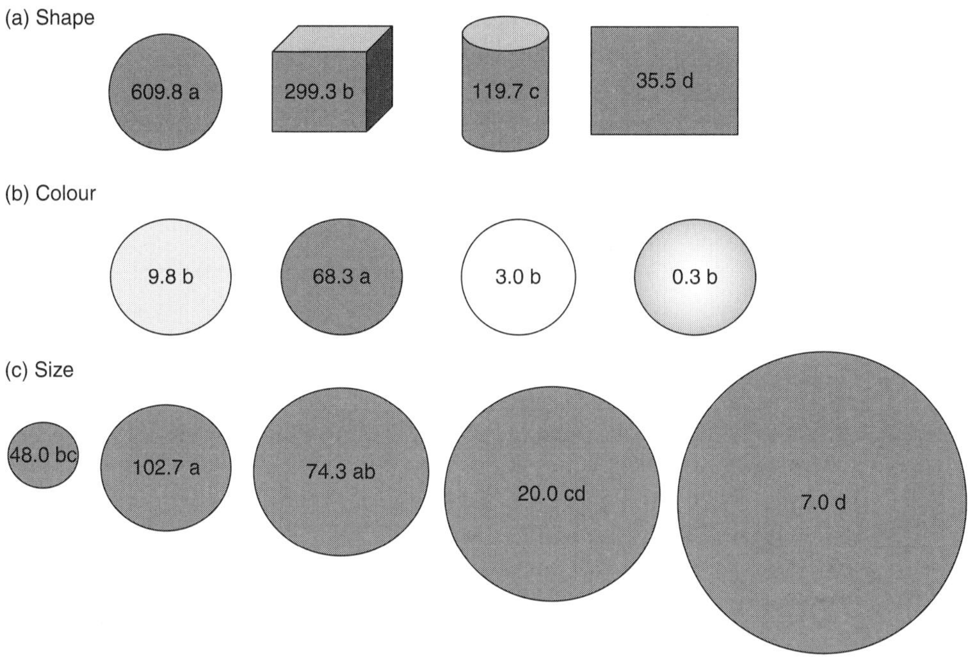

Fig. 5.1. Mean number of adult apple maggot flies captured (a) on red sticky-coated objects each with a 176-cm^2 surface area deployed from 22 July to 16 September 1967; (b) on yellow, red, white and clear spheres (respectively) 7.5 cm in diameter deployed from 14–28 August 1967; and (c) on red spheres 3.8, 7.5, 15.0, 30.0 and 45.0 cm in diameter in unsprayed apple trees in New Haven, Connecticut (Prokopy, 1968a).

fruit (Prokopy *et al.*, 1971, 1972) and early studies reveal that the visual characteristics of host apples are very attractive to foraging flies (Oatman, 1964). The shape of the host fruit itself is a key visual cue. When sticky-coated spheres, cubes, cylinders and rectangles, all with an identical surface area (176 cm^2), were suspended at heights of 1.5 m from the ground in apple trees, significantly more flies were captured on spheres (Fig. 5.1). Hue or colour also is critical. Darkly coloured objects, in particular red, appeared to be the most visually stimulating; red spheres were considered to be the most visually attractive (Fig. 5.1) (Prokopy, 1968a).

The size of the host fruit visual mimic is also a very important component of the visual cue for apple maggot fly. Several studies revealed that 7.5-cm diameter spheres were more visually stimulating than those with either a smaller or larger diameter (Fig. 5.1) (Prokopy, 1968a, 1977). Contrast of the red sphere with the surrounding background is another essential factor of the visual stimulus for foraging apple maggot flies. Adults are able to detect dark spheres reflecting low light intensity when viewed against a more brightly lit background (sky light) rather than a darker background created by foliage (Prokopy, 1968a; 1973; Owens and Prokopy, 1984).

These sticky-coated red spheres were evaluated in a number of studies for their ability to serve as a trap for the apple maggot fly. Prokopy (1968b) reasoned that these spheres could be used to estimate seasonal abundance of and as a behavioural control mechanism for the apple maggot fly within orchards, particularly since they captured large numbers of both male and female flies. The red sphere is a highly selective trap for the apple maggot fly, capturing large numbers of sexually mature adults and fewer beneficial insects (Prokopy, 1975a) and is more attractive to foraging adults than yellow rectangle traps (Prokopy and Hauschild, 1979), owing to the resource-seeking behaviour of mature flies within the canopy of host apple trees. When sticky-coated red spheres were deployed at a rate of 1 sphere/100 apples in apple trees in unsprayed orchards and compared with unsprayed orchards with no spheres, damage from the apple maggot fly in the sphere-protected orchards ranged from 1.1–2.7% compared with 97.2–98.2% for the unprotected orchards (Prokopy, 1975b).

CHEMICAL ECOLOGY

Protein-based stimuli. Immature flies forage for protein and carbohydrates. Over the years, a number of ammonium-based lures have been developed and evaluated as a potential attractant for protein-foraging apple maggot flies in Massachusetts apple orchards (Duan and Prokopy, 1992; Reynolds and Prokopy, 1997; Rull and Prokopy, 2000). However, they were never proven particularly effective at attracting apple maggot flies in commercial orchards (Reynolds and Prokopy, 1997). Their ineffectiveness had been attributed to the short lifespan of the dispensers themselves, the distance between emergence sites of immature flies near wild hosts and dispenser deployment locations in the orchard border rows and competition from natural sources of protein such as bird faeces found in and around orchards.

Rull and Prokopy (2000) demonstrated that these lures were ineffective not because of competition with natural protein sources such as bird faeces, but because most immigrating apple maggot flies that entered and penetrated orchards were sexually mature on arrival. Therefore, they would not be expected to be attracted to ammonium-based baits. If these baits were to be deployed at locations where sexually immature flies were present, they might be proven to be more attractive.

Host-fruit stimuli. Prokopy *et al.* (1973) documented that sexually mature female and male apple maggot flies responded to the odour of ripe apple fruit. In early studies, bagged, freshly picked apples or rocks were deployed in association with sticky-coated wooden rectangles; more flies were captured on the rectangles with apples than those with rocks. When bags of freshly picked apples or cut grass were deployed in fruitless host apple trees and in non-host birch and choke cherry, greater numbers of adults were captured on sticky-coated spheres in trees with bagged apples, regardless of host status.

The odour of ripening fruit is used by mature apple maggot fly as a long-range olfactory stimulus guiding them to host fruit trees. A number of studies identified volatiles released by mature apples and evaluated adult responses to these potential attractants (Fein *et al.*, 1982; Reissig *et al.*, 1982, 1985; Carle *et al.*, 1987). One of the key compounds, butyl hexanoate, was identified as being particularly behaviourally active (Carle *et al.*, 1987; Averill *et al.*, 1988). The addition of this compound in association with sticky-coated red spheres resulted in significant increases in trap captures (Duan and Prokopy, 1992; Reynolds and Prokopy, 1997) with approximately twice as many flies captured compared with unbaited spheres.

Apple maggot flies that encounter butyl hexanoate plumes exhibit oriented response to point sources of this stimulus, consistently arriving at tree canopies that contain this odour (Aluja and Prokopy, 1992). The interaction between butyl hexanoate and visual stimuli associated with host trees results in the increased capability of the apple maggot flies to locate host trees. Green *et al.* (1994) used green rectangles to simulate visual stimuli of host apple trees and found that flies were able to locate these tree models at greater distances when this olfactory stimulus was present. Apparently, butyl hexanoate is most important to foraging flies as they approach host trees, when the fruit-associated visual stimuli are less visually apparent. Once flies enter the host-tree canopy, they are able to locate host fruit based on the attractive visual stimuli of the host fruit itself, including shape, size and colour, as well as abundance (Aluja and Prokopy, 1993).

Zhang *et al.* (1999) identified an apple volatile-based blend that was more attractive to the apple maggot flies; the five-component blend consisted of butyl butanoate (10%), propyl hexanoate (4%), butyl hexanoate (37%), hexyl butanoate (44%) and pentyl hexanoate (12%). On average, sticky-coated red spheres deployed in association with the five-component blend captures twice as many flies as those deployed with the single-component butyl hexanoate and five times as many as unbaited spheres. However, there are still a number of unanswered behavioural questions associated with this new blend. One of the critical questions is over what distance flies can detect this olfactory stimulus. A number of studies indicate that deploying this olfactory stimulus can attract flies at much greater distances than the single-component butyl hexanoate, but this distance has not been quantified. Furthermore, the elegant studies investigating the interaction between visual stimuli and butyl hexanoate (Aluja and Prokopy, 1992, 1993; Green *et al.*, 1994) have not been repeated for this new blend. It would be helpful to understand if fly foraging behaviour is similar in the presence of this five-component blend compared with the single fruit volatile butyl hexanoate.

As pointed out by Berlocher and Feder (Chapter 2, this volume), the apple maggot fly is an excellent example of an insect that has undergone sympatric speciation (Bush, 1969). It has two host races (Feder *et al.*, 1988, 1990; McPheron *et al.*, 1988): one that uses native hawthorn (*Crataegus* spp.) and a second that has become an important pest of apple. As such, flies that originate from hawthorn are attracted to compounds identified from this native host. This attractive blend, based on electroantennogram activity and laboratory-based flight tunnel bioassays, includes four components, 3-methylbutan-1-ol, 4,8-dimethyl-1-3(*E*), 7-nonatriene, butyl hexanoate and dihyro-ß-ionone (Nojima *et al.*, 2003a). Butyl hexanoate is the only common component found in apple- and hawthorn-based blends, as both populations are attracted to unique volatile blends identified from their natal host fruit. An undescribed dogwood-infesting *Rhagoletis* fly, thought to be a sister species (Berlocher, 1999), also responds to a unique blend of compounds (Nojima *et al.*, 2003b). The ability of apple-, hawthorn- and dogwood-infesting populations to discriminate between these blends indicates that host-fruit odour is a key behavioural trait that establishes and maintains host fidelity (Linn *et al.*, 2005). Thus, most flies that are attracted to the five-component apple blend and those that immigrate and penetrate commercial apple orchards likely are of the apple-origin host race.

Capture mechanism and deployment strategy

One of the key components of any trapping system is the capture mechanism – the method by which a targeted insect becomes immobilized or trapped within or on a trap. As apple maggot flies alight directly on fruit, the application of a sticky coating to visually attractive red spheres has proven to be a very effective capture mechanism. Of course, the use of this material is not without shortcomings. Preservation of an efficacious sticky coating throughout the season requires frequent cleaning, which is a labour-intensive, messy and tedious process (Prokopy *et al.*, 1990).

Success or failure associated with this behaviourally based trapping system for the apple maggot fly is also dependent on the deployment strategy – the location and timing of trap installation. In the case of apple maggot fly, spheres must be deployed in such a manner as to enhance their visual apparency within the canopy of host apple trees. Spheres are deployed in the upper half of the tree canopy with a clear radius of 25–50 cm between the sphere and the surrounding fruit and foliage. If too little clear space is left between the trap and surrounding foliage and fruit, the trap will not be optimally visually apparent. Conversely, if greater than 50 cm of clear space exists between the trap and foliage and fruit, this may impede the movements made by foraging apple maggot flies in the canopy of host trees, exceeding the distance of their short hopping flights (Drummond *et al.*, 1984). Furthermore, the phenological state of the apple tree also has an impact on sphere apparency. Early in the season when apples are small and green, there is little visual competition with red spheres. However, later in the season, spheres can become less visually apparent if surrounded by large red fruit (Rull and Prokopy, 2003, 2004a). Visual competition can be reduced by leaving 30–50 cm of clear space between the sphere and the nearest foliage and fruit, as well as by removing or heavily thinning apples in the same area (Rull and Prokopy, 2004a).

Orchard architecture also is an important deployment strategy consideration and, in particular, tree size and density of plantings. Over the past several decades, the use of size-controlling rootstocks to establish high-density apple plantings has become a standard practice in most modern orchards, and these plantings have become much more prevalent due to international competition in the apple market. Rull and Prokopy (2001) reported that tree architecture found in modern orchards might, in fact, increase the efficacy of spheres. Greater numbers of marked-released flies were recovered from traps deployed in small- and medium-sized trees planted at higher densities than in larger, more standard-sized trees that were spaced more widely. Captures of immigrating flies also suggested that trapping efficiency increased in small trees compared with larger trees (Prokopy *et al.*, 2001a).

Another critical component of a behaviourally based trapping system for the apple maggot fly is deployment of the attractant, the five-component apple volatile blend (Zhang *et al.*, 1999). The behavioural response to some olfactory attractants, particularly those derived from host plants, may be weakened due to olfactory competition with the host plant itself. This behavioural effect has been demonstrated for the plum curculio in apple orchards (Leskey and Wright, 2004a); after fruit set, response to olfactory stimuli decreases. The apple maggot fly is active until harvest and olfactory competition with ripening fruit can be a concern, particularly since the five-component blend is identified from ripe apple odour (Zhang *et al.*, 1999) and because adult flies preferentially visit particular apple cultivars (Rull and Prokopy, 2004b).

However, when the five-component blend was deployed in association with sticky-coated red spheres in trees that were considered to be at low, medium and high risk of injury from apple maggot flies, i.e. those that were of high risk were the most highly preferred cultivars and vice versa, there was no evidence for olfactory competition. Instead, the deployment of the five-component blend in host apple tree canopies appeared to have an additive effect, resulting in increased captures compared with unbaited canopies. The greatest captures were recorded in preferred cultivars with the addition of the five-component blend, and similar captures were recorded in less-preferred cultivars with the addition of the five-component blend compared with unbaited canopies of preferred cultivars (Rull and Prokopy, 2005). These results point toward deployment of traps in preferred apple cultivars to increase sensitivity of the trapping system, although less-preferred hosts can be rendered more attractive with the addition of the five-component blend.

Monitoring

The visually stimulating sticky-coated red sphere has become an integral part of behaviourally based monitoring programmes for apple orchards in eastern North America. Because of the powerful visual stimulus associated with this trap type, they quickly replaced yellow rectangle traps, as they captured adult flies earlier in the season, in greater weekly and total numbers and were better correlated with fruit injury (Prokopy and Hauschild, 1979). Trials aimed at the development of a treatment-based threshold were conducted in New York apple orchards. A number of trap types were evaluated and, again, sticky-coated red spheres in combination with the olfactory attractant, butyl hexanoate, were found to be superior monitoring tools. Compared with a calendar-based spray programme, a treatment threshold of eight flies per trap resulted in a 70% decrease in insecticide applications (2.8 applications). Ultimately, a more conservative estimate of five flies per trap was recommended. Growers who used this threshold made, on average, 0.6 fewer insecticide applications against apple maggot fly, with no increase in injury (Agnello *et al.*, 1990). Unbaited spheres also can be used to monitor apple maggot fly with a treatment threshold of one or two flies per sphere, as recommended in Quebec (Chouinard *et al.*, 2001). Using a threshold of one fly per trap resulted in an almost 25% reduction in the number of insecticide treatments applied against apple maggot fly compared with a standard programme (2.0 versus 2.6 applications) (Reissig, 1988).

Interestingly, a threshold based on the five-component blend has not been established in commercial apple orchards. In general, growers who deploy an attractant use a commercially available lure comprised of the five-component blend, but rely on the threshold established for traps deployed in association with butyl hexanoate. Assuming a twofold increase in captures with the five-component blend compared with lures containing butyl hexanoate alone, the sensitivity of this monitoring tool is likely greater and the threshold for treatment may, in fact, be higher.

Management

Prokopy (1975b) recognized very early on the power of visually stimulating sticky-coated red spheres as a potential management strategy for the apple maggot fly. In an orchard comprised of 81 apple trees in which he deployed approximately 5–6 spheres per tree, he captured nearly 10,000 adults in a single season. Damage based on the presence of larvae was between 1.1–2.7%, compared with over 97% injury in trees without traps. However, he also recognized that this was not a very efficient method due to the number of spheres required and the time and labour associated with sphere maintenance. He spent the next three decades devoting himself to research questions addressing the basic biology, ecology and behaviour of the apple maggot fly. Results from these experiments charted a course for the delivery of an efficient behaviourally based management strategy for the adult fly, which eliminated the need for broad-spectrum insecticide applications to control this direct fruit-feeding pest. This is a novel accomplishment as the only other direct fruit feeders controlled in such a manner are some tortricid pests, notably the codling moth, which have the potential to be managed using only mating disruption (Cardé and Minks, 1995).

In the case of the apple maggot fly, the deployment of sticky-coated spheres in every tree within an orchard emerged as the first viable approach (Fig. 5.2a). Prokopy, in fact, deployed them in his own small commercial orchard for over 20 years at a rate of 1–3 traps per tree, depending on fruit load (Prokopy, 2003). Debris was removed from the traps and a new sticky coating applied every 2 weeks from deployment until harvest, to maintain trapping efficiency. Prokopy reported commercially acceptable levels of control for two decades, with an approximate 150-fold decrease in injury compared with nearby unmanaged

Fig. 5.2. (a) Sticky-coated red sphere; (b) biodegradable pesticide-treated sphere with internally renewable feeding stimulant; (c) pesticide-treated wood sphere with externally preserved feeding stimulant; (d) pesticide-treated plastic sphere with externally renewable feeding stimulant; and (e) pesticide-treated plastic sphere with visually integrated renewable feeding stimulant.

trees (Prokopy, 2003). However, in larger commercial orchards, it would be impractical, due to the number of spheres required and the associated labour cost of sphere maintenance.

Thus, a strategy that was more compatible with large commercial orchards was pursued. It is based on optimization of the key components of a behaviourally based trapping system and on the knowledge that most flies that damage apples in commercial orchards are sexually mature immigrants. Thus, flies that reach the border of an orchard are more likely to: (i) encounter trees on the border row; (ii) orient toward sources of olfactory attractants based on ripe apple odour, i.e. find host-tree canopies that include odour baits; (iii) respond to host fruit-associated visual stimuli, i.e. the red sphere; and (iv) be captured effectively on the sticky surface of visually apparent spheres.

This led to a perimeter-based trap out strategy that was evaluated in a 4-year study conducted in commercial orchards in Massachusetts. In this case, butyl hexanoate, and not the five-component blend, was used as the olfactory attractant. Traps and the lure were deployed in late June in border row trees approximately 5 m apart. Debris removal and new sticky applications were performed every 2 weeks (Prokopy and Mason, 1996). The results revealed an average of 1.1% fruit injury from apple maggot fly in blocks with the perimeter trapping-based strategy compared with 0.7% for blocks treated with three conventional organophosphate insecticide treatments (Prokopy *et al.*, 1996). In Quebec, a perimeter-based strategy yielded similar results (Bostanian *et al.*, 1999). In this agricultural context, the management of a key direct fruit-feeding pest is possible using an optimized behaviourally based trapping system because the pressure exerted by resident populations is not high. In the absence of insecticide sprays in a Quebec apple orchard, damage to fruit at harvest ranged from 2.3 to 61% (10-year average = 21.7 %) (Vincent and Roy, 1992).

However, as mentioned previously, the deployment and maintenance of a large number of sticky-coated spheres is a labour-intensive process. This led to concerted efforts to develop an inexpensive and effective alternative to a sticky coating to serve as the fly-killing agent for spheres. Ultimately, a unified behavioural and chemical control strategy based on a combination of insecticide and feeding stimulant in conjunction with a residue-extending agent was developed for the spheres. Three different approaches were pursued for deploying the insecticide-feeding stimulant mixture. They included: (i) an internally renewable source; (ii) an externally preserved source; and (iii) an externally renewable source of the insecticide-feeding stimulant blend.

The internally renewable system was based on the development of what was termed a biodegradable pesticide-treated sphere (Fig. 5.2b). It was envisioned that this sphere would serve as an efficacious control strategy throughout the active apple maggot fly season without a need for maintenance or replacement. Spheres would be deployed to control adults through the active season and then simply degrade over the autumn and winter months. In this case, the feeding stimulant sugar was mixed with gelatinized flour and glycerin. This mixture was shaped into spheres to create the appropriate visual stimulus and provide a large reservoir of feeding stimulant (Hu *et al.*, 1998). Dried spheres were coated with an insecticide and red latex paint mixture, which served as the killing agent and residue-extending agent, respectively. Biodegradable spheres were manufactured for several years and evaluated in New England, New York and Michigan. However appealing this strategy appeared, it was ultimately abandoned. The deployment of a starch-based structure within the context of a temperate orchard proved extremely tedious and problematic. Weaknesses included structural breakdown of the body of the sphere, fungal growth on sphere surfaces due to the sucrose-based feeding stimulant and abiotic conditions, rodent consumption of the spheres themselves and rapid loss of toxicant from the sphere surface, as described by Green and Wright (Chapter 9, this volume). However, the idea of a biodegradable sphere remains a tantalizing one in terms of a sustainable, environmentally conscious approach for the management of the apple maggot fly, if one were able to overcome some of the inherent challenges described above.

An externally dispensed system also was pursued rigorously (Fig. 5.2c). Coating plastic or wooden spheres with a mixture of insecticide, sucrose-based feeding stimulant and latex paint was found to be very effective at killing alighting flies (Duan and Prokopy, 1992, 1993), even at a very low dose of insecticide. However, these spheres had an inherent weakness. The feeding stimulant, a critical component of the system, was depleted rapidly from the mixture by less than 7 mm of rainfall (Duan and Prokopy, 1995a). In field trials conducted in commercial orchards, the feeding stimulant mixture was reapplied to spheres within 1 or 2 days following each rainfall event measuring 5 mm or more. Overall, the pesticide-treated spheres performed very well, with damage from apple maggot fly at nearly commercially acceptable levels (1.0%) compared with two foliar applications of an organophosphate (0.8%) (Duan and Prokopy, 1995b). However, the weakness associated with the loss of the feeding stimulant needed to be resolved (Bostanian and Racette, 2001) if this approach for the behavioural management of apple maggot fly were to move forward.

A project was initiated to create a slow-release, externally renewable feeding stimulant technology that would last throughout the entire 15-week activity period of the apple maggot flies without regard to environmental conditions. Rather than coating spheres with the entire mixture itself, a cap was developed that could be deployed on top of the spheres, which served to renew the sucrose-based feeding stimulant (Fig. 5.2d). The cap consisted of a 15:85 paraffin wax:sucrose matrix. The structure of the cap included shallow reservoirs formed on the upper surface to aid in the delivery of the feeding stimulant. This structure has many advantages; runoff from the cap has a high concentration of sucrose to stimulate fly feeding, and very little runoff leaves the sphere surface limiting fungal growth, the cap itself provides a season-long source of feeding stimulant and requires only a small amount of rainfall or dew (<5 mm) to recharge sphere surfaces under field conditions (Wright *et al.*, 2001).

However, a serious problem with this technology was related directly to the visual ecology of the adult fly. The cap that tops the sphere reduces the visual attractiveness of the sphere itself, as it no longer provides the visual specificity of a round silhouette within the canopy. In order to mitigate this shortcoming, Wright *et al.* (unpublished results) visually integrated the cap itself into the sphere silhouette (Fig. 5.2e). This trapping system is now being evaluated and initial results look very promising. In trials in which spinosad was used as the active ingredient and these attracticidal spheres were exposed to 30 cm of rainfall (to simulate the amount experienced throughout the entire apple maggot fly season), spheres were still extremely efficacious, killing 92% of female flies allowed to forage on their surface for only 10 min, as described by Green and Wright (Chapter 9, this volume).

Moving forward

There are still questions that need to be addressed to fulfil Ron Prokopy's goal – to develop a behaviourally based control strategy for the apple maggot fly. They include understanding how best to deploy the powerful five-component blend with the attracticidal spheres. Also important are questions surrounding specific insecticides used as the killing agent for attracticidal spheres and, in particular, how much exposure is required to kill foraging flies and if sublethal effects prevent even those females that have encountered spheres for brief periods from ovipositing successfully. The deployment strategy for use with these spheres likely will vary according to orchard architecture and fly pressure. As pointed out by Prokopy and Roitberg (2007), once the behavioural responses of flies to a particular set of ecological conditions are well understood, their environment can then be manipulated to optimize management based on the principles of behavioural ecology.

The plum curculio

Biology and pest status

The plum curculio is an insect native to North America that probably bred originally on wild plum species including *Prunus americana*, *Prunus nigra* and *Prunus mexicana*. The geographic range of the plum curculio and these wild hosts roughly correspond with one another throughout much of eastern North America. The plum curculio has expanded its host range to cultivated hosts in the family Rosaceae, including: apple, *Malus domestica*; peach, *P. persica*; sweet cherry, *Prunus avium*; tart cherry, *Prunus cerasus*; Japanese plum, *Prunus salicina*; European plum, *Prunus domestica*; quince, *Cydonia oblonga*; pear, *Pyrus communis* (Maier, 1990), nectarine, *Prunus persica* (Howitt, 1993); apricot, *Prunus armeniaca* (Howitt, 1993; Brown, 2005); and highbush blueberry, *Vaccinium corymbosum* (Beckwith, 1943; Mampe and Neunzig, 1967; Polavarapu *et al.*, 2004) in the family Ericaceae. Other recorded hosts include deerberry, *Vaccinium stamineum*, in the family Ericaceae, and muscadine grape, *Vitis rotundifolia*, in the family Vitaceae (Jenkins *et al.*, 2006). Written accounts describing plum curculio attacking cultivated stone fruit were recorded as early as 1736 and apple by 1804 (Quaintance and Jenne, 1912), although this insect also was considered a pest by native Americans, who gathered and dried wild plums (Chapman, 1938).

This insect is a key pest of both stone and pome fruit in eastern North America (Racette *et al.*, 1992; Vincent *et al.*, 1999, 2008). In a recent survey in New England, the plum curculio was cited by apple growers as the species that required the greatest pest management effort (Clifton, 2005). In the spring, adults move from overwintering sites toward orchards, where they mate in or near host fruit trees (Smith and Salkeld, 1964; Chouinard *et al.*, 1993; Piñero *et al.*, 2001).

Adults feed on and females oviposit in the developing fruit soon after petal fall (Lafleur and Hill, 1987; Chouinard *et al.*, 1993), cutting a small, crescent-shaped flap in the fruit skin and then depositing an egg (Quaintance and Jenne, 1912; Chapman, 1938). Subsequent larval feeding can lead to either fruit drop (Levine and Hall, 1977) or severe scarring (Quaintance and Jenne, 1912; Racette *et al.*, 1992). Mature larvae that leave fallen fruit burrow into the soil to pupate and emerge approximately 1 month later (Quaintance and Jenne, 1912).

In Canada, New England, New York and probably much of Pennsylvania, the plum curculio is univoltine with an obligatory diapause (Smith and Flessel, 1968), completing one generation per year (Racette *et al.*, 1992), whereas populations south of Virginia are multivoltine (Chapman, 1938). Molecular-based work to separate morphologically identical univoltine and multivoltine populations has revealed differences in the *Wolbachia* bacterium strains infecting them (McClanahan *et al.*, 2004; Zhang, 2007) and in the partial mitochondrial cytochrome oxidase gene subunit I (mtCOI) obtained from adults from different geographic populations (Zhang, 2007). These differences corresponded roughly with known univoltine and multivoltine populations in more northerly and southerly locations, respectively. By contrast, in the mid-Atlantic region including eastern Maryland, Virginia, West Virginia, New Jersey and Delaware, ovarian development data from eastern West Virginia and Maryland (Leskey and Wright, 2004b; Leskey, 2008) and genetic data from a number of states based on mtCOI (Zhang, 2007) support the presence of multivoltine populations, which is in contrast to the only map establishing the purported geographic range of univoltine and multivoltine populations (Chapman, 1938).

Adults of the univoltine populations emerge in July and August and feed on fallen fruit or fruit remaining on trees (Racette *et al.*, 1992), until they build up sufficient fat stores and emigrate to overwintering sites (Smith and Salkeld, 1964). Multivoltine populations have a facultative diapause (Smith and Flessel, 1968) and can reproduce throughout the summer months. For tree-fruit growers in the mid-Atlantic and the south-eastern USA, this means that fruit is at risk to damage by plum curculio from fruit set until near harvest. This highlights the importance of establishing an effective management programme for this destructive pest.

The adults disperse only a few metres to and from overwintering sites (Lafleur and Hill, 1987; Lafleur *et al.*, 1987). In apple orchards, most dispersal from tree to tree occurs between bloom and fruit set, but even during peak dispersal activity (fruit set), daily movements from one tree to another are seldom observed (15 movements/day/100 insects) (Chouinard *et al.*, 1994). Movements between the trees and the ground are much more frequent and are mostly the consequence of the thanatosis behaviour exhibited by adults, which explains why most of the adults can be found on the ground during the day, except around petal fall (Chouinard *et al.*, 1994). Movements occur most frequently between 3 pm and 12 pm. The insect is nocturnal at the beginning of the season and, as the season progresses, shifts towards diurnal activity (Racette *et al.*, 1990).

Conventional and alternative management

Growers have relied on broad-spectrum insecticides, in particular the organophosphates, to provide a commercially acceptable level of control. In the absence of monitoring tools, growers in the north-eastern USA typically apply three insecticide treatments to manage plum curculio (Prokopy *et al.*, 1996; Reissig *et al.*, 1998), while growers in Quebec, Canada, apply one or two. Based on the results of a 2004 survey (Clifton, 2005), 96% of New England apple growers applied insecticides to manage plum curculio, with most using phosmet and/or azinphos-methyl on the majority of their orchards. Alternatives to insecticides as a main management strategy are rare and generally costly, at least in a commercial setting. For example, Benoit *et al.* (2006) have demonstrated that the life cycle of plum curculio can be disrupted by deploying a physical barrier, a cellulose sheeting laid on the soil beneath the tree canopy. Although such a method can deliver a desirable control level over the years, it is very costly relative to insecticide sprays. The nematode, *Steinernema ribobrave*, shows promising results in controlling larvae in the soil (Shapiro-Ilan *et al.*, 2004) and further studies indicate that this approach holds promise in reducing infestations under field conditions (Shapiro-Ilan *et al.*, 2008).

Development of behaviourally based monitoring and management strategies – key components

VISUAL ECOLOGY

Toward host trees. The identification of specific visual cues used by coleopteran species as they approach host plants has led to the development of a number of visually stimulating traps. The plum curculio feeds on plants in three families, Rosaceae, Ericaceae and Vitaceae, and a polyphagous herbivore is more likely to respond to generalized rather than specific visual cues associated with host trees (Prokopy and Owens, 1978). In trials conducted to determine if adult plum curculio oriented to foliage-mimicking colours, more adults were captured on sticky-coated green rectangles meant to mimic the colour, shape and size of a small apple tree when deployed in association with apple branches (as an olfactory stimulus) compared with green rectangles with maple branches (non-host olfactory stimulus) or clear rectangles (meant to represent open space) with either apple or maple branches (Butkewich and Prokopy, 1997). These results indicated that plum curculio appeared to respond to foliar-based visual cues, although the response was not very strong or specific. These results led to other experiments aimed at finding a more powerful visual stimulus.

Some Scolytidae species respond positively to visual mimics of vertically growing stems (Schonherr, 1977). A trap based on this visual cue was developed; it consisted of a series of vertically aligned funnels (Lindgren, 1983). Similarly, a number of curculionid species have been captured by darkly coloured pyramid traps. This trap is believed to provide a foraging weevil with a visual silhouette similar in shape and size to that of a tree trunk and in visual contrast with the surrounding background. The pecan weevil, *Curculio caryae* (Tedders *et al.*, 1996), as well as the pales weevil, *Hylobius pales*, and the pitch-eating weevil, *Pachylobius picivorus* (Mizell and Tedders, 1999), have all been captured successfully by darkly coloured pyramid traps.

The use of a pyramid trap as a potentially visually stimulating trap type has been evaluated for plum curculio as well. This trap type attempts to fit their visual ecology, as adults are likely to use visual cues as they approach host trees; i.e.

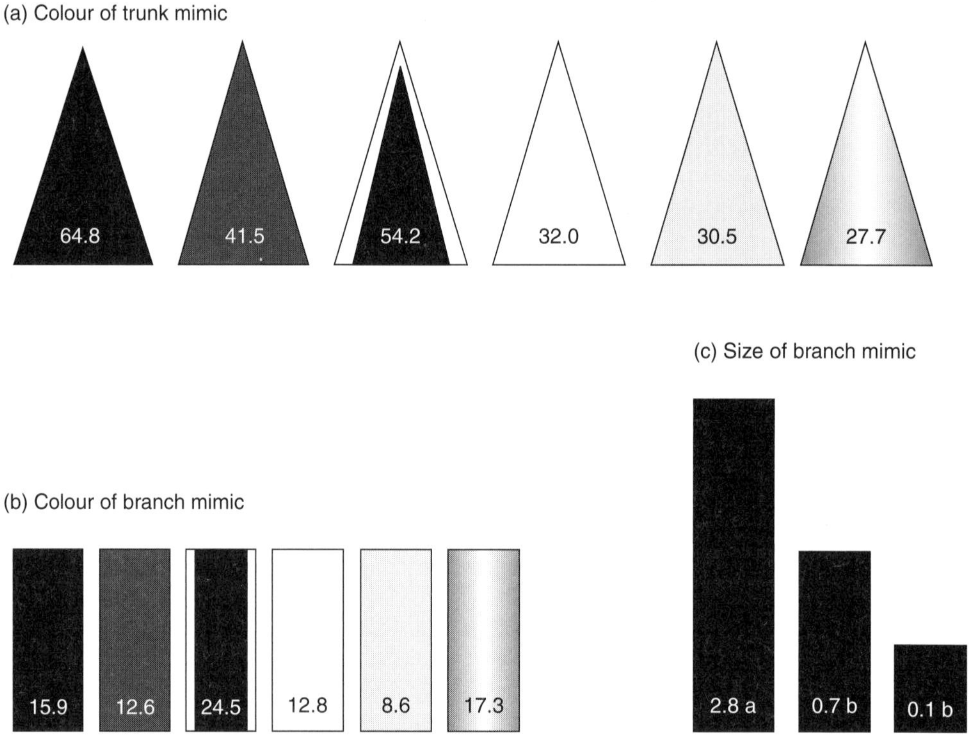

Fig. 5.3. Mean number of adult plum curculio captured in black, green, enhanced, white, yellow and clear (respectively) (a) pyramid traps and in (b) branch-mimicking traps in an unsprayed peach orchard in Summit Point, West Virginia (Leskey, 2006) and in (c) black branch-mimicking traps 50, 25 and 12.5 cm tall in unsprayed apple trees in Amherst, Massachusetts (Leskey and Prokopy, 2002).

they are able to perceive and orient toward dark silhouettes of trees against a background of sky light (Lafleur *et al.*, 1987). In one trial, pyramid traps were painted with flat latex exterior paint in the following colours: black, black with a white border along trap edges, green, yellow or white. The colours represented the following visual stimuli: standard black trunk mimic, enhanced trunk mimic (by increasing contrast between black and white (Whalon and Coombs, 2003)), foliar mimics (green and yellow) and no visual stimulus (white and clear polycarbonate). Each of these traps was capped by a conical boll weevil trap top and deployed between unsprayed peach trees to determine if plum curculio preferred pyramid traps of a specific colour. Black pyramid traps captured twice as many adults compared with white, yellow or clear pyramid traps, but not green or enhanced black (Fig. 5.3a). Plum curculio appeared not to have a preference for a specific colour, but instead approached darkly coloured traps based on their appearance as a dark silhouette in visual contrast with the surrounding background.

Observations of the propensity of adult plum curculios to walk or fly toward an apple tree or toward a pyramid trap deployed beneath an unmanaged plum tree or 2.4 m from the trunk itself revealed that adult movement was temperature-dependent. Univoltine adults walk at temperatures below 20°C, but exhibit a greater likelihood for flight at temperatures of 20°C or greater. Adults generally fly toward the tree canopy and walk toward the tree trunk or black pyramid trap when deployed next to a tree trunk. These results led Prokopy *et al.* (1999) to conclude that a black pyramid trap deployed next to the trunk of a host tree would be less effective at temperatures of 20°C or greater because adults likely would arrive in host trees by flight and bypass the pyramid trap itself. However, these results are in contrast with a study conducted in West Virginia, where pyramid traps equipped with two collection devices were deployed between host trees in an unsprayed peach orchard (Leskey, 2006). The upper collection device was intended to capture adults that flew directly to the pyramid trap and then proceeded to the top collection device, whereas the lower collection device located at the base of the trap was designed to capture adults that crawled directly from the ground to the traps. Results revealed that over six times more multivoltine adults were captured in the upper than in the lower collection device, even though the average daytime temperature throughout the trapping period was 22°C. Most of the plum curculios captured likely flew, rather than crawled, to the pyramid traps. Direct observations of plum curculio movement toward pyramid traps and trees may elucidate the nature of the differences between studies conducted with univoltine and multivoltine populations.

Within host trees. Plum curculios also appear to use visual stimuli once they arrive in the host-tree canopy. Adults are highly prone to walk rather than fly within the tree canopy after arrival, traversing the upper and lower sides of limbs (Chouinard *et al.*, 1994). Initial studies revealed that adults oriented toward visual stimuli of vertically growing limbs or branches. Therefore, a series of experiments was conducted to develop a branch-mimicking cylinder trap that could be deployed in host-tree canopies. Like pyramid traps, cylinder traps were capped with a boll weevil collection device. Cylinders constructed of polyvinyl chloride pipe and painted with black, black with contrasting white stripes, green, yellow or white flat latex exterior paint and of clear polycarbonate pipe to represent branch-mimicking (black), foliar (green and yellow) and no particular (white and clear) visual stimuli were deployed within the canopies of unsprayed peach trees. Results revealed plum curculios to be a visual generalist in terms of colours when foraging within the canopy, as there were no differences among treatments (Fig. 5.3b) (Leskey, 2006).

At close range, the dimensional qualities (size and shape) of host plants may provide important visual cues for phytophagous insects (Prokopy and Owens, 1983). Dark silhouettes that mimic closely the shape and size of spruce and pine leaders typically attacked by the white pine weevil, *Pissodes strobi*, elicit greater orientation by adults than silhouettes of other dimensions under laboratory conditions (VanderSar and Bordon, 1977). Plum curculio also appears to be a visual specialist in terms of shape and size when foraging within the canopy. When black cylinder traps of three heights (12.5, 25 and 50 cm), and each of three different diameters (3.0, 6.0 and 12.0 cm), were deployed in the canopies of unsprayed apple trees, black cylinder traps that were 50 cm tall × 6 cm in diameter captured nearly twice the number of plum curculios than other trap sizes (Fig. 5.3c). Direct observations of released adults also revealed that greater numbers arrived at taller traps compared with those of lesser height, indicating that at close range, taller cylinders might act as a stronger visual cue for foraging plum curculios seeking food and/or oviposition sites in host-tree canopies (Leskey and Prokopy, 2002). The orientation of the tall cylinder traps also influenced the attractiveness of this branch-mimicking stimulus. Like vertically oriented silhouettes for *P. strobi* (VanderSar and Bordon, 1977), cylinder traps oriented in an upright vertical position captured significantly more plum curculios in field experiments and elicited greater numbers of arrivers than cylinders positioned at other angles (Leskey and Prokopy, 2002).

CHEMICAL ECOLOGY

Aggregation pheromone. Olfactory responses of plum curculios are probably important components of host- and mate-finding as this species moves from overwintering sites at the periphery of orchards (Smith and Flessel, 1968; Lafleur and Hill, 1987; Piñero *et al.*, 2001) before fruit set (Piñero *et al.*, 2001) and mates within or beneath host fruit trees (Smith and Salkeld, 1964; Racette *et al.*, 1992; Chouinard *et al.*, 1993). Adult antennae possess sensilla types consistent with those known to be olfactory receptors in other insects, including the presence of type IV, thought to be for pheromone reception (Alm and Hall, 1986). Volatile collections from virgin multivoltine male plum curculios have yielded the identification of a single pheromone compound, (+)-(1*R*,2*S*)-1-methyl-2-(1-methylethenyl) cylcobutaneaetic acid, termed grandisoic acid. When a racemic mixture of this compound was deployed in boll weevil trap tops placed atop the cut ends of vertical branches in host fruit trees, significantly more adult female and male plum curculios were captured in traps compared with unbaited traps, indicating that this compound was an aggregation pheromone (Eller and Bartelt, 1996). Field studies revealed that both univoltine (Piñero and Prokopy, 2003) and multivoltine (Leskey *et al.*, 2005) populations were attracted to a racemic mixture of grandisoic acid when deployed in combination with host-plant volatiles. There are no published field studies that have evaluated the responses of adults to grandisoic acid when the (–) enantiomer is removed, although recent electroantennogram (EAG) studies of adult responses to purified grandisoic acid revealed an increase in magnitude of response to those formulations with decreased amounts of the (–) enantiomer (Leskey, Zhang, Chouinard, Cormier and Pichette, unpublished results). The aggregation pheromones of other related weevil species, including the strawberry blossom weevil, *Anthonomus rubi* (Innocenzi *et al.*, 2001), the pepper weevil, *A. eugeni* (Eller *et al.*, 1994), the boll weevil, *A. grandis* (Tumlinson *et al.*, 1969) and the pecan weevil, *C. caryae* (Hedin *et al.*, 1997), are comprised of multiple components. This may be the case for the plum curculio as well, but identification of other attractive compounds has proven elusive due to the difficulty in developing a reliable EAG technique for plum curculio that can be incorporated into a gas chromotography-electroantennogram detection (GC-EAD) system.

Host-plant volatiles. Behavioural studies demonstrated that plum curculios used olfactory cues to locate host fruit trees (Butkewich and Prokopy, 1993, 1997;

Leskey and Prokopy, 2001). Host apple and plum odours released by plant tissues between bloom and 2 weeks after bloom are attractive to plum curculios (Leskey and Prokopy, 2000). Compounds identified from cultivated plum collected 2 weeks after bloom (Leskey *et al.*, 2001) and apple collected 1–4 weeks after fruit set (Boevé *et al.*, 1996; Prokopy *et al.*, 2001b) were tested in the laboratory (Leskey *et al.*, 2001) and in the field (Leskey *et al.*, 2001; Prokopy *et al.*, 2001b) for plum curculio attraction. Several compounds, particularly ethyl isovalerate, limonene, benzaldehyde, benzyl alcohol, decanal, trans-2-hexenal, geranyl propionate and hexyl acetate, were found to be attractive (Leskey *et al.*, 2001; Prokopy *et al.*, 2001b). When benzaldehyde was evaluated in combination with racemic grandisoic acid in association with traps under field conditions, plum curculio captures increased significantly compared with captures in traps baited with grandisoic acid alone or in unbaited traps (Piñero *et al.*, 2001). Only benzaldehyde synergized plum curculio responses to grandisoic acid, while ethyl isovalerate and limonene did not (Piñero and Prokopy, 2003). Like other species of curculionids (Dickens, 1989; Landolt, 1997; Landolt and Phillips, 1997), the combination of synthetic host-plant volatiles and aggregation pheromone results in a synergistic response from plum curculios.

Though synergism between benzaldehyde and grandisoic acid results in increased trap captures, pyramid, screen, cylinder and panel traps baited with these attractants have failed repeatedly to monitor plum curculio activity reliably in north-eastern apple and mid-Atlantic apple and peach orchards (Prokopy *et al.*, 2003; Leskey and Wright 2004b). In fact, captures in baited traps decline quickly after fruit set (Prokopy *et al.*, 2003; Leskey and Wright, 2004b), indicating that volatiles released by rapidly developing fruit could be outcompeting synthetic attractants. Indeed, Leskey and Wright (2004a) found that the presence of host apple trees did have a negative impact on plum curculio responses to baited traps, further strengthening the hypothesis that olfactory cues produced by host trees, particularly after fruit set, were more attractive to plum curculios than synthetic attractants.

There are a number of issues associated with the approach taken toward identification of host plant-based attractants for plum curculios that could account for earlier failures associated with baited traps. Single-component, fruit-based attractants are not likely to be very competitive within the context of an orchard when the attractant itself is released by developing fruit. For example, although benzaldehyde was identified from plum (Leskey *et al.*, 2001), it also was produced by apple blossoms (Buchbauer *et al.*, 1993). Leskey and Wright (2004a) demonstrated that when traps baited with this compound were deployed in close proximity of host fruit trees, significantly fewer adults were captured compared with traps deployed in an area without competing host fruit trees. The five-component olfactory attractant for the apple maggot fly identified from ripe apple volatiles was found to be much more attractive than the single-component butyl hexanoate (Zhang *et al.*, 1999). It seems reasonable to presume that a multiple-component, fruit-based attractant for the plum curculio also would be more attractive.

Another potential issue is that volatile compounds that have been evaluated are based on the identification of compounds released by the fruit only (Leskey *et al.*, 2001; Prokopy *et al.*, 2001b). Plum curculios move from overwintering sites and locate host trees well before fruit is available (Lafleur and Hill, 1987; Piñero *et al.*, 2001; Leskey and Wright, 2004a) and therefore probably encounter and respond to host volatiles emanating from the entire tree, and not just the fruit. In fact, plum curculios are attracted to the odours of foliar and woody tissues of apple (Leskey and Prokopy, 2000) and of plum (Leskey *et al.*, 2005). When a six-component host plant volatile blend comprised of compounds identified from foliar and woody tissues of plum was deployed in combination with grandisoic acid, this blend proved to be as attractive as the standard bait combination of benzaldehyde and grandisoic acid. And when this six-component blend was combined with benzaldehyde and grandisoic acid, captures were ~1.5 times greater in traps baited with this formulation compared with benzaldehyde and grandisoic acid only, suggesting that adults might respond to volatile blends released by the entire tree and not just fruit (Leskey *et al.*, 2005).

Source material used for the identification of potential attractants for the plum curculio has included two principal hosts, European plum and apple (Leskey *et al.*, 2001, 2005). However, plum curculio has a very broad host range, using a number of cultivated hosts in the family Rosaceae. Leskey and Wright (2007) conducted an extensive behavioural study designed to elucidate the preferred rosaceous hosts of plum curculio based on host-finding behaviour of both mark-released and wild populations in order to pinpoint the best source material for the identification of potential attractants. They found that Japanese plum cultivars 'Formosa' and 'Santa Rosa' were 1.5 times more highly preferred than the European plum cultivar 'Stanley', based on a preference index. Japanese plum also was 4.0, 5.0, 11.5 and 29.7 times more highly preferred than peach, tart cherry, apple and pear, respectively. Notably, more adults were recovered from Japanese plum compared with apple or peach throughout the season, even after plums had dropped from the trees, indicating that this particular host provided the basis for a competitive attractant to be used in apple and peach orchards, the most important fruit commodities attacked by plum curculio.

The ancestral host of plum curculio is believed to be a native species of North American plum (Quaintance and Jenne, 1912; Chapman, 1938; Maier, 1990; Jenkins *et al.*, 2006). Interestingly, Japanese plum cultivars could have native plum species as part of their parentage (Jones, 1928), while European plum varieties do not (Zohary, 1992). Many Japanese plum cultivars developed in the USA were generated by Luther Burbank, who crossed imported Japanese plum with native plum species such as *Prunus americana* (Byrne, 1989). The hypothesis that Japanese plum cultivars may have native North American plum species as part of their parentage is intriguing and could explain the preference for these cultivars compared with the European plum cultivar 'Stanley'.

Capture mechanism and deployment strategy

As plum curculio exhibits temperature-dependent movement (Prokopy *et al.*, 1999), mechanisms for capturing walking and flying adults have been evaluated. They include pyramid traps of different shapes and sizes (Leskey, 2006; Lafleur *et al.*, 2007) and branch-mimicking cylinder traps; both have integrated visual cues and use a boll weevil trap top to funnel adults walking up the trap surface into a single collection device. Lindgren funnel traps, which also have a trunk-mimicking visual stimulus, use a series of vertically oriented funnels to capture flying adults that land on the funnel surface and fall to the basal collection device (Leskey, 2006). Among traps with no specific visual cues, a number of capture mechanisms have been evaluated. These include sticky-coated apples or plastic spheres hung in the canopies of host trees (Yonce *et al.*, 1995), designed to capture foraging adults walking in the host-tree canopies, and sticky-coated Plexiglas panels attached to the tops of wooden posts (Prokopy *et al.*, 2000), designed to capture flying adults entering orchards. Pitfall traps (Yonce *et al.*, 1995) and Vernon beetle traps (Leskey, 2006) deployed on the ground to passively trap adults walking beneath host trees have also been evaluated. Screen or 'circle' traps wrapped around a host tree trunk or limb are designed to intercept individuals crawling on the host (Mulder *et al.*, 1997).

Among trap types with different capture mechanisms, standard masonite pyramid and screen traps captured significantly greater numbers of adults than other trap types. Pyramid traps deployed next to (Prokopy and Wright, 1998) or between host trees, as well as screen traps deployed on the trunks of host trees (Leskey, 2006), have proven to be the most effective deployment locations. These traps probably capture more adults compared with those deployed in the host-tree canopy because as adults forage for food and oviposition resources found throughout the canopy, generally by traversing the upper and lower surfaces of limbs by walking, they can easily bypass a trap located at a particular point within the canopy. Pyramid traps and screen traps also have a greater chance of being encountered by a plum curculio compared with pitfall traps or Vernon beetle traps, which do not exploit the natural tendency of adults to walk up surfaces. Both pyramid and screen traps use a capture mechanism that guides foraging adults upwards and funnels those present over a large surface area into a single collection point (Leskey, 2006).

Monitoring

Evaluations of baited trap-based approaches incorporating these key components have yielded mixed results. In southern peach orchards, Johnson *et al.* (2002) established an economic threshold to be used as a decision-based tool for triggering insecticide applications; a threshold of 0.1 adults per trap based on captures principally in pyramid traps was established. A threshold has not been established in north-eastern and mid-Atlantic apple orchards. Various trap types including pyramid traps, Plexiglas panel traps, screen and branch-mimicking cylinder traps baited with the synthetic fruit volatile benzaldehyde in combination with grandisoic acid proved effective at capturing adults prior to fruit set. After fruit set, these traps quickly lost their ability to capture adults (Prokopy *et al.*, 2003; Leskey and Wright, 2004b), resulting in no predictive relationship between trap captures and the amount of fruit injury (Prokopy *et al.*, 2003; Leskey and Wright, 2004b). Therefore, trap-based approaches for monitoring plum curculio as a decision-making tool for growers have not moved forward in any substantial way.

Another approach evaluated in commercial apple orchards is termed a trap-tree-based approach. Like trap-based approaches, trap trees are baited with the synergistic two-component lure comprised of the fruit volatile benzaldehyde and grandisoic acid. These trap trees are located on the perimeter rows of apple orchards. Baiting selected perimeter row trees with these olfactory attractants results in an approximately 15-fold increase in aggregation of oviposition injury in these particular canopies (Prokopy *et al.*, 2003) that extends to at least 31 m along a perimeter row. After a full block petal fall insecticide application is applied, these trap trees are monitored for signs of fresh oviposition injury. If a threshold of one freshly injured fruit out of 50 sampled is found on a perimeter row trap tree, a subsequent insecticide application to the peripheral rows is recommended (Prokopy *et al.*, 2004).

Management

A number of approaches have been developed that can result in reductions in the overall amount of insecticide applied against plum curculio. These include the use of a degree-day model (Reissig *et al.*, 1998), a perimeter row treatment strategy and an insecticide-treated odour-baited trap tree (Leskey *et al.*, 2008). All of these approaches rely on a full block insecticide application at petal fall. This timing of the full block insecticide treatment at petal fall is based on the immigration of the overwintered population; at petal fall, approximately 60% of the overwintered population has moved beyond the perimeter row and penetrated the interior of the orchard by this time (Piñero and Prokopy, 2006). As the fruit are just beginning to develop, the entire orchard is vulnerable to injury and must be treated. After this treatment, the efficiency of subsequent insecticide applications can be increased by using one of the following strategies. The first method that was developed was an oviposition injury-derived degree-day model; it called for maintaining insecticide residue on trees until the accumulation of 171 degree days (base 10°C) after petal fall was reached (Reissig *et al.*, 1998).

The second method, the perimeter row strategy, is based on the risk evaluated on fresh damage caused by immigrating adults (Chouinard *et al.*, 1992; Vincent *et al.*, 1997). Starting at petal fall, fruitlets are examined two or three times a week and insecticide treatments are applied to the perimeter row only, rather than the entire orchard block, as soon as a threshold of 1% of fresh egg-laying injury is reached. This approach has been validated using Zn^{65} radiolabelled adults released at the edge of an apple orchard that received a single perimeter row application of azinphosmethyl at petal fall; 83% of recaptured insects were found dead under the sprayed trees (59%) or alive along the outside edge of the sprayed rows (24%). In commercial orchards in Quebec, this approach has proven effective in maintaining plum curculio damage at harvest below 1%, while applying only 25% of the insecticides that would have been applied using full block sprays. Although the perimeter row strategy has been validated in commercial orchards, it has not been adopted widely, presumably because the trade-off between the higher risks of fruit injury and the benefits was insufficiently rewarding. In this context, a strategy based on the evaluation of risks by trapping adults has been pursued and the perimeter row strategy is, for now, recommended and used widely for sprays needed after the first full block spray at petal fall (Chouinard *et al.*, 2001).

The odour-baited trap-tree approach is an extension of the trap-tree monitoring strategy. Insecticide applications are confined to a few perimeter row trap trees after the full block petal fall insecticide application, resulting in substantial reductions in the amount of insecticide sprayed against plum curculio. Studies conducted in New England apple orchards revealed that treating trap trees on perimeter rows separated by 50 m resulted in a reduction of approximately 70% and 90% of total trees being treated with insecticide compared with perimeter row and standard full block sprays, respectively (Leskey *et al.*, 2008). Injury in plots managed with the trap-tree treatment regime compared with perimeter row sprays were statistically identical, indicating that using this management strategy was an effective means to control immigrating plum curculio populations after petal fall and resulted in reduced insecticide inputs. One caveat for use of this strategy is that a grower must be willing to sacrifice fruit on trap trees because oviposition injury on fruit found in these trees will be high (Leskey *et al.*, 2008).

Moving forward

There are still a number of questions that need to be addressed before a behaviourally based monitoring tool, and ultimately a management strategy, will be available for commercial fruit growers. First, a better understanding of the biology, ecology and behaviour of multivoltine populations is necessary, as most of the published work to date has been with univoltine populations. Second, more competitive olfactory attractants need to be identified. With the newly developed electroantennogram technique, this goal seems to have a greater likelihood of succeeding. Third, as plum curculio are cryptic and mainly nocturnal (Racette *et al.*, 1990), novel technology such as harmonic radar (O'Neal *et al.*, 2004) will enable detailed behavioural studies of movements within the context of the orchard agroecosystem.

Conclusion

In this chapter, we summarized the body of work pioneered by the late Ron Prokopy as he chartered a new course for pest management by conducting basic behavioural studies and using those results to develop effective crop protection applications based on behavioural manipulation for the apple maggot fly and the plum curculio. His approach can be considered a model system for others seeking to develop behaviourally based management strategies for other pests and crops.

A generalized pathway for pursuing this approach is described in a recent paper by Prokopy and Roitberg (2007) and can be summarized as follows: (i) target only arthropod species that are considered to be key pests rather than secondary or sporadic pest species, because alternative management approaches such as behavioural manipulation are generally more expensive and require more labour and input and therefore are likely less economically viable in the long term for occasional pests; (ii) characterize the behaviour of the pest species in time and space by observing it in its natural habitat in order to generate a robust account of the foraging behaviour of the species; these basic behavioural observations are critical to the identification of particular resources such as preferred food and oviposition sites; (iii) conduct semi-field and laboratory studies in order to identify and quantify the effects of specific stimuli from these resources; stimuli that elicit a particular behaviour from the pest species are characterized based on the movement patterns of the pest; (iv) conduct studies of these particular stimuli under field conditions to establish the foundation for a potential behavioural management strategy; (v) replace and evaluate natural stimuli with artificial stimuli under field conditions because they provide greater flexibility and precision in behavioural manipulation approaches for pest management; and (vi) evaluate the field efficacy of this approach in the field, taking into account the genetic background, the physiological state and the current informational state of the pest, as well as the state of the current environment with respect to habitat, resources, natural enemies, etc.; all of these factors can have a profound effect on the efficacy of the system.

Ron Prokopy spent nearly 40 years working toward the development of behavioural manipulation strategies for apple maggot fly and plum curculio. Toward the end of his career, he wrote a paper detailing two decades (1981–2000) of assessments of a bottom-up IPM programme that he designed and implemented in his own small apple orchard in Massachusetts (Prokopy, 2003). As a component of this small-scale IPM demonstration, Prokopy relied exclusively on the behavioural control of apple maggot fly using unbaited, sticky-coated red spheres deployed at a rate of 1–3 spheres per tree, depending on fruit load. Through the course of this study, the damage inflicted by apple maggot fly never exceeded a 5-year mean of 1% and never differed from damage rates found in chemically managed orchards. Compared with the rate of infestation in nearby, unmanaged trees (90.8%), Prokopy demonstrated clearly the potential commercial utility of the behavioural control of apple maggot fly, affirming that sound and sustained research efforts can yield viable management strategies.

In this study, Prokopy also acknowledged that for some key pests, no biologically based management strategies were known or commercially feasible. For example, given that acceptable suppression of plum curculio was only possible through periodic treatment with a synthetic insecticide, Prokopy selected an insecticide (phosmet) based on consideration of the effectiveness, applicator safety and impact on beneficial organisms. Although Prokopy emphasized that a bottom-up ecological approach toward pest management should not rely on management strategies that might harm beneficial relationships among orchard organisms, he accepted that the development of behaviourally and ecologically based pest management strategies must fall within the bounds of economic and technical feasibility. Therefore, while conceding that plum curculio could only be controlled effectively using two applications of insecticide in this study, he pursued vigorously the research of monitoring and management techniques aimed at minimizing the impact and necessity of insecticide applications against this key pest, notably development of the trap-tree management approach.

The successes of this long-term demonstration project were remarkable achievements, but Prokopy recognized the complications of scale and context in implementing broad behaviourally and ecologically based pest management systems in commercial orchards. Given the geographic variation in the biology, ecology and severity of key pests, along with logistical and economic constraints, formidable challenges to full commercial implementation of high-level IPM growing systems still lie ahead. However, Prokopy demonstrated that through keen observation and unwavering perseverance, sustainable pest management practices could be developed and implemented successfully, bridging the gap between basic behavioural studies and crop protection applications.

References

Agnello, A.M., Spangler, S.M. and Reissig, W.H. (1990) Development and evaluation of a more efficient monitoring system for apple maggot (Diptera: Tephritidae). *Journal of Economic Entomology* 83, 539–546.

Ali Niazee, M.T. (1988) *Ecology and Management of Economically Important Fruit Flies.* Oregon State University Agriculture Station Special Report 830, Corvallis, Oregon.

Alm, S.R. and Hall, F.R. (1986) Antennal sensory structures of *Conotrachelus nenuphar* (Coleoptera: Curculionidae). *Annals of the Entomological Society of America* 79, 324–333.

Aluja, M. and Prokopy, R.J. (1992) Host search behaviour by *Rhagoletis pomonella* flies: inter-tree movement patterns in response to wind-borne volatiles under field conditions. *Physiological Entomology* 17, 1–8.

Aluja, M. and Prokopy, R.J. (1993) Host odor and visual stimulus interaction during intra-tree host-finding behavior of *Rhagoletis pomonella* flies. *Journal of Chemical Ecology* 19, 2671–2696.

Averill, A.L., Reissig, W.H. and Roelofs, W.L. (1988) Specificity of olfactory responses in the tephritid fruit fly, *Rhagoletis pomonella*. *Entomologia Experimentalis et Applicata* 47, 211–222.

Beckwith, C.S. (1943) Insects attacking blueberry fruit. New Jersey Agriculture Experiment Station Bulletin 684, New Brunswick, New Jersey.

Benoit, D.L., Vincent, C. and Chouinard, G. (2006) Management of weeds, apple sawfly (*Hoplocampa testudinea* Klug) and plum curculio (*Conotrachelus nenuphar* Herbst) with cellulose sheets. *Crop Protection* 25, 331–337.

Berlocher, S.H. (1999) Host race or species? Allozyme characterization of the 'flowering dogwood fly', a member of the *Rhagoletis pomonella* complex. *Heredity* 83, 652–662.

Bernays, E.A. and Chapman, R.F. (1994) *Host-plant Selection by Phytophagous Insects.* Chapman and Hall Inc, New York.

Boevé, J.L., Lengwiler, U., Tallsten, L., Dorn, S. and Turlings, T.J.T. (1996) Volatiles emitted by apple fruitlets infested by larvae of the European apple sawfly. *Phytochemistry* 42, 373–381.

Bostanian, N.J. and Racette, G. (2001) Attract and kill, an effective technique to manage apple maggot, *Rhagoletis pomonella* (Diptera: Tephritidae) in high density Quebec apple orchards. *Phytoprotection* 82, 25–34.

Bostanian, N.J., Vincent, C., Chouinard, G. and Racette, G. (1999) Managing apple maggot, *Rhagoletis pomonella* (Diptera: Tephritidae), by perimeter trapping. *Phytoprotection* 80, 21–33.

Brown, M.W. (2005) Host utilization and phenology of injury by plum curculio (Coleoptera: Curculionidae) in West Virginia. *Journal of Entomological Science* 40, 149–157.

Buchbauer, G., Jirovetz, L., Wasicky, M. and Nikforov, A. (1993) Headspace and essential oil analysis of apple flowers. *Journal of Agricultural and Food Chemistry* 41, 116–118.

Bush, G.L. (1969) Sympatric host race formation and speciation in frugivorous flies of the genus *Rhagoletis* (Diptera: Tephritidae). *Evolution* 23, 237–251.

Butkewich, S.C. and Prokopy, R.J. (1993) The effect of short range host odor stimuli on host fruit finding and feeding behavior of plum curculio adults (Coleoptera: Curculionidae). *Journal of Chemical Ecology* 19, 825–835.

Butkewich, S.C. and Prokopy, R.J. (1997) Attraction of adult plum curculios (Coleoptera: Curculionidae) to host-tree odor and visual stimuli in the field. *Journal of Entomological Science* 32, 1–6.

Byrne, D.H. (1989) Inbreeding, co-ancestry, and founding clones of Japanese-type plums of

California and the southeastern United States. *Journal of the American Society for Horticultural Science* 114, 699–705.

Cardé, R.T. and Minks, A.K. (1995) Control of moths by mating disruption: successes and constraints. *Annual Review of Entomology* 40, 559–585.

Carle, S.A., Averill, A.L., Rule, G.S., Reissig, W.H. and Roelofs, W.L. (1987) Variation in host fruit volatiles attractive to apple maggot fly, *Rhagoletis pomonella*. *Journal of Chemical Ecology* 14, 795–805.

Chapman, P.J. (1938) The plum curculio as an apple pest. *New York State Agricultural Experiment Station Bulletin* 684, 1–75.

Chouinard, G., Hill, S.B., Vincent, C. and Barthakur, N.N. (1992) Border-row sprays against the plum curculio (Coleoptera: Curculionidae) in apple orchards: a behavioral study. *Journal of Economic Entomology* 85, 1307–1317.

Chouinard, G., Hill, S.B. and Vincent, C. (1993) Spring behavior of the plum curculio (Coleoptera: Curculionidae) within caged dwarf apple trees. *Annals of the Entomological Society of America* 86, 333–340.

Chouinard, G., Hill, S.B. and Vincent, C. (1994) Spatial distribution and movement of plum curculio adults within caged apple trees. *Entomologia Experimentalis et Applicata* 70, 129–142.

Chouinard, G., Morin, Y. and Brodeur, C. (2001) Lutte contre les insectes et acariens. In: Chouinard, G. (coord.) *Guide de gestion intégrée des ennemis du pommier*. CRAAQ, Sainte-Foy, Quebec, Canada, pp. 99–138.

Clifton, N. (2005) New England apple pest management survey summary. Pest Resources Online New England (http://pronewengland.org/INFO/PROInfoSurvey.htm, accessed 3 November 2007).

Cook, S.M., Khan, Z.R. and Pickett, J.A. (2007) The use of push–pull strategies in integrated pest management. *Annual Review of Entomology* 52, 375–400.

Dickens, J.C. (1989) Green leaf volatiles enhance aggregation pheromone of boll weevil, *Anthonomus grandis*. *Entomologia Experimentalis et Applicata* 52, 191–203.

Drummond, F., Groden, E. and Prokopy, R.J. (1984) Comparative efficacy and optimal positioning of traps for monitoring apple maggot flies (Diptera: Tephritidae). *Environmental Entomology* 13, 232–235.

Duan, J.J. and Prokopy, R.J. (1992) Visual and odour stimuli influencing effectiveness of sticky spheres for trapping apple maggot flies *Rhagoletis pomonella* (Walsh) (Dipt., Tephritidae). *Journal of Applied Entomology* 113, 271–279.

Duan, J.J. and Prokopy, R.J. (1993) Toward development of pesticide-treated spheres for controlling apple maggot flies: carbohydrate and amino acids as feeding stimulants. *Journal of Applied Entomology* 115, 176–184.

Duan, J.J. and Prokopy, R.J. (1995a) Development of pesticide-treated spheres for controlling apple maggot flies: pesticides and residue extending agents. *Journal of Economic Entomology* 88, 117–126.

Duan, J.J. and Prokopy, R.J. (1995b) Control of apple maggot flies with pesticide-treated spheres. *Journal of Economic Entomology* 88, 700–707.

Eller, F.J. and Bartelt, R.J. (1996) Grandisoic acid, a male-produced aggregation pheromone for the plum curculio, *Conotrachelus nenuphar*. *Journal of Natural Products* 59, 451–453.

Eller, F.J., Bartelt, R.J., Shasha, B.S., Schuster, D.J., Riley, D.G., Stansly, P.A., Mueller, T.F., Shuler, K.D., Johnson, B., Davis, J.H. and Sutherland, C.A. (1994) Aggregation pheromone for the pepper weevil, *Anthonomus eugenii cano* (Coleoptera: Curculionidae): identification and field activity. *Journal of Chemical Ecology* 20, 1537–1555.

Feder, J.L., Chilcote, C.A. and Bush, G.L. (1988) Genetic differentiation between sympatric host races of the apple maggot fly, *Rhagoletis pomonella*. *Science* 336, 61–64.

Feder, J.L., Chilcote, C.A. and Bush, G.L. (1990) Geographic pattern of genetic differentiation between host associated populations of *Rhagoletis pomonella*. *Evolution* 44, 570–594.

Fein, B.L., Reissig, W.H. and Roelofs, W.L. (1982) Identification of the apple volatiles attractive to the apple maggot, *Rhagoletis pomonella. Journal of Chemical Ecology* 8, 1473–1487.

Foster, S.P. and Harris, M.O. (1997) Behavioral manipulation methods for insect pest management. *Annual Review of Entomology* 42, 123–146.

Green, T.A., Prokopy, R.J. and Hosmer, D.W. (1994) Distance of response to host tree models by female apple maggot flies, *Rhagoletis pomonella* (Walsh) (Diptera: Tephritidae): interaction of visual and olfactory stimuli. *Journal of Chemical Ecology* 20, 2393–2413.

Hedin, P.A., Dollar, D.A., Collins, J.K., Dubois, J.G., Mulder, P.G., Hedger, G.H., Smith, M.W. and Eikenbary, R.D. (1997) Identification of male pecan weevil pheromone. *Journal of Chemical Ecology* 23, 965–977.

Hendrichs, J., Lauzon, C.R., Cooley, S.S. and Prokopy, R.J. (1993) Contribution of natural food sources to adult longevity and fecundity of *Rhagoletis pomonella* (Diptera: Tephritidae). *Annals of the Entomological Society of America* 86, 250–264.

Herrick, G.W. (1920) The apple maggot in New York. *Cornell University Agricultural Experimental Station Bulletin* 102, 89–101.

Howitt, A.H. (1993) *Common Tree Fruit Pests*. Michigan State University, North Central Regional Extension Publication 63, East Lansing, Michigan.

Hu, X.P., Duan, J.J. and Prokopy, R.J. (1998) Effects of sugar/flour spheres coated with paint and insecticide on alighting female *Ceratitis capitata* (Diptera: Tephritidae) flies. *Florida Entomologist* 81, 318–325.

Innocenzi, P.J., Hall, D.R. and Cross, J.V. (2001) Components of male aggregation pheromone of strawberry blossom weevil, *Anthonomus rubi* Herbst (Coleoptera: Curculionidae). *Journal of Chemical Ecology* 27, 1203–1218.

Jenkins, D., Cottrell, T., Horton, D., Hodges, A. and Hodges, G. (2006) Hosts of plum curculio, *Conotrachelus nenuphar* (Coleoptera: Curculionidae), in Central Georgia. *Environmental Entomology* 35, 48–55.

Johnson, D.T., Mulder, P.G., McCraw, D., Lewis, B.A., Jervis, B., Carroll, B. and McLeod, P.J. (2002) Trapping plum curculio *Conotrachelus nenuphar* (Herbst) (Coleoptera: Curculionidae) in the southern United States. *Environmental Entomology* 31, 1259–1267.

Jones, D.F. (1928) Burbank's results with plums. *Journal of Heredity* 19, 359–371.

Lafleur, G. and Hill, S.B. (1987) Spring migration, within-orchard dispersal, and apple-tree preference of plum curculio (Coleoptera: Curculionidae) in southern Quebec. *Journal of Economic Entomology* 80, 1173–1187.

Lafleur, G., Hill, S.B. and Vincent, C. (1987) Fall migration, hibernation site selection, and associated winter mortality of plum curculio (Coleoptera: Curculionidae) in a Quebec apple orchard. *Journal of Economic Entomology* 80, 1152–1172.

Lafleur, G., Chouinard, G., Vincent, C. and Cormier, D. (2007) Impact of trap architecture, adjacent habitats, abiotic factors and host plant phenology on captures of plum curculio (Coleoptera: Curculionidae) adults. *Journal of Economic Entomology* 100, 737–744.

Landolt, P.J. (1997) Sex attractant and aggregation pheromones of male phytophagous insects. *American Entomologist* 43, 12–21.

Landolt, P.J. and Phillips, T.W. (1997) Host plant influences on sex pheromone behavior of phytophagous insects. *Annual Review of Entomology* 42, 371–391.

Leskey, T.C. (2006) Visual cues and capture mechanisms associated with traps for plum curculios (Coleoptera Curculionidae). *Journal of Entomological Science* 41, 97–106.

Leskey, T.C. (2008) Reproductive development of female plum curculio (Coleoptera: Curculionidae) in the mid-Atlantic: presence of multivoltine populations. *Journal of Entomological Science* 43, 208–216.

Leskey, T.C. and Prokopy, R.J. (2000) Sources of apple odor attractive to adult plum curculio. *Journal of Chemical Ecology* 26, 639–653.

Leskey, T.C. and Prokopy, R.J. (2001) Adult plum curculio (Coleoptera: Curculionidae) attraction to fruit and conspecific odors. *Annals of the Entomological Society of America* 94, 275–288.

Leskey, T.C. and Prokopy, R.J. (2002) Developing a branch-mimicking trap for adult plum curculios. *Entomologia Experimentalis et Applicata* 102, 253–259.

Leskey, T.C. and Wright, S.E. (2004a) Influence of host tree proximity on adult plum curculio (Coleoptera: Curculionidae) responses to monitoring traps. *Environmental Entomology* 33, 389–396.

Leskey, T.C. and Wright, S.E. (2004b) Monitoring plum curculio, *Conotrachelus nenuphar* (Herbst) (Coleoptera: Curculionidae) populations in apple and peach orchards in the mid-Atlantic. *Journal of Economic Entomology* 97, 79–88.

Leskey, T.C. and Wright, S.E. (2007) Host preference of the plum curculio. *Entomologia Experimentalis et Applicata* 123, 217–227.

Leskey, T.C., Prokopy, R.J., Wright, S.E., Phelan, P.L. and Haynes, L.W. (2001) Evaluation of individual components of plum odor as potential attractants for adult plum curculios. *Journal of Chemical Ecology* 27, 1–17.

Leskey, T.C., Zhang, A. and Herzog, M. (2005) Non-fruiting host tree volatile blends: novel attractants for the plum curculio (Coleoptera: Curculionidae). *Environmental Entomology* 34, 785–793.

Leskey, T.C., Piñero, J.C., Wood, S. and Prokopy, R.J. (2008) Odor baited trap trees: a potential management tool for the plum curculio. *Journal of Economic Entomology* 101, 1302–1309.

Levine, E. and Hall, F.R. (1977) Effect of feeding and oviposition by the plum curculio on apple and plum abscission. *Journal of Economic Entomology* 76, 603–607.

Lienk, S.E. (1970) Apple maggot infesting apricot. *Journal of Economic Entomology* 63, 1684.

Lindgren, B.S. (1983) A multiple funnel trap for scolytid beetles (Coleoptera). *Canadian Entomologist* 115, 299–302.

Linn, C.E., Dambroski, H., Nojima, S., Feder, J.L., Berlocher, S.H. and Roelofs, W.L. (2005) Variability in response specificity of apple, hawthorn, and flowering dogwood-infesting *Rhagoletis* flies to host fruit volatile blends: implications for sympatric host shifts. *Entomologia Experimentalis et Applicata* 116, 55–64.

McClanahan, M.E, Luckhart, S. and Pfeiffer, D.G. (2004) *Wolbachia* strains associated with univoltine and multivoltine plum curculios (Coleoptera: Curculionidae). *Journal of Entomological Science* 39, 132–135.

McPheron, B.A., Smith, D.C. and Berlocher, S.H. (1988) Genetic differences between host races of *Rhagoletis pomonella*. *Nature* 336, 64–66.

Maier, C.T. (1990) Native and exotic rosaceous hosts of apple, plum and quince curculio larvae (Coleoptera: Curculionidae) in the north-eastern United States. *Environmental Entomology* 83, 1326–1332.

Mampe, C.D. and Neunzig, H.H. (1967) The biology, parasitism and population sampling of blueberry in North Carolina. *Journal of Economic Entomology* 60, 807–812.

Maxwell, C.W. and Parsons, E.C. (1968) The recapture of marked apple maggot adults in several orchards from one release point. *Journal of Economic Entomology* 61, 1157–1159.

Mizell, R.F. and Tedders, W.L. (1999) Evaluation of trap type and color for monitoring *Hylobius pales* and *Pachylobius picivorus* in Florida. *Florida Entomologist* 82, 615–624.

Moericke, V., Prokopy, R.J., Berlocher, S. and Bush, G.L. (1975) Visual stimuli eliciting attraction of *Rhagoletis pomonella* (Diptera: Tephritidae) flies to trees. *Entomologia Experimentalis et Applicata* 18, 497–507.

Mulder, P.G., McCraw, B.D., Reid, W. and Grantham, R.A. (1997) Monitoring adult weevil populations in pecan and fruit trees in Oklahoma. *Oklahoma State University Oklahoma Cooperation Extension* Series F-7190, 1–8.

Neilson, W.T.A. and Wood. F.A. (1966) Natural source of food of the apple maggot. *Journal of Economic Entomology* 59, 997–998.

Nojima, S., Linn, C.E., Morris, B., Zhang, A. and Roelofs, W. (2003a) Identification of host fruit volatiles from hawthorn (*Crataegus* spp.) attractive to hawthorn-origin *Rhagoletis pomonella* flies. *Journal of Chemical Ecology* 29, 2321–2336.

Nojima, S., Linn, C.E. and Roelofs, W. (2003b) Identification of host fruit volatiles from flowering dogwood (*Cornus florida*) attractive to dogwood-origin *Rhagoletis pomonella* flies. *Journal of Chemical Ecology* 29, 2347–2357.

Oatman, E.R. (1964) Apple maggot trap and attractant studies. *Journal of Economic Entomology* 57, 529–531.

O'Neal, M.E., Landis, D.A., Rothwell, E., Kempel, L. and Reinhard, D. (2004) Tracking insects with harmonic radar: a case study. *American Entomologist* 50, 212–218.

Owens, E.D. and Prokopy, R.J. (1984) Habitat background characteristics influencing *Rhagoletis pomonella* (Walsh) (Dipt., Tephritidae) fly response to foliar and fruit mimics. *Zeitschrift für Angewandte Entomologie* 98, 98–103.

Piñero, J.C. and Prokopy, R.J. (2003) Field evaluation of plant odor and pheromonal combinations for attracting plum curculios. *Journal of Chemical Ecology* 12, 2735–2748.

Piñero, J.C. and Prokopy, R.J. (2006) Temporal dynamics of plum curculio, *Conotrachelus nenuphar* (Herbst) (Coleoptera: Curculionidae), immigration into an apple orchard in Massachusetts. *Environmental Entomology* 35, 413–422.

Piñero, J.C., Wright, S.E. and Prokopy, R.J. (2001) Response of plum curculio (Coleoptera: Curculionidae) to odor-baited traps near woods. *Journal of Economic Entomology* 94, 1386–1397.

Polavarapu, S., Kyryczenko-Roth, V. and Barry, J.D. (2004) Phenology and infestation patterns of plum curculio (Coleoptera: Curculionidae) on four highbush blueberry cultivars. *Journal of Economic Entomology* 97, 1899–1905.

Prokopy, R.J. (1968a) Visual responses of apple maggot flies, *Rhagoletis pomonella* (Diptera: Tephritidae). *Entomologia Experimentalis et Applicata* 11, 403–422.

Prokopy, R.J. (1968b) Sticky spheres for estimating apple maggot adult abundance. *Journal of Economic Entomology* 61, 1082–1085.

Prokopy, R.J. (1972a) Evidence for a marking pheromone deterring repeated oviposition in apple maggot flies. *Environmental Entomology* 1, 326–332.

Prokopy, R.J. (1972b) Response of apple maggot flies to rectangles of different colors and shades. *Environmental Entomology* 1, 720–726.

Prokopy, R.J. (1973) Dark enamel spheres capture as many apple maggot flies as fluorescent spheres. *Environmental Entomology* 2, 953–954.

Prokopy, R.J. (1975a) Selective new trap for *Rhagoletis cingulata* and *R. pomonella* flies. *Environmental Entomology* 4, 420–424.

Prokopy, R.J. (1975b) Apple maggot control by sticky red spheres. *Journal of Economic Entomology* 68, 197–198.

Prokopy, R.J. (1977) Attraction of *Rhagoletis* flies (Diptera: Tephritidae) to red spheres of different sizes. *Canadian Entomologist* 109, 593–596.

Prokopy, R.J. (2003) Two decades of bottom-up, ecologically based pest management in a small commercial orchard in Massachusetts. *Agriculture, Ecosystems and Environment* 94, 299–309.

Prokopy, R.J. and Bush, G.L. (1972) Apple maggot infestation of pear. *Journal of Economic Entomology* 65, 597.

Prokopy, R.J. and Hauschild, K.I. (1979) Comparative effectiveness of sticky red spheres and Pherocon® AM standard traps for monitoring apple maggot flies in commercial orchards. *Environmental Entomology* 8, 696–700.

Prokopy, R.J. and Mason, J.L. (1996) Behavioral control of apple maggot flies. In: McPheron, B. and Steck, G. (eds) *Fruit Flies*. St Lucie Press, Delray Beach, Florida, pp. 555–559.

Prokopy, R.J. and Owens, E.D. (1978) Visual generalist–visual specialist phytophagous insects: host selection behavioral and application to management. *Entomologia Experimentalis et Applicata* 24, 409–420.

Prokopy, R.J. and Owens, E.D. (1983) Visual detection of plants by herbivorous insects. *Annual Review of Entomology* 28, 337–364.

Prokopy, R. and Roitberg, B.D. (2007) Arthropod pest behavior and IPM. In: Kogan, M. and Jepson, P. (eds) *Ecological Approaches to Pest Management*. Cambridge University Press, Cambridge, UK, pp. 87–121.

Prokopy, R.J. and Wright, S.E. (1998) Plum curculio responses to unbaited pyramid and cone traps. *Journal of Economic Entomology* 91, 225–234.

Prokopy, R.J., Bennett, E.W. and Bush, G.L. (1971) Mating behavior in *Rhagoletis pomonella*

(Diptera: Tephritidae) I. Site assembly. *Canadian Entomologist* 103, 1405–1409.
Prokopy, R.J., Bennett, E.W. and Bush, G.L. (1972) Mating behavior in *Rhagoletis pomonella* (Diptera: Tephritidae) II. Temporal organization. *Canadian Entomologist* 104, 97–104.
Prokopy, R.J., Moericke, V. and Bush, G.L. (1973) Attraction of apple maggot flies to odor of apples. *Environmental Entomology* 2, 743–749.
Prokopy, R.J., Christie, M., Johnson, S.A. and O'Brien, M.T. (1990) Transitional steps toward second-stage integrated management of arthropod pests of apple in Massachusetts orchards. *Journal of Economic Entomology* 83, 2405–2410.
Prokopy, R.J., Cooley, S.S., Galarza, L., Bergweiler, C. and Lauzon, C.R. (1993) Bird droppings compete with bait sprays for *Rhagoletis pomonella* (Walsh) flies (Diptera: Tephritidae). *Canadian Entomologist* 125, 413–422.
Prokopy, R.J., Mason, J.L., Christie, M. and Wright, S.E. (1996) Arthropod and natural enemy abundance under second-level vs. first-level integrated pest management practices in apple orchards: a four-year study. *Agriculture, Ecosystems and Environment* 57, 35–47.
Prokopy, R.J., Wirth, C.B. and Leskey, T.C. (1999) Movement of plum curculios toward host trees and traps: flight versus walking. *Entomologia Experimentalis et Applicata* 91, 385–392.
Prokopy, R.J., Chandler, B.W., Leskey, T.C. and Wright, S.E. (2000) Comparison of traps for monitoring plum curculio adults (Coleoptera: Curculionidae) in apple orchards. *Journal of Entomological Science* 35, 411–420.
Prokopy, R.J., Wright, S.E., Black, J.L. and Rull, J. (2001a) Size of orchard trees as a factor affecting behavioral control of apple maggot flies (Dipt., Tephritidae). *Journal of Applied Entomology* 125, 371–375.
Prokopy, R.J., Phelan, P.L., Wright, S.E., Minalga, A.J., Barger, R. and Leskey, T. (2001b) Compounds from host odor attractive to plum curculio adults (Coleoptera: Curculionidae). *Journal of Entomological Science* 36, 122–134.
Prokopy, R.J., Chandler, B.W., Dynok, S.A. and Piñero, J. (2003) Odor-baited trees: a new approach to monitoring plum curculio (Coleoptera: Curculionidae). *Journal of Economic Entomology* 96, 826–834.
Prokopy, R.J., Jacome, I., Gray, E., Trujillo, G., Ricci, M. and Piñero, J. (2004) Using odor-baited trap trees as sentinels to monitor plum curculio (Coleoptera: Curculionidae) in apple orchards. *Journal of Economic Entomology* 97, 511–517.
Quaintance, A.L. and Jenne, E.L. (1912) The plum curculio. *USDA Bureau of Entomology Bulletin* 103, 1–250.
Racette, G., Hill, S.B. and Vincent, C. (1990) Actographs to record the daily activity of the plum curculio (Coleoptera:Curculionidae). *Journal of Economic Entomology* 83, 886–907.
Racette, G., Chouinard, G., Vincent, C. and Hill, S.B. (1992) Ecology and management of plum curculio in apple orchards. *Phytoprotection* 73, 85–100.
Reissig, W.H. (1988) Management of the apple maggot in the eastern United States. In: *Ecology and Management of Economically Important Fruit Flies. Special Report 830.* Oregon State Agricultural Experiment Station, Corvallis, Oregon, pp. 56–72.
Reissig, W.H., Fein, B.L. and Roelofs, W.L. (1982) Field tests of synthetic apple volatiles as apple maggot (Diptera: Tephritidae) attractants. *Environmental Entomology* 11, 1294–1298.
Reissig, W.H, Stanley, B.H., Roelofs, W.L. and Schwarz, M.R. (1985) Tests of synthetic apple volatiles as attractants for apple maggot flies (Diptera: Tephritidae) in commercial apple orchards. *Environmental Entomology* 14, 55–59.
Reissig, W.H., Nyrop, J.P. and Straub, R. (1998) Oviposition model for timing insecticide sprays against plum curculio (Coleoptera: Curculionidae) in New York State. *Environmental Entomology* 27, 1053–1061.
Reynolds, A.H. and Prokopy, R.J. (1997) Evaluation of odor lures for use with red sticky spheres to trap apple maggot (Diptera: Tephritidae). *Journal of Economic Entomology* 90, 1655–1660.
Roitberg, B.D., Van Lenteren, J.C., Van Alphen, J.J.M., Galis, F. and Prokopy, R.J. (1982)

Foraging behaviour of *Rhagoletis pomonella*, a parasite of hawthorn (*Crataegus virids*), in nature. *Journal of Animal Ecology* 51, 307–325.

Rull, J. and Prokopy, R.J. (2000) Attraction of apple maggot flies, *Rhagoletis pomonella* (Diptera: Tephritidae) of different physiological states to odour-baited traps in the presence and absence of food. *Bulletin of Entomological Research* 90, 77–88.

Rull, J. and Prokopy, R.J. (2001) Effect of apple-orchard structure on interception of *Rhagoletis pomonella* (Diptera: Tephritidae) flies by odor-baited traps. *Canadian Entomologist* 122, 355–363.

Rull, J. and Prokopy, R.J. (2003) Trap position and fruit presence affects visual response of apple maggot flies (Dipt., Tephritidae) to different trap types. *Journal of Applied Entomology* 127, 85–90.

Rull, J. and Prokopy, R.J. (2004a) Revisiting within-tree trap positioning for apple maggot fly (Dipt., Tephritidae) behavioural control. *Journal of Applied Entomology* 128, 195–199.

Rull, J. and Prokopy, R.J. (2004b) Host-finding and ovipositional-boring response of apple maggot (Diptera: Tephritidae) to different apple genotypes. *Environmental Entomology* 33, 1695–1702.

Rull, J. and Prokopy, R.J. (2005) Interaction between natural and synthetic fruit odor responses of apple maggot flies to visual traps. *Entomologia Experimentalis et Applicata* 114, 79–86.

Schonherr, J. (1977) Importance of visual stimuli in the host selection behavior of bark beetles (*Dendroctonus ponderosae* and *Ips montanus*). *Colloques Internationaux du CNRS* 265, 187–193.

Shapiro-Ilan, D.I., Mizell, R.F. III, Cottrell, T.E. and Horton, D.L. (2004) Measuring field efficacy of *Steinernema feltie* and *Steinernema riobrave* for suppression of plum curculio, *Conotrachelus nenuphar*, larvae. *Biological Control* 30, 496–503.

Shapiro-Ilan, D.I., Mizell, R.F. III, Cottrell, T.E. and Horton, D.L. (2008) Control of plum curculio, *Conotrachelus nenuphar*, with entomopathogenic nematodes: effects of application timing, alternate host plant, and nematode strain. *Biological Control* 44, 207–215.

Shervis, L.J., Boush, G.M. and Koval, C.F. (1970) Infestation of sour cherries by the apple maggot: confirmation of a previously uncertain host status. *Journal of Economic Entomology* 63, 294–295.

Smith, E.H. and Flessel, J.K. (1968) Hibernation of the plum curculio and its spring migration to host trees. *Journal of Economic Entomology* 61, 194–203.

Smith, E.H. and Salkeld, E.H. (1964) Ovary development and oviposition rates in the plum curculio, *Conotrachelus nenuphar* (Coleoptera: Curculionidae). *Annals of the Entomological Society of America* 57, 781–787.

Tedders, W.L., Mizell, R.F. and Wood, B. (1996) Effects of color and trunk wrap on pecan weevil catch in pyramidal traps. *Journal of Applied Entomology* 31, 414–419.

Tumlinson, J.H., Hardee, D.D., Gueldner, R.C., Thompson, A.C., Hedin, P.A. and Minyard, J.P. (1969) Sex pheromones produced by male boll weevil: isolation, identification, and synthesis. *Science* 166, 1010–1012.

VanderSar, T.J.D. and Borden, J.H. (1977) Visual orientation of *Pissodes strobi* Peck (Coleoptera: Curculionidae) in relation to host selection behavior. *Canadian Journal of Zoology* 55, 2042–2049.

Vincent, C. and Roy, M. (1992) Entomological limits to the implementation of biological programs in Quebec apple orchards. *Acta Phytopathologica et Entomologica Hungarica* 27, 649–657.

Vincent, C., Chouinard, G., Bostanian, N.J. and Morin, Y. (1997) Peripheral zone treatments for plum curculio management: validation in commercial apple orchards. *Entomologia Experimentalis et Applicata* 84, 1–8.

Vincent, C., Chouinard, G. and Hill, S.B. (1999) Progress in plum curculio management: a review. *Agriculture, Ecosystems and Environment* 73, 167–175.

Vincent, C., Chouinard, G. and Leskey, T.C. (2008) The plum curculio, *Conotrachelus nenuphar* Herbst (Coleoptera: Curculionidae). In: Capinera, J.L. (ed.) *Encyclopedia of Entomology, 2nd edn*. Springer, Dordrecht, The Netherlands, pp. 2947–2952.

Walsh, B.D. (1867) The apple-worm and the apple maggot. *Journal of Horticulture* 2, 338–343.

Whalon, M.E. and Coombs, A.B. (2003) Plum curculio monitoring in the midwest: implications for the Pacific northwest. In: *Proceedings of the 77th Annual Western Orchard Pest and Disease Management Conference*. Washington State University, Pullman, Washington, pp. 41.

Wright, S.E., Chandler, B. and Prokopy, R. (2001) Improvement of sugar deliver systems for rain-activated, pesticide-treated spheres. *Fruit Notes* 66, 25–29.

Yee, W.L. and Goughnour, R.B. (2006) New host records for the apple maggot, *Rhagoletis pomonella* (Walsh) (Diptera: Tephritidae), in Washington state. *Pan-Pacific Entomology* 82, 54–60.

Yonce, C.E., Horton, D.L. and Okie, W.R. (1995) Spring migration, reproductive behavior, monitoring procedures and host preference of plum curculio (Coleoptera: Curculionidae) in *Prunus* species in Central Georgia. *Journal of Entomological Science* 30, 82–92.

Zhang, A., Linn, C., Wright, S., Prokopy, R., Reissig, W. and Roelofs, W. (1999) Identification of a new blend of apple volatiles attractive to the apple maggot, *Rhagoletis pomonella*. *Journal of Chemical Ecology* 25, 1221–1232.

Zhang, X. (2007) Biogeography and biosystematics of plum curculio, *Conotrachelus nenuphar* (Herbst)/*Wolbachia* interactions. PhD dissertation, Virginia Polytechnic Institute and State University, Blacksburg, Virginia.

Zohary, D. (1992) Is the European plum, *Prunus domestica* L., a *P. cerasifera* EHRH × *P. spinosa* L. allopolyploid. *Euphytica* 60, 75–77.

6 Trying to Build an Ecological Orchard: a History of Apple IPM in Massachusetts

DANIEL R. COOLEY AND WILLIAM M. COLI

Department of Plant, Soil and Insect Sciences, University of Massachusetts, Amherst, Massachusetts, USA

Introduction

The way an integrated pest management (IPM) system develops depends heavily on where the system is meant to function. Even within a single crop such as apples, pest complexes and host–pest interactions vary by geographical region. For example, an apple IPM system developed in New Zealand may not function well in southern Germany. Even within a limited region such as the north-eastern USA, a system developed in New England may not function as effectively in western New York. Climate and pest complexes vary and some forecasting models and treatment thresholds are empirical, and hence less transferable to disparate regions. In addition, the social, economic and political milieu in a region can affect IPM research, development and implementation. For example, in one country or state, the government may support IPM implementation, while in another it may not.

It is through the analysis of these differences, as well as the common fundamentals in IPM programmes, that a better understanding of this applied science is built. This chapter describes the development and major components of apple IPM as developed in the north-eastern US state of Massachusetts, a case history that may be compared to the development and implementation of similar systems in other places. It is not meant to review the scientific and technical aspects of apple IPM in general, as several excellent reviews have done that already (Hull *et al.*, 1983; Whalon and Croft, 1984; Bostanian and Coulombe, 1986; Kovach and Tette, 1988; Gadoury *et al.*, 1989; Bower *et al.*, 1993; Blommers, 1994; Gurr *et al.*, 1996; Sansavini, 1997; Suckling *et al.*, 1999; Biggs *et al.*, 2000; MacHardy, 2000a; Prokopy, 2003; Ehler, 2006). Rather, it will look at the specifics of apple IPM development as they occurred in Massachusetts, in an attempt to illustrate the process of IPM development.

To understand the context of this programme history better, it is useful to know some basic facts concerning Massachusetts. It has a temperate climate with annual precipitation averaging approximately 100–120 cm. Minimum winter temperatures vary, with coastal regions rarely experiencing –20°C, while interior regions may reach –30°C. Topography is moderately hilly, with glacial drumlins that provide some of the best apple orchard land. Massachusetts is a small state, 44th of 50, (27,335 km^2) and is the third most densely populated state in the USA, behind New Jersey and Rhode Island, and just in front of Connecticut. Farmers represent only 0.15% of the total population. Agricultural product sales reached an all-time high of US$454 million in 1997 (Holm *et al.*, 2000) but, none the less, this represented a small fraction of the gross state product of US$223 billion in the same year. In short, Massachusetts is an urban state with a high-tech economy, where small-scale farming may be profitable but is not a major commercial force.

In spite of the relatively lowly economic status of agriculture in the state, the person most responsible for developing IPM in Massachusetts, Ron Prokopy, always said that it was an ideal place to do his work. For most of the period during which the apple IPM programme developed, roughly 1975 to the present, apple production in Massachusetts has been large enough to register nationally, in the top 15 states, grossing about US$10–15 million of fruit a year on approximately 1600–2000 ha of bearing trees (USDA, 2004, 2007). Yet, it is not so large that researchers could not visit and interact regularly with most growers in the state (approximately 100 significant commercial growers, none more than 2 h from the university). In many ways, Massachusetts is representative of other highly developed, densely populated parts of the world, where economic growth and change force producers to improve agricultural practices constantly in order to stay in business.

IPM is one of those improvements. It has been defined as 'a decision-based process involving coordinated use of multiple tactics for optimizing the control of all classes of pests (insects, pathogens, weeds, vertebrates) in an ecologically and economically sound manner' (Prokopy, 2003). Different classes of tactics, including chemical, cultural, genetic and biological methods, are integrated in ways that usually allow production systems to move away from traditional, chemically based management to biological and cultural strategies (MacHardy, 2000b). When chemicals are applied, the applications are guided using economic and treatment thresholds, based on the monitoring of pests, their natural enemies and environmental conditions.

Prior to the onset of the Massachusetts Apple IPM Pilot Project in 1978, apple pest management in the state and region was based largely on the prophylactic application of various pesticides (i.e. fungicides, insecticides, miticides, herbicides, rodenticides, bactericides), applied either according to the tree phenological stage or at regular intervals. At that time, alternatives to chemicals, such as the destruction of reservoir hosts, sanitation to destroy overwintering stages of pathogens and insects, planting resistant cultivars, behavioural control of insects and conservation of biological controls through the use of chemicals least toxic to beneficial natural enemies of pests (broadly classified as biological, cultural or resistance-based strategies), were employed very rarely by apple growers. Even after the initial introduction of efficient insect and pathogen monitoring, environmentally driven models, use of economically based action thresholds and other

IPM methods, most apple IPM programmes depended ultimately on regular pesticide applications, albeit in a less arbitrary, more targeted manner. This form of pest management has been termed 'chemically dependent IPM' (Frisbie and Smith, 1991), while programmes that largely substitute biological and cultural tactics have been dubbed 'bio-intensive IPM' (Benbrook, 1996).

Consistent with the concept of IPM as a 'continuum', Prokopy maintained that improving the efficiency of pesticide use in orchards was only the first step in the evolution of IPM programmes and that eventually it would be economic, social and political factors that determined the success and sustainability of IPM. He adopted Dover's concept (Dover, 1985) that, ideally, IPM consisted of these elements:

- Optimization of pest control in an ecologically and economically sound manner.
- Emphasis on coordinated use of multiple pest management tactics to assure stable crop production.
- Maintenance of pest damage below injurious levels, while minimizing hazards to humans, animals, plants and the environment in general.

Experience has shown that agroecosystems do not get to this ideal state quickly, but rather progress through recognizable levels (Prokopy, 2003) (Table 6.1). According to this concept, the first level of IPM focuses on the management of a single class of pests, for example, insects, with chemical controls selected for efficacy, low cost and ease of use. The second level integrates management tactics across all classes of pests; that is, arthropods, pathogens, weeds and vertebrates. For example, horticultural oil is commonly used as part of an IPM mite management tactic, but the fungicide, captan, must not be applied within several days of an oil application because the combination can be phytotoxic. The third level of IPM integrates pest management with the other agricultural or horticultural practices on a farm; for example, rootstock and cultivar selection, tree training and pruning, or orchard design and architecture. The first three levels of IPM are largely technical and involve the research and development of new methods to add to increasingly complex management systems. In contrast, the fourth level of IPM is highly political and social, and integrates the concerns of all those having a vital interest in pest management, including growers, researchers, Extension

Table 6.1. Levels of IPM (Prokopy, 1993; see also Kogan and Hilton, Chapter 1, this volume).

First-level IPM	Integrates chemically based and biologically based tactics for a single class of pests. It can include the full continuum of chemically based to biointensive practices (Benbrook, 1996)
Second-level IPM	Integrates multiple management tactics across all classes of pests
Third-level IPM	Integration with the entire crop production system on a farm, i.e. the interface of IPM and sustainable agriculture
Fourth-level IPM	Integration of the political and social components in which IPM works, including environmentalism, regulatory restrictions, public education and dialogue with policy makers

field staff, retailers, consumers, environmentalists and regulators. At this level of integration, consumer concerns about health and the environment, marketing and political regulation shape IPM programmes.

Two central ideas arose from the concept of IPM levels:

- IPM systems are dynamic and evolve towards increasing complexity as more pests, more crops and more aspects of crop management are included, moving away from reliance on chemically based management and towards a mix of risk forecasting and biological, cultural and resistance-based management.
- Over time, sociological and political factors become as important as biology in determining the success of IPM.

In the early levels of development, researchers can invent ways to reduce or eliminate pesticides but, to be sustainable, IPM has to be embraced by the marketplace, by consumers and by policy makers. Increased complexity in IPM systems typically brings increased expense, which ultimately must be borne by society if producers are to remain economically viable. The measure of success of IPM will be the extent to which it actually contributes to the long-term sustainability of agriculture in communities.

Early Apple IPM in Massachusetts

Beginning the IPM programme

The history of IPM shows that during the late 1960s and 1970s, agricultural scientists had begun to respond to two key pesticide issues, environmental pollution and pest resistance (Kogan, 1998; Ehler, 2006). In apples, prior to Prokopy's arrival in Massachusetts, forward-thinking scientists in Nova Scotia (Pickett *et al.*, 1958) and elsewhere began to develop 'harmonized' or 'integrated control' programmes against apple pests, principally phytophagous mites (Hoyt, 1969). By the early 1970s, integrated mite control programmes were being implemented throughout the USA and Canada (Croft and Hoyt, 1983). The fundamental IPM principle, that pesticide treatments for pest problems should depend on the risk of damage from pests rather than a preconceived schedule, was established at this time. Concurrently, private and government organizations (e.g. USDA/CSRS, Rockefeller Foundation, NSF) sponsored large scientific meetings to develop the new science, and the term integrated pest management (IPM) came into usage.

Between 1971 and 1978, the National Science Foundation (NSF), the United States Department of Agriculture (USDA) and the United States Environmental Protection Agency (US EPA) provided funding to a consortium of 19 state universities, known as the Huffaker Project (after its coordinator at the University of California, Berkeley), for a large-scale, multidisciplinary research activity. One subproject covering pome and stone fruits was led by Brian Croft and William Whitcomb of Michigan State University (Croft and Hoyt, 1983). The Huffaker Project and, subsequently, the Consortium for Integrated Pest Management, or CIPM, Project (Frisbie and Adkisson, 1985) both aimed at developing new approaches to managing pests in several important crops, including apples.

Based on these large research efforts, the USDA later provided funding on a competitive basis for establishing IPM projects throughout the USA (Ehler, 2006). After convincing University of Massachusetts (UMass) colleagues in plant pathology, horticultural science and resource economics to collaborate on a grant proposal, Prokopy used this funding to establish the first formal apple IPM project in Massachusetts in 1978. The multidisciplinary collaboration was innovative, as up to the 1980s, IPM was considered an aspect of entomology, dealing exclusively with insects and mites. The UMass project was in the vanguard of a shift that would soon consider 'pest' to include microbial pathogens and weeds, as well as arthropod pests (Jacobsen, 1997; Kogan, 1998).

Apple pests in Massachusetts

An impressive number of microbes, insects and arthropods try to take advantage of the nutrients in cultivated apples, and many of them occur in Massachusetts. Even the short list of apple pests in Massachusetts is large (Tables 6.2 and 6.3); however, insect and disease management in New England was, and in large measure still is, driven by relatively few key direct pests that have the potential to cause unacceptable damage regularly and require management every year.

Of the arthropods, the principal key pests are the plum curculio (*Conotrachelus nenuphar* (Herbst)) and the apple maggot fly (*Rhagoletis pomonella* (Walsh)). To a lesser extent, tarnished plant bugs (*Lygus lineolaris* (Palisot de

Table 6.2. Key arthropod pests in Massachusetts apple orchards.

Common name	Binomial
Apple aphid	*Aphis pomi* (DeGeer)
Spirea aphid	*A. spiraecola* (Patch)
Rosy apple aphid	*Dysaphis plantaginea* (Passerini)
Woolly apple aphid	*Eriosoma lanigerum* (Hausmann)
Apple maggot fly	*Rhagoletis pomonella* (Walsh)
Codling moth	*Cydia pomonella* (L.)
European apple sawfly	*Hoplocampa testudinea* (Klug)
European red mite	*Panonychus ulmi* (Koch)
Two-spotted spider mite	*Tetranychus urticae* (Koch)
Speckled green fruitworm	*Orthosia hibisci* (Guenée)
Humped green fruitworm	*Amphipyra pyramidoides* (Guenée)
White apple leaf hopper	*Typhlocyba pomaria* (McAtee)
Potato leaf hoppers	*Empoasca fabae* (Harris)
Rose leaf hoppers	*Edwardsiana rosae* (L.)
Plum curculio	*Conotrachelus nenuphar* (Herbst)
Red-banded leaf roller	*Argyrotaenia velutinana* (Walker)
San José scale	*Quadraspidiotus perniciosus* (Comstock)
Spotted tentiform leaf miner	*Phyllonorycter blancardella* (F.)
Apple blotch leaf miner	*P. crataegella* (Clemens)
Tarnished plant bug	*Lygus lineolaris* (Palisot de Beauvois)

Table 6.3. Key microbial pests in Massachusetts apple orchards.

Disease name	Pathogen binomial
Apple scab	*Venturia inaequalis* (Cke.) G. Wint.
Black rot	*Botryosphaeria obtusa* (Schwein.) Shoemaker
Cedar apple rust	*Gymnosporamgium juniperi-virginianae* Schwein.
Fire blight	*Erwinia amylovora* (Burr.) Winslow *et al.*
Flyspeck	*Schizothyrium pomi* (Mont. & Fr.) Arx
Powdery mildew	*Podosphaeria leucotricha (*Ell. & Ev.) E.S. Salmon
Sooty blotch	*Peltaster fructicola* Johnson
	Leptodontidium elatius (G. Mangenot) De Hoog
	Geastrumia polystigmatis Batista & M.L. Farr

Beauvoir)), European apple sawfly (*Hoplocampa testudinea* (Klug)) and the codling moth (*Cydia pomonella* L.) require treatments. In the latter case, the timing of first-generation emergence and activity coincides with the application of petal fall and subsequent sprays of organophosphate insecticides for plum curculio, so that specific sprays typically are not needed against the moth. At the programme's onset, key indirect pests such as spider mites (e.g. European red mite, *Panonychus ulmi* (Koch), and the two-spotted spider mite, *Tetranychus urticae* (Koch)) and *Phyllonorycter* leaf miners also drove pesticide applications.

Other arthropod pests present less potential risk to New England apples, but are still important enough to require assessment. These include leaf rollers (red-banded leaf roller, *Argyrotaenia velutinana* (Walker), and oblique-banded leaf roller, *Choristoneura rosaceana* (Harris)), fruitworms (speckled green fruitworm, *Orthosia hibisci* (Guenée), and humped green fruitworm, *Amphipyra pyramidoides*)), aphids (green apple aphid, *Aphis pomi* (DeGeer), and woolly apple aphid, *Eriosoma lanigerum* (Hausman)), leaf hoppers (white apple leaf hopper, *Typhlocyba pomaria* (McAtee) and potato leaf hopper, *Empoasca fabae* (Harris)) and scale insects (e.g. *Quadraspidiotus perniciosus* (Comstock)).

Of pathogens, *Venturia inaequalis* Cke., which causes apple scab, is the focus of most attention. The disease presents the highest risk on a regular basis and drives the application of fungicides for the first 3 months of the growing season. Fungicides targeting scab can be selected and applications timed to control other diseases that may infect during this time, such as cedar apple rust (*Gymnosporangium juniperi-virginianae* (Schwein.)) and powdery mildew (*Podosphaeria leucotricha* (Ell. & Ev.) E.S. Salmon). The summer blemish diseases, sooty blotch and flyspeck, a disease complex caused by several fungi including, but probably not limited to, *Peltaster fructicola* Johnson *et al.*, *Leptodontidium elatius* (G. Mangenot) De Hoog, *Geastrumia polystigmatis* Batista & M.L. Farr, *Gloeodes pomigena* (Schwein.) Colby and *Schizothyrium pomi* (Mont. & Fr.) Arx, did not present significant problems in the early years of the programme. The highly effective ethylene bisdithiocarbamate (EBDC) fungicides were used widely through the summer and climatic conditions did not appear to encourage disease development to the extent seen in the south-eastern USA. Fire blight (*Erwinia amylovora* (Burr.) Winslow *et al.*) occurred very sporadically and growers seldom applied streptomycin to control it. Since the late 1980s, these diseases have become increasingly damaging.

The initial apple IPM approach for insects and mites in Massachusetts was outlined at the outset of the programme (Prokopy *et al.*, 1980). Disease management targeted scab and was based largely on methods developed at the University of New Hampshire (MacHardy and Sondej, 1981; Gadoury and MacHardy, 1982a). This apple IPM approach was similar in many ways to that being developed in other parts of the country (Asquith and Hull, 1979; Tette *et al.*, 1979; Gruys, 1982; Whalon and Croft, 1984; Gadoury *et al.*, 1989). The focus in these early years was on reducing pesticides in apple orchards, not eliminating them. The primary intent was to use selective pesticides that allowed the natural enemies of pests to survive and to limit the application of pesticides to instances where monitoring the weather, the first appearance of or the abundance of pests indicated the need for an application. The first operating manual for this system was developed at the end of the first 5 years of the programme (Coli, 1984).

Entomology scouts relied on a number of techniques for monitoring pests and attempting to determine how many of each constituted a treatment threshold. For plant bugs and sawflies, sticky coated, non-UV-reflecting white rectangles were deployed when apple trees were at the silver tip or pink phenological stage, respectively, to monitor activity and assess the need for treatment (Owens and Prokopy, 1978; Prokopy *et al.*, 1979, 1982; Coli *et al.*, 1985). For *Phyllonorycter* leaf miners, monitoring and spray decision making from silver tip through 2 weeks past petal fall relied on tent-shaped, red enamel traps (Coli *et al.*, 1985; Green and Prokopy, 1986). Visual examination of leaves for the presence of mines was also used to confirm whether populations were high enough to require controls. Sticky red sphere visual traps (Prokopy, 1968) were deployed to monitor the emergence and activity of apple maggot flies from late July to harvest in mid-September. Commercially available pheromone traps were used to monitor the emergence of male codling moth, San José scale, red-banded and oblique-banded leaf rollers and leaf miners. Plum curculio, spider mites and their predators, aphids and their predators, fruitworms, leaf hoppers and scale insects were monitored by direct examination of plant parts (e.g. terminals, leaves, flower clusters, fruit, pruning cuts) with a 10× hand lens on ten appropriate sampling units in each of three tree locations (low inside, low outside and top of tree canopies).

Monitoring the development of fungal spores and measuring the environmental conditions needed to cause infection were used to determine the application timing of fungicides to arrest or eradicate apple scab infections (MacHardy and Sondej, 1981; Gadoury and MacHardy, 1982a,b). Scab-infected leaves were collected in October and November and placed to overwinter in the same locale as cooperating orchards. Beginning in March of the next year, a sample of overwintered, scab-infested leaves would be examined weekly to determine ascospore maturity and thereby determine the beginning and end of the period during which primary apple scab infection might occur (Gadoury and MacHardy, 1982a). During this time, wetting periods were monitored using modified hygrothermographs and the Mills periods determined (Zuck and MacHardy, 1981). Because the fungicides available at the time had no more than 48 h of post-infection activity, growers generally did not wait until after Mills infections occurred to make fungicide applications. Any reductions in the number of fungicide applications made in IPM blocks were the result of the elimination of an application at the very beginning or end of the primary scab season, a function of more accurate determination of the availability of primary inoculum.

Apple IPM as participatory research – research and education

A 1948 pest control guide for Massachusetts's apple growers (Bourne *et al.*, 1948) recommended a series of 12 pesticide spray applications, starting before the trees began to grow in the spring and continuing at intervals determined by growth stage or date through August. Over the next 30 years, the chemicals changed, from lime sulfur, iron carbamate, lead arsenate and DDT to new types of chlorinated hydrocarbons, carbamates and other synthetic organic pesticides. Yet, by 1978, the methods of and guidelines for applying these chemicals had changed only slightly and IPM was virtually unknown in the orchards of Massachusetts. For decades, university and county Extension had been telling apple growers to apply chemicals religiously if they had any hope of growing a 'clean crop', a term for the unblemished fruit that brought the highest prices. Rather than asking how to use less pesticide, apple growers expected recommendations on using chemicals that would minimize the risk of fruit damage. Most growers had little interest in why mites were becoming resistant to miticides, but simply wanted to know which new miticides would work. The new IPM approach sounded needlessly complicated and risky to growers, who had been told for years that the key to success was regular applications of chemicals. Growers had had many years of success with this approach. Not only were they skeptical, but also many regional Extension specialists shared their skepticism. Not surprisingly, pesticide salesmen fought the new ideas, fearing they threatened sales. In the face of this hostility, simply recommending change would not work, particularly when the recommendations were coming from people who had yet to build trust with the grower community. With this in mind, the IPM programme asked growers to participate in IPM research, using a small block of trees in each of their orchards to test the new methods. Team members, particularly Prokopy, spent many, many hours at field sites talking with growers about the new methods and the particular results on their farms, building relationships and also confidence in IPM.

After a growing season, a few influential growers began to adopt IPM methods beyond the 'test blocks', notably using the sticky red croquet ball traps developed by Prokopy to monitor apple maggot fly. Perhaps more importantly, the orchard advisor for the largest apple packer and distributor in Massachusetts, William Pearse of J.P. Sullivan Co, saw an opportunity to increase fruit quality. In his rounds checking on crop quality in various orchards, he suggested quietly that growers try IPM and, at times, he travelled with the IPM team members, learning first-hand the purpose of the various traps and monitoring methods and how the research would establish a threshold count for each insect. In his own discussions with growers, Pearse emphasized fruit quality and pesticide savings in monetary terms. In his discussions with researchers, he emphasized the constraints faced by people trying to make their living by growing apples. The 'early adopter' growers saw a way to solve the burgeoning problem of pesticide resistance, reduce pesticide costs and produce a high-quality crop. It was also becoming apparent to apple growers that environmental and food safety issues associated with pesticides had begun to attract public attention, particularly in the politically liberal state of Massachusetts, and that a proactive approach such as IPM could ease public concerns.

Regular orchard visits from researchers and scouts, usually on a weekly basis during the growing season, have always been a critically important part of apple IPM work. As noted above, one aspect of this is sampling and monitoring pest populations and recommending treatment actions in grower test blocks. Just as importantly, the regular interactions allow researchers to obtain grower input regarding the practicality of the IPM methods being tested. While researchers did not refer to the process as such, the apple IPM programme was using participatory research methods: experiments planned by scientists, but employed by growers and modified according to grower input (Park, 1993; Riley and Alexander, 1997). On-farm demonstration of proven methods was a powerful learning tool, first described formally and developed by Seaman Knapp, the founder of the US Extension system (Martin, 1921). The apple IPM programmes asked growers to do more than demonstrate proven methods and to help plan and execute experiments on their farms. The regular interactions during the growing season allowed for, and sometimes demanded, relatively rapid adjustment to field-study protocols. Annual meetings between researchers and participating growers, usually held prior to the beginning of a field season, discussed the previous year's results and what should be done in the coming year. It is our opinion that engaging growers in the planning and execution of IPM research is instrumental in the rapid development and adoption of apple IPM in Massachusetts.

The programme used other tools, both old and new, to educate growers. Grower meetings were held regularly at IPM sites to enable people not involved directly in the project to see IPM in action. Results were published in the long-established periodical publication, *Massachusetts Fruit Notes*, and *The Proceedings of the Annual Meeting of the Massachusetts Fruit Growers*, the annual conference for New England apple growers. Two new methods were developed for getting information to the orchardists, the annual *March Message to Massachusetts Fruit Growers* and the weekly *Pest Message*. The *March Message* reviewed the current status of apple IPM on an annual basis in applied terms designed to stimulate growers to think about and try the latest IPM methods. It provided growers access to newly developed methods and ideas relevant to apple IPM, information that was not included in the established *New England Apple Pest Control Guide*. The *Pest Message* was an adaptation that provided apple IPM information to regional Extension specialists, who would then distribute it in their newsletters during the growing season.

Extension newsletters were a long-standing method for disseminating information to growers; the innovation of the *Pest Message* involved getting time-critical information collected, analysed and disseminated to growers within a useful time frame. Obtaining and interpreting the ever-changing data relevant to pest management decisions is both central to the application of IPM and one of its major costs. Ultimately, IPM decisions should be based on data obtained at an individual farm, but collating and analysing data from several sites in a region can provide a valuable context for management decisions and, in the absence of individual, on-site data, regional data can still guide treatment decisions. For example, if information in a *pest message* indicated that plum curculio activity at monitored sites exceeded the threshold, those orchards that did not monitor plum curculio would be wise to treat. In the early years of the Massachusetts programme, cooperating growers received scouting and management recommendations that were subsidized heavily by government funds. The rationale for this was that it gave growers an incentive to learn and use IPM methods, while giving researchers field sites to test new methods. In addition, the pest messages made information available to all growers in the state. Initially, the process of gathering information, analysing it, preparing it for growers and getting it to them took several days, largely because the final delivery step involved the mail. This was soon augmented by recorded phone, fax and Internet technology.

Early results

By 1979, support for the apple IPM programme was widespread. The Massachusetts Fruit Growers Association was contributing significant money to the project (US$16,000 annually, adjusted for inflation), complimenting an USDA grant and base funding for research and Extension supplied by Federal Smith-Lever funds and state funds. In addition, one grower in the eastern part of the state donated an apartment to house field staff. At the end of 1982, the federal IPM grant in Massachusetts ended, with the basic components of the apple IPM system in place. The results of the programme were encouraging, reducing insecticide use by about one-third and fungicide use by about one-fifth (Becker *et al.*, 1982; Coli *et al.*, 1982) (Fig. 6.1; Table 6.4). Field staff formed one private IPM consulting business in 1983 and, a few years later, another member of the field staff formed a second business in the region. Both businesses still provide apple IPM consulting advice and other services. At that stage, by any reasonable measure, the apple IPM programme in Massachusetts could be considered a success. Several factors contributed to this success: an apple industry of a size that allowed frequent, direct interaction with researchers; sufficient number of faculty and professional staff to undertake meaningful research and Extension in multiple aspects of apple IPM; a zealous academic leader with extensive experience in apple production and IPM; a supportive industry leader; and a rich network of collaborative researchers and educators throughout the north-eastern USA and eastern Canada (Becker *et al.*, 1983; Coli *et al.*, 1983).

Evolving: Second-level IPM

The primary question facing the programme in 1983 was, 'Where should Massachusetts apple IPM go now?' One option would be to declare the apple IPM project 'done', in that the majority of growers around the state had adopted at least some of the recommended IPM practices. Alternatively, the University of

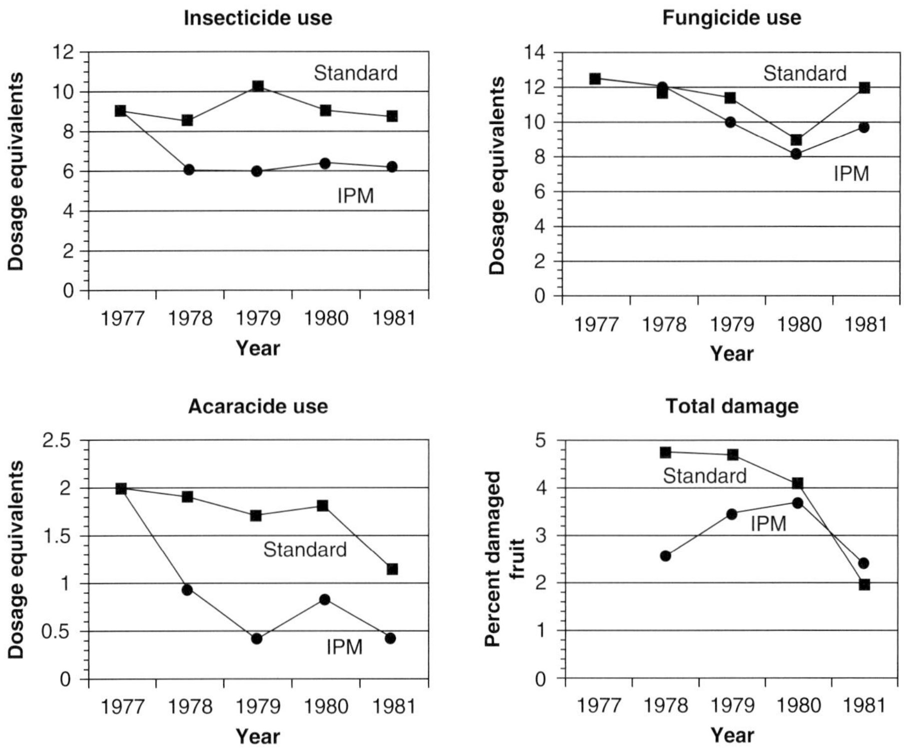

Fig. 6.1. Results of the Massachusetts apple IPM programme, 1977–1981.

Table 6.4. Results of Massachusetts apple IPM programme, 1979.

	Number of applications		Average DE		Per cent damage at harvest		Cost (US$/ha)	
Block type	Fung	In/Aca	Fung	In/Aca	Dis	In/Aca	Fung	In/Aca
IPM	10.6	6.0	9.8	6.4	1.0	3.6	US$269	US$171
Control	13.0	11.0	11.1	12.4	0.9	4.7	US$324	US$334

DE = dosage equivalent: amount of pesticide used divided by the average recommended rate of pesticide; Fung: fungicide; Dis: disease; In/Aca: insecticide and acaricide.

Massachusetts (UMass) group had begun to explore how the apple IPM system might be pushed to enable further pesticide reductions, using traps to remove insects rather than simply monitoring their presence, introducing mite predators and predator-friendly pesticides or growing scab-resistant apple cultivars (SRCs). At the same time, it had become apparent that changing pest management practices had gradually and subtly altered the pest complex in the apple orchards of Massachusetts. The ecology of an apple orchard is intricate, and indiscriminate use of pesticides destroys that intricacy. As chemical use declines and the types of chemicals used changes, ecological complexity returns and, inevitably, the pest community changes. Prokopy also noted that non-chemical changes, such as modification of the groundcover habitat orchards, would change arthropod communities and that such changes might enhance or hinder pest management (Prokopy, 1994). It was at this point that the ideas that eventually would be formalized as levels of IPM were first discussed (Hill *et al.*, 1999) (Table 6.1). It was apparent that within a single class of pests, such as insects, not only could the timing of pesticide applications be made more efficient by using traps, thresholds and weather-based models, but that more environmentally benign pesticides could be selected as well. In addition, biologically and culturally based methods could be integrated with more benign pesticides, moving the whole IPM system to be less dependent on broadly toxic chemicals and thus presumably more sustainable, a concept also described by others (Benbrook, 1996; Hill *et al.*, 1999). Inevitably, this means increasing the integration of management across multiple pest classes, for example, insects and diseases, the integration of pest management with horticultural practices and shifts to more ecologically benign alternatives. Following each significant management change, the agroecosystem changes, sometimes in a matter of days and sometimes more slowly over months or years, and then the IPM system needs to be adjusted, in a cycle that hopefully moves the agroecosystem towards sustainability. The UMass apple IPM programme had reached a point of major readjustment in the IPM system.

The UMass team decided to pursue a research programme to develop an apple IPM approach that would be more integrated, particularly across the management of arthropods, diseases and horticultural practices. Extension programming focused on the established IPM methodologies described earlier, while research was started on more advanced IPM methods. During this time, the established IPM system was referred to as first-stage IPM, while the more advanced IPM being developed was called second-stage and, eventually, second-level, IPM. Second-level IPM employed increased behavioural and biological approaches to pest management as substitutes for most pesticide treatments (Prokopy, 1987; O'Brien and Prokopy, 1988; Mason *et al.*, 1995).

In second-level IPM, early-season insects such as plant bugs, sawflies and plum curculios were managed with three to four insecticide sprays, while apple scab and rusts were managed with three to five fungicide applications. Mites were managed with one or two oil applications and by building up mite predator populations through one of several options, including planting alternate plant hosts, introducing predators and using pesticides that would not sterilize or kill mite predators. Unmanaged and wild *Malus* and *Pyrus* trees around the orchard borders were removed to reduce codling moth and other pest immigration (Prokopy *et al.*, 1990). After mid-June, reliance was as much on biologically based tactics as on practical tactics. Apple maggot flies were trapped using sphere traps baited with apple odour volatiles and placed around the border of a block. The application of fungicides for the management of sooty blotch and flyspeck was based on the amount of rain and degradation of fungicide over time (Rosenberger *et al.*, 1990).

Based on the understanding that the predator mite *Typhlodromus pyri* (Scheuten) is a more effective biocontrol than the more common *Amblyseius fallacis*

(Garman) (Nyrop *et al.*, 1997), the programme introduced *T. pyri* to experimental blocks. This introduction led to established populations in these orchards (Hu *et al.*, 1995). To encourage establishment and spread, it was necessary to use a regime of pesticides that would not kill or sterilize mites, 'mite-friendly pesticides' (Prokopy *et al.*, 1997, 1999). The concept of mite-friendly pesticides demanded integration with disease management because some fungicides used in apples can have a negative effect on mite predators (Van Driesche *et al.*, 1988; Hardman *et al.*, 1991) and captan can have phytotoxic interactions with horticultural oil. For example, benzimidazole and ethylene bisdithiocarbamate fungicides can sterilize *A. fallacis* and other mite predators. Hence, the most effective fungicides against sooty blotch and flyspeck were of limited use in second-level IPM. Early-season horticultural oils used to suppress mites constrain use of the fungicide captan because oil and captan applied within several days of each other can interact to cause a phytotoxic reaction on apples.

The major effort in scab management was a test of the potential ascospore dose (PAD) method to evaluate initial inoculum in order to determine whether to delay the first fungicide application in a growing season (Gadoury and MacHardy, 1986). By measuring the amount of scab present in an orchard in late autumn, one can estimate how long it would take a scab epidemic to start in that orchard the following spring if no fungicides were applied. It is generally assumed that apple scab epidemics are not limited by inoculum but begin with the availability of the first green-leaf tissue in spring; however, Gadoury and MacHardy showed that amounts of mature *V. inaequalis* inoculum sufficient to initiate an epidemic were not always available during the early growth stages of the host. A very low proportion of the season's total ascospore inoculum matures during this early period and, in orchards where there is little inoculum from the previous year, it is insufficient to cause infections. In such orchards, enough inoculum to cause an epidemic may develop as late as the tight cluster phenophase, allowing growers to refrain from applying fungicides until up to three Mills infection periods have occurred. To estimate PAD, a systematic survey of foliar scab in an orchard must be made in the autumn (MacHardy *et al.*, 1993).

In 1987, field tests of individual components of the second-stage approach in commercial orchards began. That is, in six commercial blocks, red baited spheres were used to intercept apple maggot fly; in six different blocks, perimeter row sprays were used against plum curculio after early June; and in another six blocks, PAD was used to direct when the first fungicide application should be made. The approaches were integrated in a single programme and field-tested from 1991 to 1994.

At the end of 4 years of the second-level IPM project, while the second-level approach was not a failure, it was not a clear success (Prokopy *et al.*, 1996; Cooley and Autio, 1997). Insecticide and miticide use was lower in the second-level IPM blocks than in the first-level blocks, but damage was higher, particularly in later years (Fig. 6.2; Table 6.5). Similarly, while second-level disease management used less fungicide, summer disease in the second-level blocks was nearly twice that in the first-level blocks, though the difference was not significant. It was clear that the system would have to be developed further before it could be recommended on a commercial basis. The project also indicated that the growers'

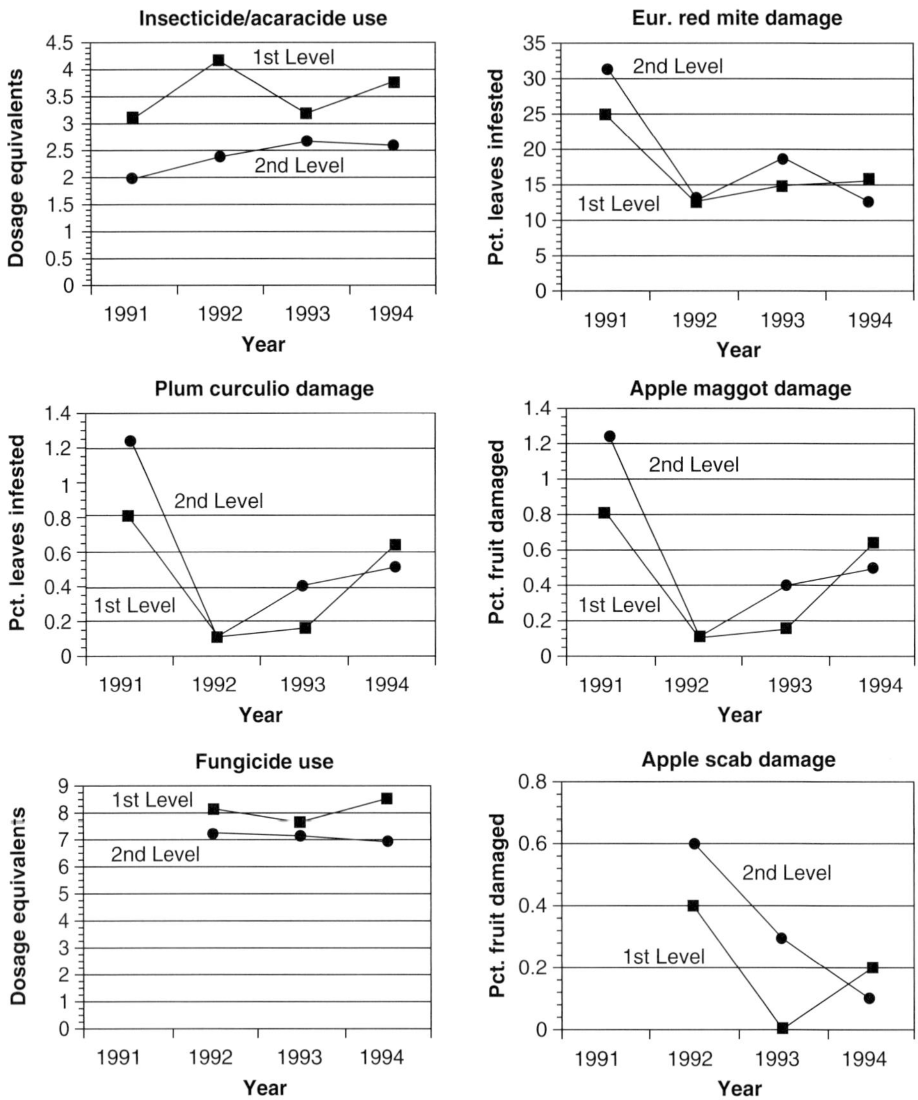

Fig. 6.2. Results of the second-level apple IPM programme, 1991–1994.

willingness to cooperate in innovative IPM development was dropping. For example, in nearly half the cases where it was recommended that the first scab fungicide application be delayed, the growers failed to do it, largely because they were unwilling to take what they perceived to be unacceptable risk (Cooley and Autio, 1997).

Behind the growers' unwillingness lay the fact that a huge increase in global production had caused apple prices to stagnate or drop, while production costs continued to rise. Growers faced increasing pressure to eliminate as much fruit

Table 6.5. Results of Massachusetts apple IPM programme, 1991–1994.

	Avg. DE before mid-June		Avg. DE after mid-June		Damage early-season pests (%)		Damage late-season pests (%)	
Block type	Fung	In/Aca	Fung	In/Aca	Dis	In/Aca	Dis	In/Aca
Second-level IPM	4.7	2.6	2.3	0.0	0.3	2.0	4.8	2.6
First-level IPM	4.6	2.5	3.4	1.6	0.2	1.9	2.5	1.2

DE = dosage equivalent: amount of pesticide used divided by the average recommended rate of pesticide; Fung: fungicide; Dis: disease; In/Aca: insecticide and acaricide.

damage as possible as cheaply as possible in order to maximize returns; yet, the second-level IPM approach demanded more time and risk tolerance from them. For some labour-intensive jobs, such as placing and cleaning red sphere traps for monitoring apple maggot, the University team and scouts carried the burden, while for others, such as mixing special loads of fungicide and applying it to a small block of trees, the growers were asked to do the work. In general, it was clear that new, advanced management methods would take more time and effort than the first-level methods had. Second-level IPM also appeared to be riskier, allowing marginally higher damage in some cases. Increased management and increased damage ultimately would cost money and global competition in wholesale apples would not allow growers to recoup extra costs by charging more.

Alar and 'Partners with Nature'

The integration of pest management with horticulture was pushed ultimately by a largely political event, demonstrating that politics and society have a very real impact on IPM systems. The episode also demonstrates the power that the cosmetic qualities of fresh food can have in driving chemical applications during production. The growth regulator, daminozide (Alar®), was an integral part of New England apple production, largely because the cultivar McIntosh represented approximately 75% of the apples grown in the region at the time, the late 1980s. McIntosh fruit drops easily from trees as the fruit ripens. To maximize value, growers try to delay the harvest of McIntosh for as long as possible in order to allow the fruit to develop as much red colour as possible. If fruit drops to the ground before it is picked, it inevitably bruises, discolours and loses most of its commercial value. Alar inhibits fruit dropping, allowing the development of optimal colour before harvest. In 1986, Massachusetts proposed to eliminate daminozide use by 1988, based on reports that daminozide presented a cancer risk to children (Anonymous, 1986; Rosenberger *et al.*, 1996). Ultimately, a long and bitter debate over whether daminozide presented a cancer risk became irrelevant, as the manufacturer removed it from the market in 1989.

Alar presented a dilemma for IPM scientists, since reasonable analysis suggested that it represented little, if any, actual risk to people or the environment, but it did allow McIntosh apple trees to tolerate better the foliar damage from indirect pests such as mites, leaf hoppers or leaf miners. Foliar damage exacerbates early apple drop, especially in the presence of other stresses such as heat, drought or winds. Without daminozide, the risk represented by foliar-feeding damage increased and entomologists needed to reduce treatment thresholds for foliar damage, causing growers to spray pesticides more often to control leaf damage that formerly, using daminozide, was tolerable (Prokopy, 1988).

Horticulturists responded to the loss of daminozide by recommending a horticultural practice called summer pruning, which removed small branches from dense trees during the summer in order to increase light penetration, thereby increasing red colour and fruit value and allowing earlier harvest than would otherwise be possible (Autio and Greene, 1991). Improved quality and return on fruit more than paid for the increased expense of carrying out the pruning. Entomologists concluded that summer pruning had no effect, positive or negative, on foliar-feeding insects and mites (Butkewich *et al.*, 1989). Later research would indicate that summer pruning reduced fruit damage from oblique-banded leaf rollers in large trees (Lawson *et al.*, 1998). In addition, summer pruning reduces sooty blotch and flyspeck by causing a slight reduction in the relative humidity in the tree canopy and concurrently improving the coverage of summer fungicide applications (Cooley *et al.*, 1997). The same benefits in disease and insect management do not occur in small trees and, as large vegetative trees are replaced by high-density trees, the benefits of summer pruning in both horticultural and pest management terms have decreased.

The Alar debate also led the IPM team to ask how the public might be better informed about IPM and pesticides. In terms of IPM levels, this was what Prokopy considered fourth level IPM, involving consumers and the public in developing and maintaining apple IPM. In the wake of bad publicity surrounding Alar, growers wanted to be able to demonstrate to the public that they were good stewards, growing apples without excessive pesticide use. They wanted to communicate their use of IPM to consumers and the public, and the IPM team responded by developing a programme that would document and recognize IPM use by apple producers. The first task was to determine whether a grower was, in fact, using IPM because, in a few instances, growers claiming to use IPM clearly were not. The programme assembled a list of all tactics used in the apple IPM system, assigning each tactic a point value. If growers accumulated 70% or more of the possible points, they were certified as using IPM. This allowed them to use signs and stickers for packaging that identified them as part of the ‘Partners with Nature’ programme, a very small attempt at using IPM to market the fruit. This was an early example of an ‘eco-label’ (Hollingsworth *et al.*, 1992; Van Zee, 1992).

As ‘Partners with Nature’ took shape, a number of growers signed up to participate; however, for some growers, the idea was controversial. These growers, including some involved in IPM for years, were concerned that the adoption of a certification programme would bring to the public’s attention the fact that pesticides were used in apple production and would, therefore, be detrimental to sales and make regulations more restrictive. A few growers resented the implication that producing apples with conventional pest management programmes, rather than IPM, was somehow unsafe. Wholesalers, and the retailers they sold to, did not use the 'Partners with Nature' material, either because it did not fit existing marketing programmes or was not perceived to offer any market advantage. In general, wholesale growers did not like IPM certification, while direct market growers did. Unfortunately, from 1993 to 1999, of all the fruit and vegetable farms that might have participated in the 'Partners with Nature' programme, at most, only 53% did in any one year. The Massachusetts Department of Food and Agriculture decided that the 'low participation' did not warrant further funding and, without the money, the IPM programme was forced to end 'Partners with Nature' in 1999, though the IPM certification standards have been updated since (Hollingsworth and Coli, 2007).

Sustainable apple production

A significant part of the second-level IPM effort had been coordinated with a regional USDA Sustainable Agriculture Research and Education (SARE) project designed to examine disease-resistant apple cultivars (sometimes referred to as scab-resistant cultivars, or SRCs) and their potential for commercial use. Conceptually, it seemed obvious that using genetic resistance to eliminate one or more significant diseases would provide a sound foundation to a biointensive apple-production system. To test this, Prokopy planted a small orchard of SRCs in 1977 (Prokopy and Cooley, 1991; Prokopy, 2003) and, later, the SARE project planted SRCs in commercial orchards in Massachusetts.

Experience using SRCs in an advanced IPM system confirmed that disease resistance alone could not provide the basis for sustainable apple production systems (Merwin *et al.*, 1994; Ellis *et al.*, 1998). The SRCs could eliminate the need for many fungicide applications, but not all. For example, summer blemish diseases can become epidemic in orchards where no fungicides have been sprayed (Rosenberger *et al.*, 1996). In addition, leaves on SRC trees that had not been treated with fungicides became chlorotic and necrotic and then dropped prematurely, evidently the result of putative but unidentified fungal infections that otherwise would be controlled by summer fungicides. Over 2 or more years, this foliar damage reduced yield. As well as unmanaged disease, other factors such as market demand, adequate supply, production efficiency and storability have, to date, made large-scale wholesale production of even relatively high-quality SRCs such as Liberty untenable (Murphy and Willett, 1991; Penrose, 1995). By 2002, the few large commercial SRC blocks in Massachusetts had been replanted or renovated to other, more profitable but non-resistant cultivars.

Development of Biointensive Methods for Key Apple Pests

In the late 1990s and early 2000s, research in the apple IPM programme continued to focus on solving the major roadblocks to more biointensive IPM in Massachusetts: mites, plum curculio, apple maggot, apple scab and sooty blotch/flyspeck. The challenge was to develop biointensive management methods that would be acceptable within the economic constraints of a global apple market.

It looked at orchard and border architecture: the make-up of cultivars, the size of trees and the topographic location of blocks.

For mite management, work emphasizing the introduction of the predator mite, *T. pyri*, and consequently using a mite-friendly pesticide regimen, continued (Prokopy *et al.*, 1999). The mite-friendly approach used no pyrethroid fungicides, no EBDC fungicides after mid-June, with no insecticides and only limited fungicides thereafter. After 3 years, the study recommended that *T. pyri* be released no further apart than every sixth tree in high-density plantings or every third tree in low-density plantings (Prokopy *et al.*, 2000a). Given that, and appropriate pesticide selection, *T. pyri* could be established successfully and used to manage phytophagous mites.

For plum curculio, the focus in 2001 was on traps and 'trap trees' that were baited with odours attractive to plum curculio, benzaldehyde and grandisoic acid (Prokopy, 2002; Piñero and Prokopy, 2003). The baited traps were designed to signal a treatment threshold and, while they trapped curculio effectively, orchard populations of curculio and damage did not correlate with trap captures (Prokopy *et al.*, 2002). This may have been the result of odour competition from nearby fruiting hosts (Leskey and Wright, 2004), the finding that most curculio migrate into orchards well before most damage occurs, or temperature effects on ovipositing activity (Piñero and Prokopy, 2006). Baiting a tree with the same odours promised a better monitoring tool, which would promote aggregation and allow direct monitoring of early damage (Prokopy, 2002). This trap-tree approach was developed to establish a threshold for treatment following a full orchard insecticide targeting curculio applied at petal fall (Prokopy *et al.*, 2004b,c; Piñero *et al.*, 2006). Work also was carried out on a so-called 'bomb tree' approach that would attract immigrating curculios to a single tree, which was then treated with pesticide, eliminating the need for a more general orchard pesticide application.

Based on the same attract-and-kill strategy, work also began on using red sphere traps for apple maggot as direct controls, rather than monitoring tools, by treating them with an insecticide and feeding stimulant (Duan and Prokopy, 1995; Wright *et al.*, 1997). While odour-baited sticky traps had performed well (Reynolds and Prokopy, 1997), the labour and expense of maintaining or replacing the traps, necessitated by the build-up of insects on the trap surface, limited their acceptability as a direct control tactic (Duan *et al.*, 1994). The development of non-sticky, pesticide-treated spheres (PTS) for apple maggot control has gone through a series of steps, beginning with an attempt to sustain the effectiveness of the traps using chemical and physical barriers to keep toxicants and feeding stimulants in place (Hu *et al.*, 1996, 1997). This was followed by studies of spheres that had different forms of internal or external renewal of feeding stimulants and different toxicants (Prokopy *et al.*, 2000a,b, 2003, 2004a). Most recently, PTS development has centred on a visually integrated system named the 'CurveBall' (S. Wright and T. Leskey, Massachusetts, 2008, personal communication), which maximizes the availability and lethality of the toxicant. In addition to PTS technology, an index to determine optimal placement of the traps on orchard perimeters has been developed using tree size, tree density and pruning, cultivar(s) and border habitat (Prokopy *et al.*, 2005).

Decreased use of scab fungicides in apples during the late 1980s and early 1990s depended largely on demethylation-inhibiting (DMI) fungicides that allowed relatively long intervals of approximately 10 days between applications and that complemented the PAD-based delay of the first fungicide application (Gadoury *et al.*, 1989; Wilcox *et al.*, 1992). By 2000, resistant strains of *V. inaequalis* had developed in commercial orchards. The alternatives to DMI fungicides generally must be applied prophylactically and more frequently (Köller *et al.*, 1997, 2005), and both reducing chemical use for scab management and moving beyond chemically based tools for scab management has proven very difficult. Using PAD and inoculum destruction remains the most promising non-chemical tactic, but growers remain reluctant to accept the risk and cost.

Both threshold-based fungicide applications and non-fungicide alternatives hold promise for more biointensive management of sooty blotch and flyspeck. A model based on the accumulated hours of leaf wetting has been developed and, in parts of the USA, it can save three to four fungicide applications relative to calendar-based timing (Williamson and Sutton, 2000; Babadoost *et al.*, 2004), though performance has been erratic in the north-eastern USA and efforts are being made to improve it. In Massachusetts, the physical characteristics of orchard sites have been evaluated in an attempt to develop a risk index that would allow growers to reduce fungicide applications or try biointensive methods in low-risk blocks (Tuttle *et al.*, 2002). It may be possible to reduce sooty blotch and flyspeck incidence by modification of the orchard borders, either by increasing the distance from apples to reservoir hosts of the blemish fungi or by removing these hosts entirely. Summer sprays of calcium salts, applied to increase apple firmness and storability, can also suppress sooty blotch and flyspeck, though sprays must be augmented with a conventional fungicide such as captan to reach commercial levels of control (Biggs *et al.*, 2000; Cooley *et al.*, 2006).

The biointensive approaches for the management of mites, plum curculio, apple maggot, apple scab and the sooty blotch/flyspeck complex have been used on a limited basis in trials in commercial blocks and have yet to be tested as an integrated programme. Ultimately, IPM operates within the constraints of profitability and biointensive methods generally will increase input costs. To date, there is also some increased risk with these tactics, though further development may decrease them. While public policy will almost certainly push for decreased use of the most toxic pesticides, apple growers in Massachusetts will not be anxious to adopt biointensive IPM unless they can increase the value of their crop by earning more per apple sold, selling more apples, or both.

Marketing Biointensive IPM

Making biointensive IPM profitable engages consumers and marketing in the IPM process. In 2003, the apple IPM team looked again at eco-labelling, this time cooperating with a non-profit wholesale marketing organization based in Massachusetts, Red Tomato, Inc. Red Tomato was concerned not only with biointensive IPM but with local production, as summarized in their mission statement: 'Red Tomato's mission is connecting farmers and consumers through marketing, trade, and education, and through a passionate belief that a family-farm, locally-based, ecological, fair trade food system is the way to a better tomato.'

The pilot project addressed the issue that the most problematic in terms of toxicity, the organophosphate (OP) insecticides, the EBDC fungicides and captan (potential B_2 carcinogens), were also commonly used pesticides. In designing an advanced IPM system, generally accepted second-level IPM methods were combined to minimize the use of these materials. In 2004, the Eco Apple™ protocol of advanced IPM methods was formalized by UMass and Cornell working with the IPM Institute of North America (Madison, Wisconsin). The protocol classified pesticides as either 'green', 'yellow' or 'red', based on several types of toxicity criteria and the potential for resistance development (Red Tomato, 2007):

- Green – use with justification, e.g. trap captures or weather-based thresholds.
- Yellow – use with justification when green list or other alternatives are not adequate.
- Red – do not use.

IPM specialists were hired to examine production records from growers in the Eco Apple programme and ensure compliance with the protocol.

Eco Apple marketing was supported with a branding campaign and shipping and sales were coordinated by Red Tomato. Much of Eco Apple sales have gone through stores that emphasize environmentalism, such as Whole Foods Market. In addition to a focus on environmentally sound production, part of the Eco Apple marketing plan is an emphasis on local production. The amount of fruit sold through the programme grew by a factor of ten from 2004 to 2007. Growers reported higher prices gained through the Eco Apple programme but rate expertise in the partnership and access to new markets as even more important advantages.

In 2007, the programme involved nine farms and 227 ha in New England and New York, a very small fraction of the apple producers there. It is not clear how large the market for Eco Apples can grow and whether the market advantages in the programme can be maintained if a significant proportion of apple growers in the region adopt a similar production and marketing approach. In addition, the most advanced IPM methods, such as the PTS, have not been brought into the Eco Apple protocol.

In many ways, the Eco Apple project is a continuation of the basic paradigm used in the early days of the UMass apple IPM programme. The direct engagement and shared leadership roles played by the non-profit wholesaler, Red Tomato, and the IPM Institute go well beyond previous involvement with non-academic institutions and businesses but, in the early years of the programme, apple IPM worked closely, if informally, with the wholesaler, J.P. Sullivan, to promote IPM adoption. Grower visits are less frequent and the growers are spread around several states in the Eco Apple programme, but they are still involved closely in developing the project agenda. The key difference is that, at this point in the Eco Apple project, the emphasis is on using established IPM tactics outlined in a defined protocol and marketing fruit based, in part, on that protocol. The emphasis on field biology and ecology, which ran through most of the UMass programme, has taken a secondary role, though in the future, continued research will be needed to improve the reliability of new biointensive IPM tactics and to develop acceptable alternatives to commonly used pesticides. It remains unclear whether public research universities in the USA will hire research scientists to work in the area of IPM. To a large extent, the decreased field research on apple IPM in Massachusetts and New England is attributable to the loss of a key scientist, Prokopy, and with him, the position charged with applied entomology in tree fruit. As traditional relationships between land grant universities in the USA, state and federal governments and growers change, the methods for funding and pursuing IPM research must change too. Assuming advanced IPM methods are developed, at least in the short- to midterm, these new methods probably will cost more to use than present pest management practices. These increased costs will need to be subsidized by government incentives or paid for by consumers. Increasingly, attempts in Massachusetts to build an ecological orchard will depend as much on policy makers and consumers as on researchers and producers.

References

Anonymous (1986) Alar-treated apples: a potential health threat, especially for children. *Child Health Alert* April, 1–2.

Asquith, D. and Hull, L.A. (1979) Integrated pest management systems in Pennsylvania apple orchards. In: Bethel, D. and Eikenbary, R. (eds) *Pest Management Programs For Deciduous Tree Fruits and Nuts.* Plenum, New York, pp. 203–222.

Autio, W.R. and Greene, D.W. (1991) Summer pruning, a continuing look at the benefits. *Fruit Notes* 56(1), 27–28.

Babadoost, M., McManus, P.S., Helland, S.N. and Gleason, M.L. (2004) Evaluating a wetness-based warning system and reduced-risk fungicides to manage sooty blotch and flyspeck of apple. *HortTechnology* 14, 51–57.

Becker, C.M., Pategas, K. and Manning, W.J. (1982) Apple disease management in Massachusetts: 1981 results and four-year summary. *Fruit Notes* 47(2), 12–19.

Becker, C.M., Bardinelli, T., Cooley, D.R., Pategas, K. and Manning, W.J. (1983) Disease management of apples in Massachusetts: 1982 results and summary of the five-year program. *Fruit Notes* 48(1), 10–16.

Benbrook, C.M. (1996) *Pest Management at the Crossroads.* Consumers Union of United States, Inc, Yonkers, New York.

Biggs, A.R., Hogmire, H.W. and Collins, A.R. (2000) Assessment of an alternative IPM program for the production of apples for processing. *Plant Disease* 84, 1140–1146.

Blommers, L.H.M. (1994) Integrated pest management in European apple orchards. *Annual Review of Entomology* 39, 213–241.

Bostanian, N.J. and Coulombe, L.J. (1986) An integrated pest management program for apple orchards in southwestern Quebec. *Canadian Entomologist* 118, 1131–1142.

Bourne, A.I., Boyd, O.C., Roberts, R.G. and Whitcomb, W.D. (1948) Apple pests and their control. Leaflet No. 189. University of Massachusetts Cooperative Extension, Amherst, Massachusetts.

Bower, C.C., Penrose, L.J. and Dodds, K. (1993) A practical approach to pesticide reduction on apple crops using supervised pest and disease control – preliminary results and problems. *Plant Protection Quarterly* 8, 57–62.

Butkewich, S.L., Prokopy, R.J., Pyne, W., Autio, W.R. and Greene, D.W. (1989) The effects of summer pruning on insect and mite populations in apple trees. *Fruit Notes* 52(2), 6–7.

Coli, W.M. (1984) *Integrated Pest Management of Apple Pests in Massachusetts and New England.* University of Massachusetts Cooperative Extension, Amherst, Massachusetts.

Coli, W.M., Morin, G., Mancino, L., Gordon, D., Green, T., Sienkiewicz, C., Spitko, R. and Prokopy, R.J. (1982) Integrated management of apple pests in Massachusetts, 1981 results: insects. *Fruit Notes* 47(1), 15–25.

Coli, W.M., Zahnleuter, R., Gordon, D., Leahy, K., Parella, J., Roberts, D. and Prokopy, R.J. (1983) Integrated management of apple pests in Massachusetts, 1982 results: insects. *Fruit Notes* 48(1), 24–34.

Coli, W.M., Green, T.A., Hosmer, T.A. and Prokopy, R.J. (1985) Use of visual traps for monitoring insect pests in the Massachusetts apple IPM program. *Agriculture Ecosystems and Environment* 14, 251–265.

Cooley, D.R. and Autio, W.R. (1997) Disease-management components of advanced integrated pest management in apple orchards. *Agriculture Ecosystems and Environment* 66, 31–40.

Cooley, D.R., Gamble, J.W. and Autio, W.R. (1997) Summer pruning as a method for reducing flyspeck disease on apple fruit. *Plant Disease* 81, 1123–1126.

Cooley, D.R., Autio, W.R., Tuttle, A.F. and Krupa, J. (2006) Alternative fungicides for management of sooty blotch and flyspeck. *Fruit Notes* 72(3), 1–3.

Croft, B.A. and Hoyt, S.C. (1983) *Integrated Management of Insect and Mite Pests of Pome and Stone Fruits.* Wiley Interscience, New York.

Dover, M.J. (1985) *A Better Mousetrap. Improving Pest Management for Agriculture. Study 4.* World Resources Institute, Washington, D.C.

Duan, J.J. and Prokopy, R.J. (1995) Control of apple maggot flies (Diptera, Tephritidae) with pesticide-treated red spheres. *Journal of Economic Entomology* 88, 700–707.

Duan, J.J., Hu, X., Prokopy, M. and Prokopy, R.J. (1994) Build-up of bugs causes decline in effectiveness of sticky for capturing apple maggot flies on red sphere traps. *Fruit Notes* 59(2), 8–9.

Ehler, L.E. (2006) Integrated pest management (IPM): definition, historical development and implementation, and the other IPM. *Pest Management Science* 62, 787–789.

Ellis, M.A., Ferree, D.C., Funt, R.C. and Madden, L.V. (1998) Effects of an apple scab-resistant cultivar on use patterns of inorganic and organic fungicides and economics of disease control. *Plant Disease* 82, 428–433.

Frisbie, R.A. and Adkisson, P.L. (1985) *Integrated Pest Management on Major Agricultural Systems.* Texas A&M University, Texas Agricultural Experimental Station, College Station, Texas.

Frisbie, R.A. and Smith, J.W. (1991) Biologically intensive integrated pest management: the future. In: Menn, J.J. and Steinhauer, A.L. (eds) *Progress and Perspectives for the 21st Century: Entomological Society of America National Symposium.* Entomological Society of America, Lanham, Maryland, pp. 151–164.

Gadoury, D.M. and MacHardy, W.E. (1982a) Preparation and interpretation of squash mounts of pseudothecia of *Venturia inaequalis*. *Phytopathology* 72, 92–95.

Gadoury, D.M. and MacHardy, W.E. (1982b) A model to estimate the maturity of ascospores of *Venturia inaequalis*. *Phytopathology* 72, 901–904.

Gadoury, D.M. and MacHardy, W.E. (1986) Forecasting ascospore dose of *Venturia inaequalis* in commercial apple orchards. *Phytopathology* 76, 112–118.

Gadoury, D.M., MacHardy, W.E. and Rosenberger, D.A. (1989) Integration of pesticide application schedules for disease and insect control in apple orchards of the northeastern United States. *Plant Disease* 73, 98–105.

Green, T.A. and Prokopy, R.J. (1986) Visual monitoring trap for the apple blotch leafminer moth, *Phyllonorycter crataegella* (Lepidoptera, Gracillariidae). *Environmental Entomology* 15, 562–566.

Gruys, P. (1982) Hits and misses – the ecological approach to pest control in orchards. *Entomologia Experimentalis et Applicata* 31, 70–87.

Gurr, G.M., Valentine, B.J., Azam, M.N.G. and Thwaite, W.G. (1996) Evolution of arthropod pest management in apples. *Agricultural Zoology Reviews* 7, 35–69.

Hardman, J.M., Rogers, R.E.L., Nyrop, J.P. and Frisch, T. (1991) Effect of pesticide applications on abundance of European red mite (Acari, Tetranychidae) and *Typhlodromus pyri* (Acari, Phytoseiidae) in Nova Scotian apple orchards. *Journal of Economic Entomology* 84, 570–580.

Hill, S.B., Vincent, C. and Chouinard, G. (1999) Evolving ecosystems approaches to fruit insect pest management. *Agriculture Ecosystems and Environment* 73, 107–110.

Hollingsworth, C. and Coli, W.M. (2007) IPM guidelines: apple (http://www.umass.edu/

umext/ipm/guidelines/ipm_guidelines_apple.html, accessed 21 March 2008).

Hollingsworth, C., Coli, W.M., Cooley, D.R. and Prokopy, R.J. (1992) Massachusetts integrated pest management guidelines for apple. *Fruit Notes* 57(4), 12–16.

Holm, D., Lass, D. and Rogers, R. (2000) The changing landscape of Massachusetts agriculture. *Massachusetts Benchmarks* 3, 12–18.

Hoyt, S.C. (1969) Integrated chemical control of insects and biological control of mites on apple in Washington. *Journal of Economic Entomology* 62, 74–86.

Hu, X., Prokopy, R.J., Wright, S.E. and Mason, J.L. (1995) Released *Typhlodromus pyri* show success in colonization and dispersion in Massachusetts apple orchards. *Fruit Notes* 60(2), 6–7.

Hu, X., Clark, J. and Prokopy, R.J. (1996) Progress in 1995 toward development of toxicant-treated spheres for control of apple maggot flies. *Fruit Notes* 61(2), 10–13.

Hu, X., Kaknes, A. and Prokopy, R.J. (1997) Improved pesticide-treated wooden spheres for controlling apple maggot flies. *Fruit Notes* 62(2), 3–5.

Hull, L.A., Hickey, K.D. and Kanour, W.W. (1983) Pesticide usage patterns and associated pest damage in commercial apple orchards in Pennsylvania. *Journal of Economic Entomology* 76, 577–583.

Jacobsen, B.J. (1997) Role of plant pathology in integrated pest management. *Annual Review of Phytopathology* 35, 373–391.

Kogan, M. (1998) Integrated pest management: historical perspectives and contemporary developments. *Annual Review of Entomology* 43, 243–270.

Köller, W., Wilcox, W.F., Barnard, J., Jones, A.L. and Braun, P.G. (1997) Detection and quantification of resistance of *Venturia inaequalis* populations to sterol demethylation inhibitors. *Phytopathology* 87, 184–190.

Köller, W., Parker, D.M., Turechek, W.W., Rosenberger, D.A., Wilcox, W.F., Caroll, J., Agnello, A.M. and Reissig, W.H. (2005) Fungicide resistance of apple scab: status quo and management options. *New York Fruit Quarterly* 13(1), 9–17.

Kovach, J. and Tette, J. (1988) A survey of the use of IPM by New York apple producers. *Agriculture Ecosystems and Environment* 20, 101–108.

Lawson, D.S., Reissig, W.H. and Agnello, A.M. (1998) Effects of summer pruning and hand fruit thinning on obliquebanded leafroller (Lepidoptera: Tortricidae) fruit damage in New York state apple orchards. *Journal of Agricultural Entomology* 15, 113–123.

Leskey, T.C. and Wright, S.E. (2004) Influence of host tree proximity on adult plum curculio (Coleoptera: Curculionidae) responses to monitoring traps. *Environmental Entomology* 33, 389–396.

MacHardy, W.E. (2000a) Action threshold for managing apple scab with fungicides and sanitation. *Acta Horticulturae* 525, 123–131.

MacHardy, W.E. (2000b) Current status of IPM in apple orchards. *Crop Protection* 19, 801–806.

MacHardy, W.E. and Sondej, J. (1981) Weather monitoring instrumentation for plant disease management programs and epidemiological studies. *New Hampshire Agricultural Experiment Station Bulletin* 519, 40 pp.

MacHardy, W.E., Gadoury, D.M. and Rosenberger, D.A. (1993) Delaying the onset of fungicide programs for control of apple scab in orchards with low potential ascospore dose of *Venturia inaequalis*. *Plant Disease* 77, 372–375.

Martin, O.B. (1921) *The Demonstration Work: Dr Seaman A. Knapp's Contribution to Civilization.* The Stratford Co, Boston, Massachusetts.

Mason, J.L., Prokopy, R.J., Wright, S.E., Black, J., Chang, C., Cook, J., Goodall, S. and Ma, Y. (1995) Apple integrated pest management in 1994: insects and mites in second-level orchard blocks. *Fruit Notes* 60(1), 1–7.

Merwin, I.A., Brown, S.K., Rosenberger, D.A., Cooley, D.R. and Berkett, L.P. (1994) Scab-resistant apples for the Northeastern United States – new prospects and old problems. *Plant Disease* 78, 4–10.

Murphy, C. and Willett, L.S. (1991) Issues in the development and marketing of reduced chemical agricultural products: a look at disease-resistant apple cultivars. *Agricultural Economics Extension Bulletin 91-34.* Cornell University, Ithaca, New York.

Nyrop, J.P., English-Loeb, G. and Roda, A. (1997) Conservation biological control of spider mites in perennial cropping systems. In: Barbosa, P. (ed.) *Perspectives on the Conservation of Natural Enemies of Pests*. Academic Press, New York, pp. 307–333.

O'Brien, M. and Prokopy, R.J. (1988) Results of the first year of second-stage apple IPM practices. *Fruit Notes* 53(2), 8–11.

Owens, E.D. and Prokopy, R.J. (1978) Visual monitoring trap for European apple sawfly (Hymenoptera-Tenthredinidae). *Journal of Economic Entomology* 71, 576–578.

Park, P. (1993) What is participatory research? A theoretical and methodological perspective. In: Jackson, T. and Park, P. (eds) *Voices of Change: Participatory Research in the United States and Canada*. Bergin and Garvey, Westport, Connecticut, pp. 1–20.

Penrose, L.J. (1995) Fungicide use reduction in apple production – potentials or pipe dreams. *Agriculture Ecosystems and Environment* 53, 231–242.

Pickett, A.D., Putnam, W.L. and Le Roux, E.J. (1958) Progress in harmonizing biological and chemical control of orchard pests in eastern Canada. 10th International Congress of Entomology. Montréal, Québec, Canada.

Piñero, J.C. and Prokopy, R.J. (2003) Field evaluation of plant odor and pheromonal combinations for attracting plum curculios. *Journal of Chemical Ecology* 29, 2735–2748.

Piñero, J.C. and Prokopy, R.J. (2006) Temporal dynamics of plum curculio, *Conotrachelus nenuphar* (Herbst.) (Coleoptera: Curculionidae), immigration into an apple orchard in Massachusetts. *Environmental Entomology* 35, 413–422.

Piñero, J.C., Tuttle, A.F., Agnello, A.M., Faubert, H., Koehler, G., Morin, G., Leahy, K., Los, L., Cooley, D.R. and Prokopy, R.J. (2006) Using odor-baited trap trees as indicators of need and timing of insecticide spray against plum curculio in the northeast: validation studies 2004–2005. *Fruit Notes* 72(4), 1–11.

Prokopy, R.J. (1968) Sticky spheres for estimating apple maggot adult abundance. *Journal of Economic Entomology* 61, 1082–1085.

Prokopy, R.J. (1987) The second stage of apple IPM in Massachusetts. *Fruit Notes* 52 (3), 9–12.

Prokopy, R.J. (1988) Benefits of Alar to apple IPM programs. *Fruit Notes* 53(2), 7–8.

Prokopy, R.J. (1993) Stepwise progress toward IPM and sustainable agriculture. *The IPM Practitioner* 15, 1–4.

Prokopy, R.J. (1994) Integration in orchard pest and habitat management – a review. *Agriculture Ecosystems and Environment* 50, 1–10.

Prokopy, R.J. (2002) An odor-baited 'trap-tree' approach to monitoring plum curculio. *Fruit Notes* 67(1), 23–24.

Prokopy, R.J. (2003) Two decades of bottom-up, ecologically based pest management in a small commercial apple orchard in Massachusetts. *Agriculture Ecosystems and Environment* 94, 299–309.

Prokopy, R.J. and Cooley, D.R. (1991) Very low spray apple production. *Fine Gardening* 22, 46–49.

Prokopy, R.J., Adams, R.G. and Hauschild, K.I. (1979) Visual responses of tarnished plant bug (Hemiptera, Miridae) adults on apple. *Environmental Entomology* 8, 202–205.

Prokopy, R.J., Coli, W.M., Hislop, R.G. and Hauschild, K.I. (1980) Integrated management of insect and mite pests in commercial apple orchards in Massachusetts. *Journal of Economic Entomology* 73, 529–535.

Prokopy, R.J., Hubbell, G.L., Adams, R.G. and Hauschild, K.I. (1982) Visual monitoring trap for tarnished plant bug adults on apple. *Environmental Entomology* 11, 200–203.

Prokopy, R.J., Johnson, S.A. and O'Brien, M. (1990) Second-stage integrated management of apple arthropod pests. *Entomologia Experimentalis et Applicata* 54, 9–19.

Prokopy, R.J., Mason, J.L., Christie, M. and Wright, S.E. (1996) Arthropod pest and natural enemy abundance under second-level versus first-level integrated pest management practices in apple orchards: a 4-year study. *Agricultural Ecosystems and Environment* 57, 35–47.

Prokopy, R.J., Wright, S.E., Mason, J.L., Nyrop, J.P., Wentworth, K. and Herring, C. (1997) Comparative level of establishment of released *Typhlodromus pyri* predatory mites in first-level and second-level IPM apple orchard blocks. *Fruit Notes* 62(2), 6–8.

Prokopy, R.J., Wright, S.E., Black, J., Nyrop, J.P., Wentworth, K. and Hering, C. (1999) Establishment and spread of released *Typhlodromus pyri* predator mites in apple orchard blocks of different tree size: 1998 results. *Fruit Notes* 64(1), 10–15.

Prokopy, R.J., Wright, S.E., Black, J., Nyrop, J.P., Wentworth, K. and Hering, C. (2000a) Establishment and spread of released *Typhlodromus pyri* predator mites in apple orchard blocks of different tree size: 1999 and final results. *Fruit Notes* 65, 6–16.

Prokopy, R.J., Wright, S.E., Black, J.L., Hu, X.P. and McGuire, M.R. (2000b) Attracticidal spheres for controlling apple maggot flies: commercial-orchard trials. *Entomologia Experimentalis et Applicata* 97, 293–299.

Prokopy, R.J., Chandler, B.W. and Piñero, J.C. (2002) Commercial-orchard evaluation of traps monitoring plum curculio: 2001 results. *Fruit Notes* 67(1), 17–22.

Prokopy, R.J., Chandler, B.W. and Wright, S.E. (2003) Improved sugar delivery onto pesticide-treated spheres for controlling *Rhagoletis pomonella* (Diptera: Tephritidae). *Canadian Entomologist* 135, 909–918.

Prokopy, R.J., Jácome, I., Bigurra, E., Blanco, M., Chandler, B.W. and Wright, S.E. (2004a) Evaluation of pesticide-treated spheres for control of apple maggot flies in 2003. *Fruit Notes* 69(2), 21–25.

Prokopy, R.J., Jácome, I., Gray, E., Trujillo, G., Ricci, M. and Pinero, J.C. (2004b) Using odor-baited trap trees as sentinels to monitor plum curculio (Coleoptera: Curculionidae) in apple orchards. *Journal of Economic Entomology* 97, 511–517.

Prokopy, R.J., Jácome, I. and Piñero, J.C. (2004c) A threshold for spraying against plum curculio using odor-baited trap trees. *Fruit Notes* 69(1), 14–15.

Prokopy, R.J., Jácome, I. and Bigurra, E. (2005) An index for assigning distances between odor-baited spheres on perimeter trees of orchards for control of apple maggot flies. *Entomologia Experimentalis et Applicata* 115, 371–377.

Red Tomato (2007) Eco-Apple protocol and grower self-assessment. IPM Institute of North America (http://www.ipminstitute.org/pdf/Eco-Apple%20Protocol%20for%20 2007%20v3-2%20Revised%20081007.pdf, accessed 1 May 2008).

Reynolds, A.H. and Prokopy, R.J. (1997) Evaluation of odor lures for use with red sticky spheres to trap apple maggot (Diptera: Tephritidae). *Journal of Economic Entomology* 90, 1655–1660.

Riley, J. and Alexander, C.J. (1997) Statistical literature for participatory on-farm research. *Experimental Agriculture* 33, 73–82.

Rosenberger, D.A., Meyer, F.W. and Rist, D.L. (1990) Effect of spray timing on incidence and severity of sooty blotch and flyspeck on Liberty apple. *Fungicide and Nematicide Tests* 46, 31.

Rosenberger, D.A., Engle, C.A. and Meyer, F.W. (1996) Effects of management practices and fungicides on sooty blotch and flyspeck diseases and productivity of liberty apples. *Plant Disease* 80, 798–803.

Sansavini, S. (1997) Integrated fruit production in Europe: research and strategies for a sustainable industry. *Scientia Horticulturae* 68, 25–36.

Suckling, D.M., Walker, J.T.S. and Wearing, C.H. (1999) Ecological impact of three pest management systems in New Zealand apple orchards. *Agriculture Ecosystems and Environment* 73, 129–140.

Tette, J., Glass, E.H., Bruno, D. and Way, D. (1979) New York Tree Fruit Pest Management Project 1973–1978. Food Life Science. Bulletin 81. New York.

Tuttle, A.F., Bergweiler, C., Hall, J., Reisner, L., Christle, S., Autio, W.R. and Cooley, D.R. (2002) Development of a model for predicting flyspeck risks in blocks of apple trees. *Fruit Notes* 67(1), 5–8.

USDA (2004) The 2002 Census of agriculture. National Agricultural Statistics Service. United States Department of Agriculture, Washington, DC.

USDA (2007) Noncitrus fruits and nuts 2006 summary. National Agricultural Statistics Service, United States Department of Agriculture, Washington, DC.

Van Driesche, R.G., Prokopy, R.J. and Coli, W.M. (1988) Controlling spider mites in Massachusetts apple orchards through conservation of predator mites. *Fruit Notes* 53(4), 8–12.

Van Zee, V. (1992) Partners with nature: a Massachusetts integrated pest management recognition program. *Fruit Notes* 57(4), 23–24.

Whalon, M.E. and Croft, B.A. (1984) Apple IPM implementation in North America. *Annual Review of Entomology* 29, 435–470.

Wilcox, W.F., Wasson, D.I. and Kovach, J. (1992) Development and evaluation of an integrated, reduced-spray program using sterol demethylation inhibitor fungicides for control of primary apple scab. *Plant Disease* 76, 669–677.

Williamson, S.M. and Sutton, T.B. (2000) Sooty blotch and flyspeck of apple: etiology, biology, and control. *Plant Disease* 84, 714–724.

Wright, S.E., Hu, X. and Prokopy, R.J. (1997) Tests of imidacloprid-treated spheres for controlling apple maggot fly. *Fruit Notes* 62(4), 1–4.

Zuck, M.G. and MacHardy, W.E. (1981) Recent experience in timing sprays for control of apple scab – equipment and test results. *Plant Disease* 65, 995–998.

7 Managing Pestiferous Fruit Flies (Diptera: Tephritidae) through Environmental Manipulation

Martín Aluja and Juan Rull

Instituto de Ecología, Xalapa, Veracruz, México

Introduction

Overview and aims

One of the most humbling challenges a fruit-fly biologist can face is to try to understand what drives the biology, ecology and behaviour of fruit flies and, based on this knowledge, design practical and cost-effective biorational management schemes aimed at minimizing the damage inflicted by the pestiferous species to owners of commercial orchards and backyard gardens through environmental manipulation. Our objective here is to offer the first compendium on this topic as a tribute to the late Ronald J. Prokopy, as he pioneered many of the environmentally friendly fruit-fly management schemes currently applied in many temperate regions of the world.

We define the environment as all biotic and abiotic elements with which an organism interacts and, as a consequence, understand environmental manipulation in the context of pest management as a series of measures aimed at manipulating such elements with the purpose of eliminating or reducing damage to a valued crop. Because the pest organism itself is also part of the environment and alterations to environmental elements will influence its behaviour, centring pest management tactics/practices on careful behavioural observations will increase their long-term effectiveness greatly. This was one of Ron Prokopy's dreams and we therefore find it appropriate to review this complex topic in a book dedicated to him. In addition, and quoting Southwood and Way (1970), 'The choice of the pest management procedure depends on four characteristics of the agroecosystem: 1) the diversity of vegetation within the agroecosystem, 2) the permanence of the various crops within the agroecosystem, 3) the extent of the isolation of the agroecosystem from natural vegetation and 4) the stability of the climate.' One of the fundamental themes of this chapter is that some of these ecological characteristics of the agroecosystem (environment) should not only be taken into account when designing fruit-fly management schemes but also that they can be manipulated directly.

Successful cases of environmental manipulation for pest control have been developed almost exclusively in the case of annual crops or short-lived trees such as papayas (Altieri *et al.*, 1983; Aluja *et al.*, 1997a; Lewis *et al.*, 1997; Verkerk *et al.*, 1998; Banks and Ekbom, 1999; Thies and Tscharntke, 1999; Van den Berg *et al.*, 2001; Gurr *et al.*, 2004; Landis *et al.*, 2005; Opdam *et al.*, 2005; Shelton and Badenes-Pérez, 2006). In our opinion, the main reason for this is that annuals are amenable to experimentation (Shea *et al.*, 2002). By contrast, environmental manipulation represents a daunting challenge when dealing with perennial crops such as fruit trees, which can live productively for over 30 years. Under such a scenario, mistakes can be costly and decisions have permanent, or at least long-term, consequences. Furthermore, biological processes are highly dynamic and, in the case of the fruit-fly/fruit-tree interaction, it is likely that during the lifetime of a long-lived tree, strong selection on fly populations exposed to new environmental challenges will result in individuals overcoming the hurdles presented to them artificially through environmental manipulation schemes.

Additionally, trying to manage pestiferous fruit flies through environmental manipulation is a complex task because even after more than 110 years of uninterrupted research worldwide, our knowledge on fruit-fly natural history, basic biology, ecology and behaviour is still surprisingly scant (Aluja, 1999). For example, we are still trying to understand fully basic aspects such as the dynamics of habitat use by highly polyphagous species such as *Ceratitis capitata* (Wiedemann), *Anastrepha ludens* (Loew), *Bactrocera dorsalis* (Hendel), *B. tryoni* (Froggatt) and *B. zonata* (Saunders). Although much has been learned on the sequential use of habitats/hosts over an entire season or over several seasons (e.g. Bateman, 1977; Nishida *et al.*, 1985; Harris and Lee, 1986, 1987; Hendrichs and Hendrichs, 1990; Hendrichs *et al.*, 1991; Aluja and Birke, 1993; Harris *et al.*, 1993; Celedonio-Hurtado *et al.*, 1995; Aluja *et al.*, 1996, 2000; Papadopoulos *et al.*, 1996, 1998, 2001, 2003; Israely *et al.*, 1997; Katsoyannos *et al.*, 1998; Clarke *et al.*, 2001; Ovruski *et al.*, 2005), little is known with respect to variability within hosts (i.e. same plant) or between hosts of the same species over time in the presence/concentration of toxic secondary metabolites or nutritional quality (Sivinski *et al.*, 2004). Similarly, we are also still trying to determine why recently emerged adults prefer certain trees over others as shelter sites or how adult feeding history will influence individual responses to the suite of volatiles encountered by a foraging female in search of an oviposition substrate or a male in search of a mate (e.g. Prokopy *et al.*, 1994a,b).

Despite the above-mentioned gaps in our knowledge on the basics of fruit-fly biology and ecology, research efforts all over the world, past and present, have nevertheless yielded valuable information and insights into daily activity and host-use patterns (e.g. preference of certain cultivars), including host resistance, nutritional, chemical and behavioural ecology and relationship with natural enemies (e.g. Metcalf, 1990; Hendrichs *et al.*, 1991; Aluja and Birke, 1993; Jang and Light, 1996; Light and Jang, 1996; Craig *et al.*, 1999; Díaz-Fleischer and Aluja, 2000; Drew and Yuval, 2000; Papaj, 2000; Raghu *et al.*, 2004). Much of the work published since the late 1970s on these topics was influenced by Ron

Prokopy in one way or the other, or produced by him, his students and his collaborators. Ron, in turn influenced by the thinking of Pickett *et al.* (1946), Geier and Clark (1961), Hoyt (1969), Asquith and Colburn (1971), Stern (1973), Croft (1978) and Kogan (1988), translated the most important insights on fly behaviour, particularly in temperate regions, into management tools that he integrated skillfully into programmes that were constantly fine-tuned under commercial orchard settings (second and third levels of integration) (Prokopy, 1994).

In this chapter, we first address three issues tied closely to the management of highly mobile insects such as fruit flies through environmental manipulation: (i) how do fruit flies perceive the environment we intend to manipulate? (ii) how do these insects use the resources available to them in variable environments? and (iii) at what spatial scale do we want to manipulate the environment? We then review the most relevant publications on topics such as the biotic and abiotic factors that influence female and male behaviour in natural habitats and agroecosystems (fruit orchards), the effect of fruiting phenology on fruit-fly attack (and the possibility of manipulating it), and the variability in susceptibility to the attack by fruit flies of various hosts and cultivars thereof. We will also cursorily review fly responses to visual and chemical cues, including traps, as these monitoring and control tools end up forming a critical part of the environment in which adult flies live, frequently altering it in significant ways. Based on this information, we will finish by reviewing the few published cases of fruit-fly management through environmental manipulation and by sharing ideas and future scenarios on this topic in the areas of agroecosystem and orchard design (e.g. the possibility of fostering colonial spider webs, an old idea of one of Ron's students (Jorge Hendrichs)), trap cropping, interception using traps, possible use of repellent plants, combined strategies such as push–pull and bait stations and chemical environment manipulation, as this strategy has been applied successfully to control other tree-fruit pests (Greenblatt and Lewis, 1983). In keeping with one of Ron's strongly held beliefs, we always try to keep in mind the ultimate user of the information: the grower. In this respect, and building upon ideas proposed originally by one of us in the mid-1980s (Aluja and Liedo, 1986), we discuss environmental manipulation schemes tailored to the needs of contrasting types of growers: small or big, rich or poor, technologically advanced or subsistence and conventional or organic.

Conceptual Framework

How do fruit flies perceive the environment we intend to manipulate and how do they use the resources available to them in variable environments?

As noted earlier, before being able to manipulate the environment for management purposes, we must understand first how fruit flies perceive the environment we intend to manipulate and how they use the resources available to them in variable environments.

In recent reviews, Bruce *et al.* (2005) and Aluja and Mangan (2008) discuss the most critical factors that lead a foraging insect to a particular tree and that, after arrival, help it locate individual fruit which can be used as an oviposition substrate if all suitable signals are present (i.e. host acceptance). For further insight into this topic, we refer the reader to two chapters in this book (see Roitberg *et al.*, Chapter 3, this volume; Dorn and Piñero, Chapter 4, this volume). Hassel and Southwood (1978) proposed that insects followed a step-by-step (catenary) process during which they perceived the environment at three hierarchical levels: habitat, patch and food item (a fruit in the case of a foraging, sexually mature/gravid female fruit fly). Doutt (1959) previously had proposed a similar approach in the case of parasitic Hymenoptera: host-habitat location, host finding, host acceptance and host suitability. These hierarchical levels at which insects perceive the environment can be fit into categories of spatial scale, as we will discuss later.

There are three classes of chemicals of interest when trying to understand how fruit flies perceive their environment: (i) general 'host-plant habitat volatiles' (*sensu* Light and Jang, 1996; additional details in the later section on fruit volatiles); (ii) more specific host-plant volatiles (i.e. associated with a specific species of tree); and (iii) highly specific fruit volatiles that help a foraging, host-seeking female locate a suitable oviposition substrate. There are also non-volatile chemicals found on the surface of the fruit (e.g. host-marking pheromones, cuticular waxes) and secondary metabolites inside the fruit that allow the same female to reach a final decision on whether to accept or reject a particular fruit as an oviposition substrate (Aluja and Díaz-Fleischer, 2006; Aluja and Mangan, 2008). Some of these chemicals can be incorporated into management schemes based on environmental manipulation (details follow).

For a review of the critical components of environment-shaping resource use by fruit flies and their parasitoids, we refer the reader to Roitberg and Prokopy (1984), Prokopy *et al.* (1984), Averill and Prokopy (1987), Fitt (1990), Messina and Jones (1990), Aluja and Birke (1993), Aluja *et al.* (1997a,b, 2003a), Roitberg and Mangel (1997), Sivinski *et al.* (1997, 2000a, 2004, 2007), Aluja (1999), Dalby-Ball and Meats (2000, 2002), Cronin *et al.* (2001), Rull and Prokopy (2001), Papadopoulos *et al.* (2003), Raghu *et al.* (2004) and Balagawi *et al.* (2005).

Environmental manipulation and the issue of scale

In our opinion, environmental manipulation as a management tool for controlling highly mobile insects such as fruit flies needs to be viewed from a flexible perspective spanning all the way from an entire geographic region to a single fruit (Fig. 7.1). At every spatial unit, there are opportunities to manipulate the environment to help us reduce the populations of pestiferous fruit flies. For the purposes of this chapter, and borrowing from Doutt (1959), Hassel and Southwood (1978), Forman (1995), Pearson (2002) and Fischer and Lindenmayer (2007), we will consider from now on the following scales: region, landscape (which henceforth will be equated with agroecological matrix), agroecosystem/single orchard (which henceforth will be equated with patch), single tree and single fruit (Fig. 7.1).

Region/area from an 'area-wide' pest management perspective has been defined by Lindquist (2000) as 'a combination of geography and the range of

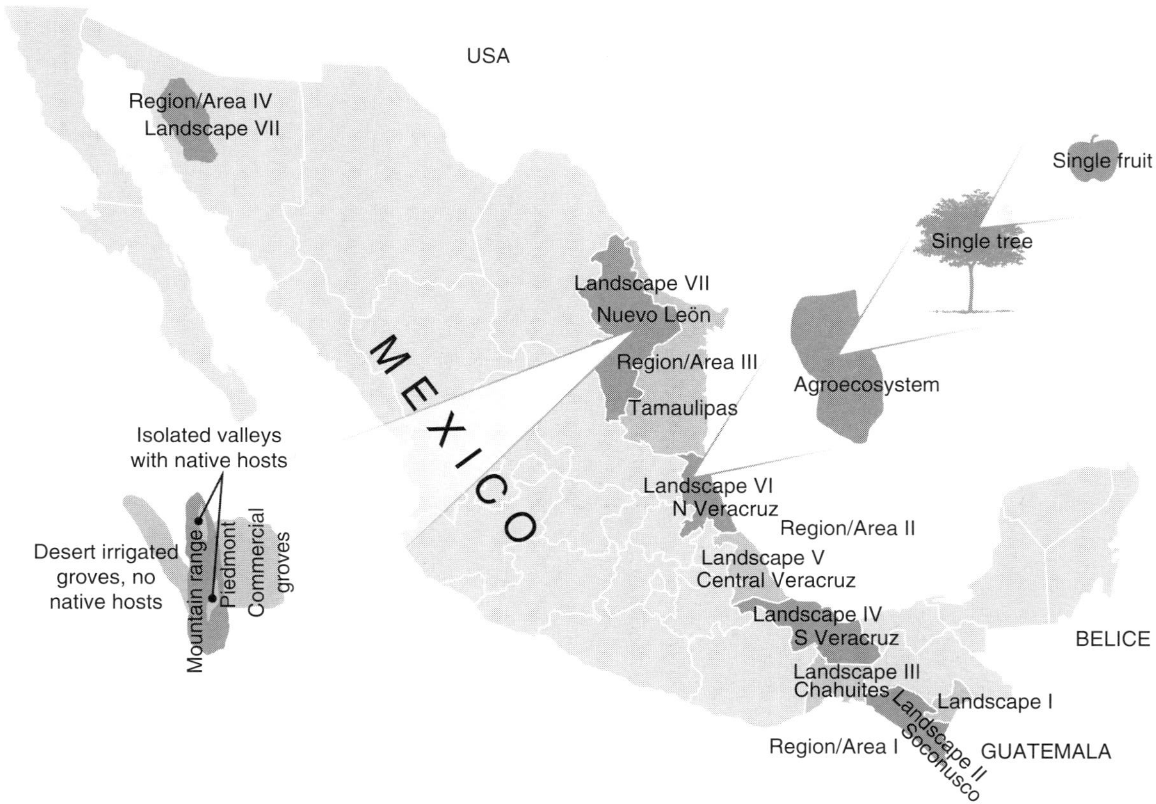

Fig. 7.1. Different levels of spatial scale at which environmental manipulation schemes can be implemented to manage pestiferous fruit flies. Four scales are considered: region/area, landscape (= agroecological matrix), agroecosystem/single orchard (= patch), single tree/single fruit. To illustrate, four regions and eight landscapes within those regions are depicted in Mexico. Region/area I is located in the humid tropics where many fruit-fly hosts can be found generating ideal ecological conditions for population growth (details in Fig. 7.2). In sharp contrast, region/area IV is located in one of the driest parts of the country where hosts are found only in isolated agroecosystems (patches) and flies have a hard time surviving.

hosts of the target insect pests which are found in this area'. The same author indicates that the term 'area' in 'area-wide' 'refers to the area where the target insect population survives'. Most importantly, 'the area is not limited to production of the major crop(s) to be protected'.

In the case of landscape, we adopt the definitions provided by Forman (1995), '. . . a kilometer-wide mosaic over which local ecosystems recur . . .', and more recently by Fischer and Lindenmayer (2007), 'a human-defined area ranging in size from *c*.3 km^2 to *c*.300 km^2'. Based on the latter definitions, we will equate landscape with agroecological matrix (i.e. area encompassing many agroecosystems) and define region/area (the largest scale we are using) as any area bigger than 3 km^2. Our next hierarchical level (patch or single orchard) was defined by Pearson (2002) as '. . . a continuous region of the same type of habitat . . .'. In the case of habitat, we adopt the definitions of Pearson (2002), '. . . sites having appropriate levels of the biotic . . . and abiotic . . . features required by a given species . . .' and Fischer and Lindenmayer (2007), 'the range of environments suitable for a particular species'.

From the above, it follows that any of these spatial units can provide suitable conditions for pestiferous insects, which potentially could be manipulated.

Region/area

In the case of highly mobile insects such as fruit flies, the only effective approach to long-term successful control/management is to view the problem from an 'area-wide' perspective (Klassen, 2000; Lindquist, 2000). For example, in the case of fruit flies, it has been shown that they are able to move between landscapes (agroecosystems), patches (orchards) or single trees on a seasonal or even daily basis (Aluja and Birke, 1993; Aluja *et al.*, 1997b; Nestel *et al.*, 2004). Given the large spatial scale involved, environmental manipulation schemes are difficult to apply, but there are some options. For example, through the strategic planting of trees that are hosts of non-pestiferous flies (e.g. *Ximenia americana* L., a host of *A. alveata* (Stone); Piedra *et al.*, 1993), one can enhance generalist parasitoids (e.g. *Doryctobracon areolatus* (Szepligeti)) that are also able to reduce populations of pestiferous species (e.g. *A. obliqua* (Macquart)). For fruit-fly species amenable to sterile insect technique (SIT) programmes, such a scheme can be well integrated with sterile male releases over large areas where pest population suppression is achieved by targeting two developmental stages (larvae and adults) (see Aluja, 1999, for further details on this environmentally friendly fruit-fly management scheme, or Verkerk *et al.*, 1998, and Tscharntke *et al.*, 2007, for a treatment of the topic on a broader scale).

On occasion, large-scale host-plant eradication schemes have been followed successfully, particularly in very dry environments suitable for the establishment of 'fruit-fly-free areas' (USDA, 1999). Such a scheme, although effective, may be questioned from an environmental perspective, particularly if the trees being eradicated are native and, as a consequence, serve as resources to local fauna. Additionally, and as discussed by Aluja and Mangan (2008), if females of highly polyphagous species such as *A. ludens*, *B. dorsalis* or *C. capitata* are confronted with a scarcity of hosts, they will lay eggs readily into fruit/vegetables outside their host range. Therefore, the strategy of stripping all native hosts can backfire, as females will be forced to lay eggs in the remaining fruit or vegetables. We believe that it would be wiser to use native hosts as population sinks. For example, large numbers of parasitoids can be released strategically in strips of native vegetation, or at the peak of the fruiting period. In addition, targeted sprays with environmentally friendly insecticides such as Spinosad® (Burns *et al.*, 2001; Stark *et al.*, 2004) can control entire pestiferous fruit-fly populations effectively without the need to eradicate native vegetation of value to other types of insects and animals (e.g. birds, mammals).

Landscape/agroecological matrix

The critical role of landscape structure/fragmentation in driving the dynamics of biological processes, including the presence/absence and abundance of insect pests and their natural enemies, has been reviewed by Turner (1989, 2005), Saunders *et al.* (1991), Forman (1995), Marino and Landis (1996), Khan *et al.* (1997), Fahrig (1998, 2003), Weibull *et al.* (2003), Cronin and Reeve (2005), Thies *et al.* (2005), Bianchi *et al.* (2006), Klein *et al.* (2006), Marino *et al.* (2006), Anton *et al.* (2007), Chust *et al.* (2007), Elzinga *et al.* (2007), Rauch and Weisser (2007) and Tscharntke *et al.* (2007). Among the most important landscape characteristics shaping these dynamics, the following stand out: degree of landscape patchiness and patch distribution, as well as patch size, shape and structure (Andrén, 1994; Mazerolle and Villard, 1999; Fahrig, 2003; Fischer and Lindenmayer, 2007); landscape connectivity (distance, isolation and degree of connectivity) (Beier and Noss, 1998; Fahrig, 2003; Haddad *et al.*, 2003; Chetkiewicz *et al.*, 2006; Falcy and Estades, 2007; Fischer and Lindenmayer, 2007); type of vegetation present in any given matrix (area in which the patches are embedded) or in the patch itself (Fahrig, 2003; Turner, 2005; Fischer and Lindenmayer, 2007).

There are two life-history attributes that need to be considered when trying to manipulate environmental conditions at the landscape level (and also at the patch (orchard) levels) to lower the damage of fruit flies: number of generations per year (univoltine versus multivoltine species; details in Zwölfer, 1983) and host breadth (monophagous, stenophagous, oligophagous and polyphagous species; details in Aluja and Mangan, 2008). There are many pestiferous fruit flies representing the first two categories: univoltine versus multivoltine. In general terms, species living in temperate environments are functionally univoltine, even though a few second-generation individuals emerge that are usually unable to reproduce (Phipps and Dirks, 1933; Feder *et al.*, 1997). The most notorious example is the apple maggot fly, *Rhagoletis pomonella* (Walsh). Other examples are *R. completa* (Cresson) (walnut husk fly), *R. mendax* (Curran) (blueberry maggot) and *R. tomatis* (Foote) (South American tomato fly). As noted by Zwölfer (1983), these flies exhibit long diapause periods and their emergence schedules are synchronized closely with the fruiting phenology of their hosts. As we will discuss later, this offers opportunities for environmental manipulation, for example, through the strategic timing of fruiting phenology (Yokoyama and Miller, 1993, 1994).

The large majority of pestiferous, polyphagous species are multivoltine and highly mobile (Zwölfer, 1983). The most notorious example is represented by the

Medfly, *C. capitata*, which, under particular conditions, can complete up to 13 generations in a year (Copeland *et al.*, 2002). This situation complicates management schemes based on environmental manipulation greatly, as individuals are able to exploit a wide variety of resources over the season and survive harsh conditions thanks to their mobility and physiological/behavioural plasticity. Nevertheless, detailed knowledge on population dynamics, seasonal dynamics of host-use patterns and the influence of seasonal climatic conditions on the life cycle of multivoltine polyphagous fruit-fly species could offer the opportunity to identify critical periods and key locations as targets for the application of area-wide suppression techniques that will impact pest populations at the landscape level in their most fragile stage.

With respect to host breadth, with a few exceptions (e.g. *A. grandis* (Macquart), *Toxotrypana curvicauda* Gerstaecker), most tropical pestiferous fruit flies are either oligophagous or polyphagous (attacking plants within a single family or of many families, respectively) (Aluja and Mangan, 2008). In general terms, the lower the number of host plants attacked, the easier it is to apply environmental management schemes.

Agroecosystem (single orchard or patch)

The next level in our hierarchical spatial scale is represented by the agroecosystem, which is inserted artificially into the landscape. It can be a little perturbed area covered with native vegetation or an agroecological matrix consisting of many agroecological units (single orchards/patches) that can be located next to each other or distributed patchily. We note that, functionally, an agroecosystem can be equated with a patch and we will therefore treat them as the same ecological unit. Importantly, in the case of highly mobile, polyphagous fruit flies, most of the factors driving their population dynamics and behaviour are found outside the agroecosystem. Many resources such as shelter, adult food and, most importantly, native hosts (or feral fruit) are found in the native vegetation surrounding orchards (details follow). The latter both complicates and facilitates environmental manipulation. On the one hand, the fact that flies move between the native vegetation (or/and abandoned orchards) and semi-commercial and commercial orchards can be used to our advantage, as individuals can be intercepted/trapped out during the process (Fig. 7.2). The species that are more amenable to the application of this strategy are those in which adults move in and out of orchards (e.g. papayas, mangos) or plantations (e.g. cucurbits, peppers) on a daily basis: *A. grandis* (Silva and Malavasi, 1993), *A. obliqua* (Aluja and Birke, 1993), *B. cucurbitae* (Coquillett) (Ebeling *et al.*, 1953), *Dacus* (*Didacus*) *frontalis* (Becker) (Steffens, 1983), *T. curvicauda* (Aluja *et al.*, 1997a,b), *Zonosemata electa* (Say) (Boucher *et al.*, 2001). But, it can also be applied in the case of species in which adults move from native vegetation surrounding orchards into orchards and then stay there for variable amounts of time: *R. pomonella* (Aluja and Prokopy, 1992; Rull and Prokopy, 2003, 2004a,b), various *Anastrepha* species (Aluja *et al.*, 1996; Kovaleski *et al.*, 1999), *B. dorsalis* (Chen *et al.*, 2006) and other insects (Bhar and Fahrig, 1998; Landis *et al.*, 2000; Nault and Kennedy, 2000).

There are two key characteristics of agroecosystems/patches that either complicate or facilitate environmental management schemes: size and structural

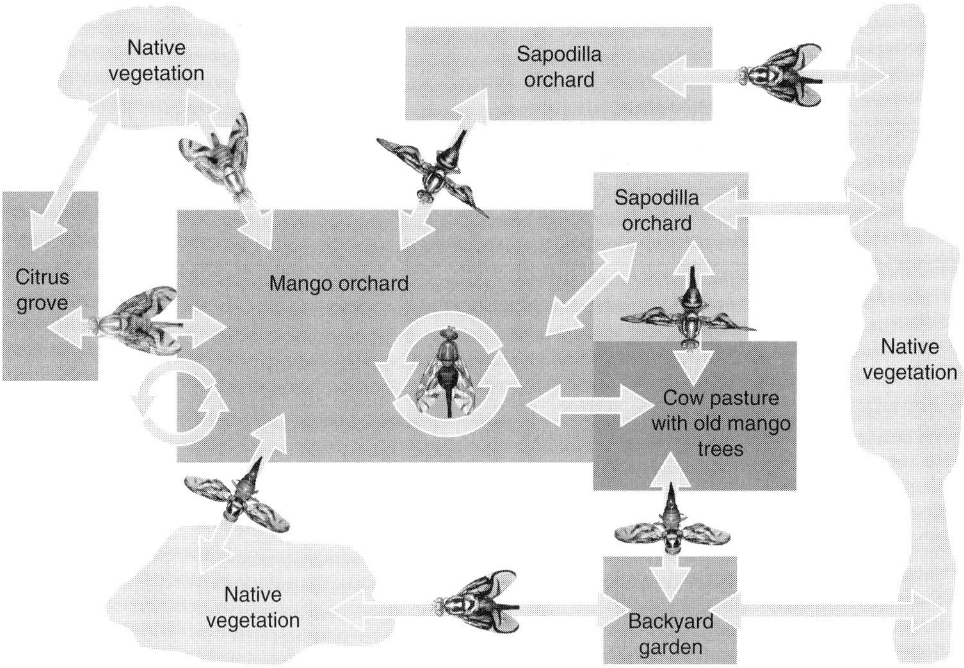

Fig. 7.2. Movement of individual fruit flies among the different ecological/agroecological units composing an agroecological matrix. Flies move between units seeking food, shelter, mating sites and oviposition substrates.

diversity or heterogeneity, *sensu* Gould and Stinner (1984) and Addicott *et al.* (1987) (also see Forman, 1995; Turner, 1989; Estevez *et al.*, 2000; Fischer and Lindenmayer, 2007). Agroecosystem size and shape can be highly variable and agroecosystems can be separated by regions/areas with unsuitable environmental conditions or be interconnected by 'benign' corridors that allow insects to move among/between agricultural areas of various sizes (Forman, 1995; Chetkiewicz *et al.*, 2006) (Fig 7.1). Following Gould and Stinner (1984), heterogeneity can be of two types: presence–absence (temporal variability with respect to areas with and areas without a given resource) and qualitative heterogeneity (temporal or spatial variability in the quality of a given resource, for example, trees exhibiting variable levels of toxic allelochemicals). For the purposes of this chapter, we distinguish further between three functionally distinct types of orchards: monocultures, polycultures containing only trees of commercial value, and polycultures that contain not only trees of commercial value but also other types of tree, some of them non-fruiting ones, such as timber of various types. Such polycultures are the norm among resource-poor farmers in Latin America, Africa and Asia, who favour highly diverse mixtures of fruit trees interspersed with annual crops, herbs and medicinal plants for cultural and economic reasons (Michon *et al.*, 1983; Fernandes and Nair, 1986; Lambin *et al.*, 2003). For example, it is common to find a papaya grove with mango (*Mangifera indica* L.),

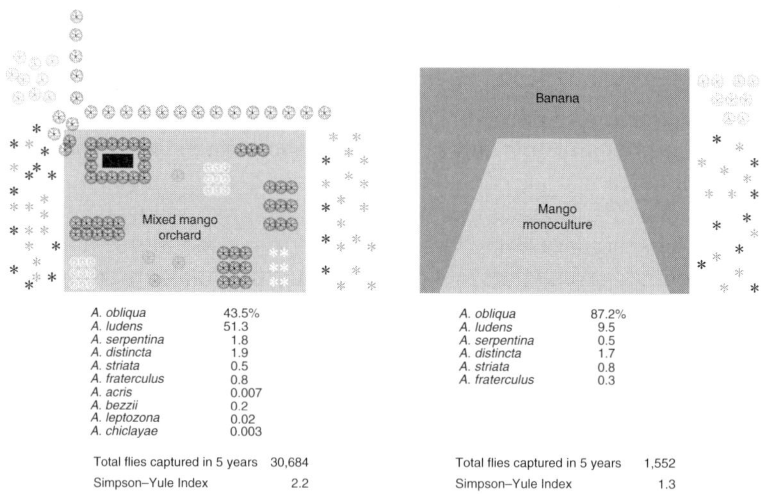

Fig. 7.3. Effect of type of orchard (i.e. mono- and polyculture) on total number of flies captured over a 5-year period and guild composition (diversity) (information published originally in Aluja *et al.*, 1996). Trees adjacent to orchards are either alternative hosts (e.g. *Spondias purpurea*, *S. mombin*, *Citrus* spp., *Psidium guajava*) or non-hosts.

avocado (*Persea americana* Mill) and guanabana (*Annona muricata* L.) trees interplanted or forming mini-patches of non-host vegetation inserted into the grove (examples in Aluja and Liedo, 1986; and Aluja *et al.*, 1997a). Such structure will have an impact on fruit-fly guild composition (Aluja *et al.*, 1996; Fig. 7.3) and influence host use by generalist species. Consequently, host-plant preference hierarchies and seasonality will influence seasonal trap efficiency and crop damage directly (Papadopoulous *et al.*, 1998, 2001).

What sorts of resources do flies seek outside single-species patches (i.e. monocultures) of cultivated trees? As documented by Back and Pemberton (1918a,b), Baker *et al.* (1944), Dean and Chapman (1973) and more recently by Aluja and Birke (1993), Aluja *et al.* (1997a), Boucher *et al.* (2001) and Shelly and Kennelly (2007), flies seek refuge from the elements (i.e. a more benign microclimate during the hottest hours of the day), food and mating sites among hosts and non-hosts. For example, Boucher *et al.* (2001) documented the fact that *Z. electa* adults used sugar maples (which they preferred over choke cherry trees) as shelter sites or possibly as sites where they found food (e.g. insect honeydew, bird droppings or other sources of essential nutrients) (Hendrichs and Prokopy, 1990). Earlier, Aluja and Birke (1993) showed that during the hottest hours of the day, *A. obliqua* females and males were found in mango trees, where the dense canopy offered ideal microclimatic conditions. But early in the morning and late in the afternoon, flies were found mainly in neighbouring tropical plum (*Spondias purpurea* L.) trees.

Single tree/single fruit

The last hierarchical level of interest to us is a single tree containing many fruit or a single fruit within such a tree (Fig. 7.1). Both have functional significance when it comes to analysing ways of manipulating environmental conditions to manage pestiferous fly populations or also individuals; for example, fruit quality in the internal or external parts of the tree canopy, as described by Sivinski *et al.* (2004, 2007). This hierarchical level is important from a management perspective because it has been found that within-tree trap positioning can be optimized by managing the abundance of visually competing fruit surrounding traps (Rull and Prokopy, 2004b).

Biotic and abiotic factors that influence fruit-fly behaviour in nature and in a fruit orchard

We note that any environmental manipulation scheme for the management of pestiferous and non-pestiferous fruit flies hinges on a thorough understanding of their behavioural ecology. We refer the reader to Potting *et al.* (2005) and Roitberg *et al.* (Chapter 3, this volume) for an erudite treatment of this topic. To avoid overlap, we will concentrate only on the particular environmental elements that can influence fly behaviour in unmodified environments or modified agricultural settings.

Effects of seasonality on fly appearance and abundance

Tephritid females rely on ephemeral resources (fruit, flower heads or perennial plants) that vary along the season in terms of suitability for oviposition or larval development (Sivinski *et al.*, 2007; and references therein). Fruit-fly life strategies have therefore evolved to match host-plant phenology. Species inhabiting temperate habitats with cold winters undergo diapause. It has been found that diapause is influenced by a number of factors. Dirks (1935) first studied the effect of apple variety on larval production and adult emergence of apple maggot flies. The effects of temperature, relative humidity, photoperiod and food on diapause regulation have been studied by Neilson (1962, 1964), Prokopy (1968a) and Dean and Chapman (1973). Pre-winter and winter duration effects on diapause of apple maggot fly host races were studied by Feder *et al.* (1993, 1997). Teixeira and Polavarapu (2004) provided evidence of heat-induced quiescence during pupal development of the blueberry maggot. Interest in such factors arises due to their implications for forecasting fly pest species emergence (Reissig *et al.*, 1979; Teixeira and Polavarapu, 2001) and on their effects as pre-mating barriers during host-race formation (Filchak *et al.*, 2000). An accurate forecast of adult fruit-fly appearance, coupled with the determination of economic thresholds, contributes to optimizing the timing of chemical treatments or the deployment of traps for the interception of adults moving into orchards, significantly lowering damage.

Tropical and subtropical fruit-fly species are often multivoltine but experience population peaks during the fruiting period of their main hosts (Jirón and Hedström, 1991; Celedonio-Hurtado *et al.*, 1995; Aluja *et al.*, 1996). In areas where several species of fruit flies cohabit, species composition in orchards is governed by the relative abundance of host-tree species and variety within the orchard (Aluja *et al.*, 1996; Fig. 7.3). For example, the availability of fruit of different host species and its fruiting succession affects Medfly abundance (Israely *et al.*, 1997; Papadopoulous *et al.*, 2001, 2003). In tropical regions with monsoon weather events, population peaks often occur during the onset of the rainy season, although this phenomenon cannot be uncoupled from the fact that many fruit-tree species also produce fruit during this period (Celedonio-Hurtado *et al.*, 1995; Ronchi-Teles and da Silva, 2005). Adult tephritids in such climates may undergo aestivation or dormancy during periods of fruit scarcity and adverse climatic conditions (Bressan-Nascimento, 2001), particularly during periods of drought, although no diapause has been reported (Aluja *et al.*, 1998). Furthermore, mean daily temperature has been found to be an important factor in explaining melon fly population density (Inayatullah *et al.*, 1991). Temperature, humidity and precipitation have also been found to influence Medfly abundance in California (Barry *et al.*, 2004), although in dry climates, such factors could have been influencing the response of flies to dry and wet traps and therefore trap capture would not necessarily be an accurate estimate of population size. Some subtropical and tropical fruit flies such as the Medfly and the South American fruit fly (*A. fraterculus*) remain in fruit as larvae in cold periods during winter, developing at a very slow rate (Papadopoulous *et al.*, 1996; Israely *et al.*, 1997; Kovaleski *et al.*, 1999).

Factors influencing daily adult activity patterns

Fruit-fly daily activity patterns are shaped by a large number of biotic and abiotic factors that often interact in complex ways. A thorough understanding of how such factors shape fruit-fly and natural enemy behaviour and their interactions is crucial in developing efficient integrated biorational management schemes, particularly those based on environmental manipulation.

Daily adult activity is often reduced at the time of greatest ambient temperature (Aluja and Birke, 1993). Flies of different species have been found to seek shelter during such periods (Aluja and Birke, 1993; Aluja *et al.*, 1997b). Mating activity often takes place at dusk or early morning, perhaps to avoid the hottest part of the day (Aluja *et al.*, 2000). Male calling behaviour in *Anastrepha* and *Toxotrypana* can be influenced by weather, with the onset being delayed during cool days (Aluja *et al.*, 1997b, 2000). Weather has to be taken into account when performing releases for SIT, as high levels of mortality and/or poor mating performance are to be expected during periods of extreme temperature. Tephritid larvae have been found to exit fruit subject to solar radiation before abscission (Aluja and Birke, 1993; Sivinski *et al.*, 2007), and high levels of larval mortality have been observed on soil devoid of cover and exposed to radiation at midday (Aluja *et al.*, 2005). Due to such factors, larvae appear to exit fruit during the early hours of the day (Hodgson *et al.*, 1998; Aluja *et al.*, 2000).

Factors influencing immature stages

Fruit temperature can influence the survival of larvae inside fruit (Thomas, 1993; Sivinski *et al.*, 2007) and weather, ground cover and soil compaction can affect ant predation of tephritid larvae as they exit the fruit and burrow into the ground in tropical climates (Aluja *et al.*, 2005). Pupation depth is affected by factors such as soil temperature, soil type, compaction and moisture (Guillén *et al.*, 2002; Dimou *et al.*, 2003) and exposure to predators (Aluja *et al.*, 2005). It has been found that pupal mortality can be affected by soil moisture, with dry and saturated soils causing greater mortality (Hulthen and Clarke, 2006). Therefore, larvae may actively seek shaded pupation areas or areas with higher soil moisture to avoid dessication (Thomas, 1995; Alyokhin *et al.*, 2001; Hulthen and Clarke, 2006). Although no records of ploughing as a management technique for fruit flies exist, turning the top soil under the canopy of host trees at the end of the fruiting season could expose pupae to ground predators and desiccation. Nevertheless, such a procedure could lead to horticultural problems if ploughing damages the root system, facilitating the penetration of disease agents.

Effects of food sources

Adult food availability and acquisition affects different aspects of fly behaviour. For example, the acquisition of protein by adult tephritids results in sexual maturation and influences sexual behaviour/success (Hendrichs and Prokopy, 1990; Drew and Yuval, 2000; Aluja *et al.*, 2001a,b). Mating success, sperm transfer and duration of female mating refractory periods are greater for protein-fed than for protein-deprived male Medflies (Blay and Yuval, 1997) and in various *Anastrepha* species (Aluja *et al.*, 2001b; Pérez-Staples and Aluja, 2006; Pérez-Staples *et al.*, 2008), a fact that may influence the effectiveness of sterile males used in SIT programmes (Kaspi and Yuval, 2000). Male Caribbean fruit flies, *A. suspensa* (Loew), deprived of food do not engage in calling activity (Landolt and Sivinski, 1992), while protein-fed male and female apple maggot flies are more responsive to synthetic fruit volatiles than protein-starved flies (Rull and Prokopy, 2000). Apple maggot fly adults may be arrested in areas containing important amounts of bird faeces (Rull and Prokopy, 2000). Furthermore, females of species that do not mate on host fruit become responsive to host-fruit odour once they have acquired protein and matured sexually (Prokopy *et al.*, 1991). Importantly, natural sources of food have been found to outcompete protein baits in attracting flies (Prokopy *et al.*, 1992, 1993a,b).

For almost a century, glass McPhail traps baited with an aqueous slurry of torula yeast (or other liquid proteins of various types) have been used to monitor flies in the genus *Anastrepha* and *Ceratitis*. The McPhail trap (McPhail, 1939) is an excellent example of how monitoring mechanisms can modify the environment in which fruit flies live. Cunningham *et al.* (1978), working in low, intermediate and high rainfall areas in Hawaii, reported that this liquid-based trap captured significantly more flies in the low rainfall areas. The authors speculated that such a phenomenon was related to the fact that, in dry areas, the McPhail trap represented a water and food source for flies. Since, as noted above, females need protein for ovary development, it is not surprising that McPhail traps capture greater proportions of immature females than males.

Biotic and abiotic mortality factors at the orchard level (e.g. sunshine, ants, spiders, parasitoids)

The role of biotic and abiotic factors as fruit-fly mortality agents was last reviewed by Debouzie (1989) and Meats (1989). Partial updates have been provided by Thomas (1993, 1995), Hodgson *et al.* (1998) and Aluja *et al.* (2005). An overview of the predators related to an entire pest complex in fruit orchards can be found in Solomon *et al.* (2000). A thorough understanding of such mortality factors and their potential interactions can lead to the successful development of biorational management schemes (examples in the section 'Managing fruit flies through environmental manipulation', this chapter). One of the clearest and most widely adopted examples in the case of annual crops is the use of predators in regulating secondary pests in rice plantations in South-east Asia (Matteson, 2000).

Chemicals produced by plants and fruit flies

These two critical components of the environment in which flies live will be discussed in the next section, as they serve as the basis for the development of monitoring tools which, in turn, become part of environmental manipulation schemes.

Response to visual and chemical stimuli and the development of monitoring/environmental manipulation tools based on this information

Chemical and visual cues are two of the most important environmental stimuli that influence/modulate fruit-fly behaviour in nature that are prone to manipulation. But, the response to such stimuli is complex as flies often respond to the interaction between chemical and visual cues. For example, once a host-seeking female has landed on a fruit-bearing tree based on orientation toward the chemical cues emanating from such a tree (e.g. Aluja and Prokopy, 1992), she resorts principally (but not exclusively) to chemical cues if the visual stimuli emanating from the host are weak (green, unripe fruit), but relies mainly on vision if the host fruit is fully ripe and, therefore, the visual stimulus is strong (Aluja and Prokopy, 1993).

In two thought-provoking publications, Jang and Light (1996) and Light and Jang (1996) discuss the role of 'tephritid olfactory semiochemicals'. They note that:

> Olfactory semiochemicals initiate, control or modify various tephritid searching/foraging behaviors, including host-plant- and habitat-finding via plant kairomones, food foraging via food odours, mate finding, mate selection and courtship via sex pheromones, male aggregations via plant kairomones/parapheromones, ovipositional site foraging and host selection via fruit kairomones and dispersal and avoidance of intraspecific competition via epideictic host-marking pheromones.
>
> (Jang and Light, 1996)

These authors coined the term 'host-plant habitat volatiles' when discussing the highly ubiquitous 'green-leaf volatiles' (GLVs), principally six-carbon aliphatic aldehydes and alcohols (e.g. monoterpenes, sesquiterpenes) emitted by leaves and immature unripe fruit that most likely lead adult flies to particular patches of vegetation (Light and Jang, 1996). It would be completely unreasonable to think of ways to use these types of chemicals in environmental manipulation schemes against fruit flies, as they are so widespread and overwhelmingly abundant in nature. Once fruit ripen, another group of chemicals are emitted (mainly six- and eight-carbon esters such as ethyl hexanoate), allowing gravid females to locate them (Aluja and Prokopy, 1993; Light and Jang, 1996; and references therein). These types of chemicals, much more specific and usually associated with a particular plant species, are tractable by a fruit fly and therefore potentially prone to manipulation. But, despite the latter, it must be noted that in habitats/patches/individual plants, many other organisms can be found that contribute to the chemical environment in which flies live and to which they respond. In particular, bacteria, insects and birds produce exudates/faeces emitting attractive odours to fruit-fly adults. Frequently, it is the combination of fruit volatiles and these other types of chemicals that attract flies to a patch of trees or to a particular tree (Epsky *et al.*, 1997; Robacker *et al.*, 1998, 2004; Robacker and Lauzon, 2002; Robacker, 2007).

In what follows, we will provide a brief overview of the most influential studies exploring responses to visual and chemical stimuli by fruit flies, and on follow-up research incorporating such knowledge into the development of monitoring/control tools that modify/manipulate the environment in which flies live. In doing so, we place special emphasis on recognizing the pioneers who inspired R.J. Prokopy and his students and many other scientists throughout the world. For historical and in-depth technical reviews on this work, we suggest Prokopy and Owens (1983), Owens and Prokopy (1986), Metcalf (1990), Aluja and Prokopy (1993), Green *et al.* (1994), Epsky and Heath (1998) and Díaz-Fleischer and Aluja (2000).

Fruit volatiles

We commence with chemical cues, as these are the ones a foraging fly perceives from a distance. Early work by Prokopy *et al.* (1973) showed that apple maggot flies appeared to be attracted by volatiles emitted by ripe fruit placed in inconspicuous bags hung on fruitless host trees. Later, Fein *et al.* (1982), Carle *et al.* (1987), Aluja and Prokopy (1993), Aluja *et al.* (1993) and Zhang *et al.* (1999), among others, documented the response of apple maggot fly adults to different apple volatiles and blends thereof. Furthermore, it has been found recently that non-host volatiles mixed with host volatiles might have antagonistic effects (Linn *et al.*, 2005), suggesting that adults not only may be attracted from long distances by host volatiles, but also may be repelled from sources emitting non-host volatiles (Feder and Forbes, 2007). This finding opens up an interesting avenue of research related to environmental manipulation schemes aimed at managing pestiferous fruit flies.

Prokopy and Vargas (1993) showed that Medflies of both sexes were attracted to the odour of ripe crushed coffee (protein-fed females to a greater degree), although attraction in the field was not greater than that of conventional protein baits. Such a finding was also reported by Robacker and Heath (1996) for irradiated, sugar-fed *A. ludens* in response to volatiles of yellow chapote, and for *B. dorsalis* in field cages and in the field (Cornelius *et al.*, 2000; Sidehurst and Jang, 2006). It would appear from these results that the response of generalist fruit flies to host-fruit volatiles is not as specific and strong as that of specialist, monophagous *Rhagoletis* species. In this respect, Light *et al.* (1988) recorded electroantennogram responses of Medfly to compounds typically emanating from fruit and foliage, reporting greater response to 'general green-leaf volatiles'.

Behaviour-modifying chemicals produced by fruit flies (e.g. sexual pheromones, host-marking pheromones)

The behaviour and nature of the response to sexual pheromones has been reviewed by Metcalf (1990), Landolt and Phillips (1997), Heath *et al.* (2000) and Sivinski *et al.* (2000b). Detailed information on various aspects related to host-marking behaviour and host-marking pheromones (HMPs) can be found in a number of reviews or research articles (Prokopy, 1981; Roitberg and Prokopy, 1987; Boller and Aluja, 1992; Aluja *et al.*, 2000; Díaz-Fleischer *et al.*, 2000; Nufio and Papaj, 2001; and Aluja and Díaz-Fleischer, 2006). The phenomenon was described originally by Martelli (1909) and later redescribed by Cirio (1971).

Prokopy (1972) reported the host-marking behaviour of *R. pomonella*, suggesting that females deposited a pheromone deterring other females from ovipositing in occupied fruit. Prokopy later explored the existence of similar HMPs in other species of *Rhagoletis* (Prokopy *et al.*, 1976) and Medflies (Prokopy *et al.*, 1978). Studies on the function in nature of these types of chemicals (Averill and Prokopy, 1987) and on the behavioural reactions of foraging females to marked fruit (e.g. Roitberg *et al.*, 1984; Aluja and Díaz-Fleischer, 2006) followed. Katsoyannos and Boller (1976, 1980) and Aluja and Boller (1992) studied the possible applications of this behaviour-modifying chemical by evaluating the response of *R. cerasi* L. females to synthetic isomers (details in the section 'Push–pull strategy, this chapter).

Variability in the susceptibility to attack by fruit flies of various hosts and host cultivars

Susceptibility to different cultivars of apples to apple maggot flies was the subject of much attention early in the 20th century (Dirks, 1935; Dean and Chapman, 1973). The idea that early cultivars of apples were more attractive to apple maggot flies than mid- and late-ripening cultivars arose as a generally held principle (Murphy *et al.*, 1991) because early cultivars were susceptible to fly attack when apple maggot fly populations peaked. However, such a view was later questioned, based on the fact that early, mid- and late-ripening cultivars could be both attractive and unattractive to adults and susceptible and unsusceptible to oviposition by females (Rull and Prokopy, 2005). For a recent review on the topic, see Aluja and Mangan (2008).

Several studies have dealt with listing and ranking hosts based on the degree of susceptibility to fruit-fly attack. Unfortunately, some of this literature is mired with anecdotal accounts, requiring substantial effort to clarify true host range for species of quarantine importance (Aluja and Mangan, 2008). Competitive displacement of fruit flies resulting in host-range modifications has been reviewed by Duyck *et al.* (2004). Establishing host and cultivar preferences is important from a management perspective because it enables predictions on fly distribution in orchards, which can be exploited for optimal trap deployment, directioning bait sprays or augmenting predators in areas of adult aggregation (details follow).

Management Possibilities and Ideas: From Theory to Practice

Managing fruit flies through environmental manipulation

Having reviewed most of the factors that influence fly behaviour in nature or that influence responses to monitoring tools, we will discuss novel methods to potentially manipulate some of these factors to manage pestiferous fruit flies (for a broader treatment of the topic, we refer the reader to the recent review by Foster and Harris, 1997). In congruence with our discussion on the need to consider the issue of scale or spatial units when applying environmental manipulation schemes for fruit-fly management, we will start with the broadest spatial scale (region) and finish with the smallest (single fruit).

The sterile insect technique (SIT) and mating disruption

Methods affecting the reproductive capacity of pests directly have been used on a number of tephritid species. The most common method is SIT, which consists of rearing, sterilizing and releasing large numbers of males of the target pest species in areas with large wild populations to reduce their reproductive capacity (Knipling, 1955). Because sterile males have to survive, move, attract, court and mate with wild females, behaviour is a critical component of the success of SIT (Hendrichs *et al.*, 2002). When such factors are taken into account, SIT has proven to be an outstandingly effective area-wide approach with minimal negative environmental impact (Klassen and Curtis, 2005).

One could ask, 'How does SIT relate to environmental manipulation?' By releasing millions of sterile males to suppress populations on a large spatial scale, one is altering the environment. For example, in extremely dry areas, the effect of such large numbers of insects can be such that food sources for wild flies may dwindle. In general terms, we believe that SIT represents the management technique that is most amenable to the largest spatial scale being considered here: the region. Actually, for SIT to achieve its highest potential, it must be applied over entire regions, often encompassing two or more countries (Hendrichs *et al.*, 2002; Vreysen *et al.*, 2007).

An additional strategy that can be applied over entire regions, in this case, fruit-growing regions, is mating disruption through the use of artificial pheromone sources. Mating disruption, as applied against many moth species (Cardé and Minks, 1995) has, to our knowledge, only been tested successfully in the case of *B. oleae* (Montiel-Bueno and Mata, 1985). But recently, a novel twist was proposed: cross-species disruption. In this case, interrupting the mating activity of one species is achieved by using another species as a mercenary agent (Suckling *et al.*, 2007). The method consists of treating one species with a sufficiently powerful attractant for a second species, interfering with mating of the second species by leading males astray in pursuit of false trails created by dosing individuals of the first species. Reciprocal test systems used were: (i) methyl eugenol, an attractant for male Oriental fruit flies (*B. dorsalis*), applied to melon flies (*B. cucurbitae*); and (ii) cue lure, a lure for male melon flies, applied to *B. dorsalis*. The experiment achieved a significant reduction in the mating of melon flies (Suckling *et al.*, 2007).

Bait stations/male annihilation

Bait stations have been defined as 'discrete containers of attractants and toxins targeting specific pests' (Mangan and Moreno, 2007). Bait stations have been employed extensively in area-wide fruit-fly eradication programmes, based on the male annihilation technique for flies in the genus *Bactrocera* in the form of wooden or fibreboard blocks impregnated with methyl eugenol and a toxicant (see Hancock *et al.*, 2000, for a recent example). In SIT programmes against fruit flies in other genera, bait stations have been used for some time for population suppression before sterile male release. However, the only formal test of the efficiency of this method is very recent. Mangan and Moreno (2007) found that a piece of sponge material soaked in protein hydrolysate, sugar, adjuvants and a toxicant attached to a plastic cover suppressed 70–90% of the adult *A. ludens* population compared to a control. The main limitation of this technique is the loss of effectiveness of the station over time, a problem solved elegantly in the case of the apple maggot fly by developing attracticidal spheres (see Green and Wright, Chapter 9, this volume). Bait stations may become an important area-wide management tool alternative due to citizen resistance to pesticide cover sprays in urban areas (Moreno and Mangan, 2000; Mangan and Moreno 2007). We note, too, that this technique is not only amenable with the regional and landscape spatial scales proposed here, but can also be applied at the smaller patch scale.

Conservation biological control

Conservation biological control has been defined by Hajek (2004) as 'modification of the environment or existing practices to protect and enhance specific natural enemies or other organisms to reduce the effect of pests' (also see Barbosa, 1998; Pickett and Bugg, 1998; Eilenberg *et al.*, 2001). Broadly speaking, conservation biological control deals with habitat management to enhance/promote natural enemies by providing refuges in areas surrounding annual crops/orchards or within fields/orchards, enhancing cover crops, or by providing food and shelter for natural enemies (Barbosa, 1998; Picket and Bugg, 1998; Hajek, 2004). This pest management strategy can be applied at the landscape (i.e. agroecological matrix) and patch (= agroecosystem/single orchard) spatial scales.

In the case of fruit flies, there are several modalities of conservation biological control that involve parasitoids as well as predators. The case of parasitoids has been discussed by Aluja (1999) and Aluja *et al.* (2003b) and the case of predators, particularly wasps and ants, was addressed by Hendrichs *et al.* (1991, 1994) and van Mele *et al.* (2007).

PARASITOIDS General reviews on the use of parasitoids for fruit-fly control can be found in Wharton (1989), Sivinski (1996) and Ovruski *et al.* (2000). Historically, the most common approaches have been classical biological control (i.e. the introduction of exotic parasitoids for the long-term control of exotic pests) and augmentative (inundative)/inoculative biological control (i.e. short- to mid-term control of pests through repeated releases of large numbers of natural enemies). Suppression on a large spatial scale, over an entire region in southern Florida, was reported for the Caribbean fruit fly, *A. suspensa* (Sivinski *et al.*, 1996). Similarly, suppression of *A. ludens* and *A. obliqua* populations was demonstrated in large (i.e. 1600 ha) mango-producing areas of Chiapas, Mexico, through augmentative releases of the parasitoid, *Diachasmimorpha longicaudata* (Ashmead) (Montoya *et al.*, 2000). As a consequence, augmentative biological control in conjunction with SIT currently is applied over extensive areas in Mexico (e.g. mango-growing regions in the state of Sinaloa).

Real-world examples of conservation biological control in the case of fruit flies are nowhere to be found in the literature. But, Aluja (1999) proposed a theoretical scheme through which two goals could be achieved simultaneously: conservation of tropical trees and natural enemies (mainly native parasitoids) and enhancement/multiplication of native parasitoids. As noted before, the basic idea is that, in many rural areas of Latin America, there are certain host trees of *Anastrepha* species considered not to be economically important pests that are attacked by generalist parasitoids. After life cycle completion in larvae of, for example, the rare *A. alveata*, these parasitoids move to wild hosts of economically important *Anastrepha* species such as *A. obliqua*, exerting considerable pressure on local populations. In addition, some trees serve as reservoirs of as many as five parasitoid species. Many of these trees produce highly valuable timber and are disappearing locally at alarming rates. Therefore, the idea is to implement dual tree conservation–fruit-fly management programmes over entire regions that would help suppress pestiferous fly populations and, at the same time, conserve/recover populations of ecologically/economically 'useful' trees (Aluja, 1999).

WEAVER ANTS Recently, Peng and Christian (2005, 2006), van Mele and Cuc (2000, 2003), van Mele and Chien (2004) and van Mele *et al.* (2001, 2007) have documented the effectiveness of weaver ants of the genus *Oecophylla* in lowering the damage of flies within *Ceratitis* and *Bactrocera* in commercial mango orchards in Africa, Asia and Australia. For example, high abundance of the African weaver ant, *O. longinoda* (Latreille) significantly lowered the damage inflicted to mangos by *C. cosyra* (Walker), *C. quinaria* (Bezzi), *C. silvestrii* Bezzi and *B. invadens* Drew in Benin (van Mele *et al.*, 2007). Interestingly, they detected a cultivar effect on ant abundance, with 'Governeur' exhibiting the highest. The latter opens up the door for the selective planting of these types of trees to attract ants into the orchard, a mechanism that is amenable with the third level in our spatial-scale categorization: the patch (= agroecosystem/single orchard). Van Mele *et al.* (2007) report that 'although predation on adult fruit flies took place, deterrence and disturbance by ants during fruit oviposition seem to be the most important cause of reducing fruit-fly damage.' In the case of organic mango orchards, one could apply HMP (e.g. Aluja and Boller, 1992; Aluja and Díaz-Fleischer, 2006) which, in addition to the ant activity, most probably would reduce significantly oviposition activity by flies. Given that flies would be searching for additional oviposition sites, they could be trapped out (push–pull strategy; Miller and Cowles, 1990; Cook *et al.*, 2007) or killed in strategically placed bait stations.

WASPS Another scheme that potentially could be applied at the patch or single-tree spatial scales is the enhancement/local release of predatory wasps. Hendrichs *et al.* (1991, 1994) and Hendrichs and Hendrichs (1998) documented strong wasp predation on male Medflies mediated by wasp response to male sexual pheromone emission. In the case of females, they were preyed upon heavily while ovipositing or visiting leks. It is known that individual wasp foragers are influenced by past foraging experience and by the presence of other foragers on resources. A forager's ability to learn odours and landmarks, which direct its return to foraging sites, and to associate cues such as odour or leaf damage with resource availability, provide the behavioural foundation for facultative specialization by individual foragers (Richter, 2000). Under appropriate environmental conditions, wasp nest establishment could be favoured in fruit orchards and exposed to odours such as male pheromone, HMP and damaged fruit in order to foster specific predation and suppress pestiferous fruit-fly populations.

COLONIAL SPIDER WEBS The idea of using social spider webs to lower the number of ovipositing females was proposed by Hendrichs (1990) but has never been tested formally in the field. Nevertheless, we believe that it has potential in the case of low-income growers who maintain a few trees in their backyard gardens or in the case of small groves that potentially could be ringed by spider webs.

Social spiders (Arachnida: Araneae) are restricted to the subtropics and tropics, and are particularly common in the Neotropics and Africa (Burgess, 1978; Buskirk, 1981; Agnarsson and Kuntner, 2005; Agnarsson, 2006). In contrast to the majority of spiders, which are aggressive and solitary, most *Anelosimus* species are social, including all studied New World species. Species are placed in two social categories, both non-territorial (web-sharing): (i) subsociality, in which cooperative siblings and their providing mother temporarily share a nest; and (ii) quasisociality (sociality), with multiple adult spiders in semi-permanent communal webs (Agnarsson, 2004). Subsocial *Anelosimus* have nests usually containing between 20 and 100 individuals, whereas social nests can contain thousands of individuals (Agnarsson, 2006).

As noted above, we envision enhancing social spider webs in small backyard gardens in tropical areas of Latin America and Africa. Once established, they could work in conjunction with weaver ants and non-social spiders to exert pressure on fruit flies entering these backyard gardens or living there temporarily. Perhaps, colonial spider webs in border vegetation surrounding small orchards could act in a manner similar to interception traps for frugivorous tephritids entering orchards (see the section 'Trap cropping', this chapter, for further details), while within the orchards, ants and parasitoids could suppress immature stages and other predators (e.g. wasps) would act on adults bypassing peripheric webs.

Orchard design and within-orchard vegetation management

The possibility of designing orchards prior to being planted, and to manipulate vegetation structure within them to help reduce fruit-fly damage to fruit of commercial value was first addressed formally by Aluja and Liedo (1986). The idea is amenable with the agroecosystem/single orchard spatial scale (= patch) and applies to orchards in both temperate and tropical environments, in which growers are interested in planting several cultivars of one species (e.g. apples, mangos) to spread the harvest season or in planting several species of fruit trees. On one hand, such diversity can be viewed as a nuisance since it provides flies with ideal conditions throughout the year. If unchecked, populations can explode and reach unsustainable levels. On the other, it opens up the possibility of developing a refined management scheme that takes advantage of the diversity to manipulate the environment, generating suboptimal or negative conditions for both immatures and adults. For example, Aluja and Liedo (1986) discussed the possibility of designing orchards using blocks of certain tree species/cultivars thereof instead of a more haphazard location of trees. They used the example of a commercial mango orchard dominated by one mango cultivar. Between rows of these trees, or in orchard edges or certain sectors of the orchards, one could find trees of other mango cultivars or tree species (e.g. guavas, tropical plums, sapodilla and citrus). In the case of *A. ludens*, females prefer citrus over mangos when available (Aluja *et al.*, 1996). In the case of *A. obliqua*, they prefer tropical plums to mangos (Aluja and Birke, 1993). Females of this species also prefer certain mango cultivars to others, usually ones that have little commercial value but are nevertheless sought after by local people for cultural reasons. When these preferred hosts have finished fruiting, females move to the commercially valuable mango trees, inflicting significant damage to the crop. If such preferred trees could be planted strategically (Fig. 7.4), they could serve as population sinks if the grower was able to decimate populations by timing insecticide applications properly or by releasing parasitoids augmentatively. But, if such actions are not implemented or are applied late, the cost can be very high, not only to the owner of the orchard but also to the neighbours, as adults will move within the orchard or between orchards. Such an ability to move over large distances also raises another issue. If environmental manipulation schemes, or if fruit-fly management actions are not applied by all growers on a regional or landscape scale, the long-term result is that the overall effectiveness will be minimal. Nevertheless, we believe that if a small-scale grower is, indeed, able to maintain large populations of predators (e.g. colonial spiders, harvester ants, wasps) in his orchard, adult population levels will be kept at low enough levels to reduce to a maximum ovipositional activity by females. As noted by Aluja (1996) and Aluja and Piñero (2004), these growers can also resort to bagging their fruit once the critical fruit maturity level is reached. Bagging is used widely in Asia for high-quality, high-value fruit for lucrative export markets (Jia *et al.*, 2005), but also can be very cost effective in the case of resource-poor growers who are interested in the fruit for self-consumption only and have ready access to children who can climb trees to place bags over the fruit (Mosqueda-Vázquez *et al.*, 1996).

Another interesting alternative with respect to vegetation management is to plant trees that serve as shelter sites for flies and where a large number of adults will accumulate, facilitating their control with bait stations, traps or through the application of insecticides accepted for organic groves, such as Spinosad® (Burns *et al.*, 2001; Stark *et al.*, 2004). A good example was provided by Aluja *et al.* (1997a) in the case of the papaya fruit, *T. curvicauda*. In a mixed papaya orchard (as opposed to a monoculture), adults spent the night in densely foliated trees planted in the middle of the grove. In contrast, in a grove comprised of a monoculture, adults flew in and out on a daily basis between patches of native vegetation and the grove.

Finally, another alternative would be to plant repellent trees/bushes, but such a possibility is theoretical at this stage as there are no quantitative reports available in the literature on any non-host tree/plant that is effectively repellent against fruit-fly adults.

Trap cropping

At the landscape (= agroecological matrix) and patch (= agroecosystem/single orchard) scales, there are several environmental manipulation schemes that can be applied successfully. Among them, trap cropping, interception using traps, the selective

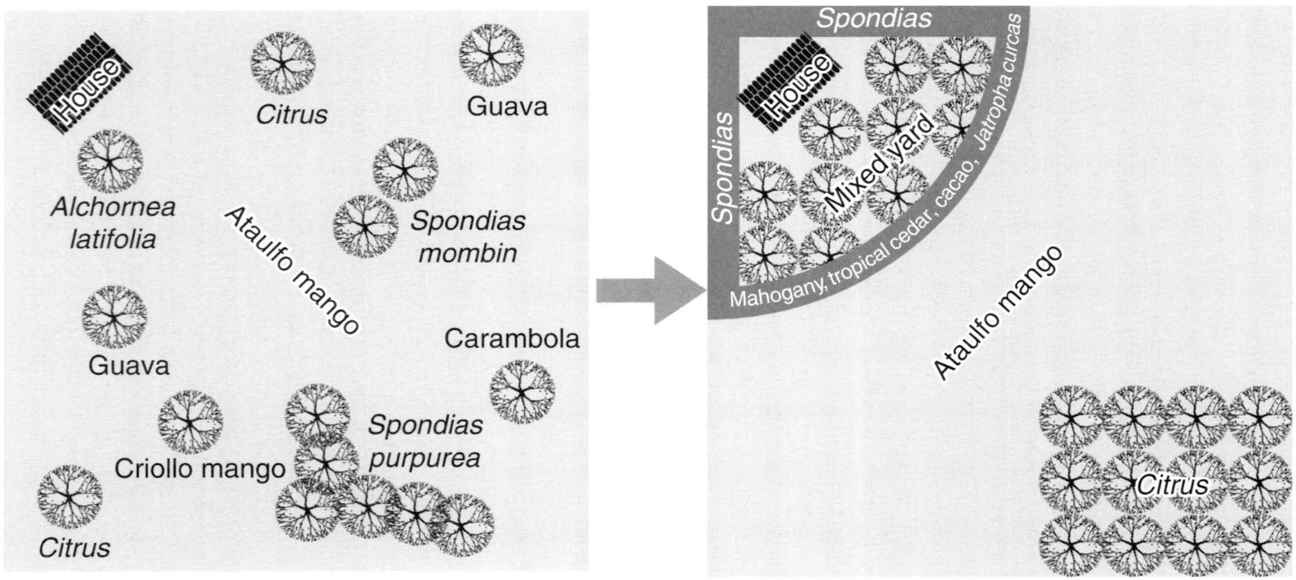

Fig. 7.4. Design of an orchard with the aim of maintaining high fruit-tree species diversity but facilitating management of pestiferous fruit flies (originally proposed by Aluja and Liedo, 1986). Two main commercially valuable fruit-tree species are planted on a large scale, while additional species of particular value to growers are compacted in a mixed yard next to the house and surrounded by a living fence containing valuable hardwoods or other plants valued locally.

application of insecticides, the push–pull strategy and manipulation of fruiting phenology hold strong potential and will be discussed in the following sections.

Trap cropping exploits hierarchical host-plant preferences by attracting and arresting pestiferous fruit flies in patches of cultivars or plants planted especially for that purpose. Trap cropping, as a means of lowering populations of pestiferous flies entering orchards from adjacent native vegetation, was first suggested by Steffens (1983) and formally tested by Aluja *et al.* (1997a) in the case of the papaya fruit fly, *T. curvicauda*. These authors concluded that designing papaya groves in such a way that a trap crop consisting of plants located 10 m from the main block of trees and entirely surrounding it, could reduce fly damage significantly. If pheromone-baited fly traps are hung in these peripheral rows, the trap crop effect can be enhanced and, as a result, damage to the commercial block of trees can be reduced even further. Such a fruit-fly management scheme should allow growers to avoid or reduce pesticide applications drastically and has been implemented successfully for the pepper maggot (Boucher *et al.*, 2003), and suggested in conjunction with interception trapping as a management tool for the apple maggot fly (Rull and Prokopy, 2004a, 2005).

VISUAL MANIPULATION Coloured traps placed in tree canopies can be viewed as a form of environmental manipulation at the tree level if we consider that supernormal visual stimuli alter fruit flies' perception of the environment. Such alterations, if well understood in the context of fruit-fly behaviour, can be used to direct the movement of flies to objects designed to capture them for monitoring or controlling during foraging bouts. In some instances, visual traps constitute the control agent per se (e.g. interception trapping, push–pull strategy) and therefore a thorough understanding of the visual ecology of target pests becomes the cornerstone of efficient control.

Visual responses of fruit flies to different stimuli were studied thoroughly by Prokopy, inspired by early work by Oatman (1964), who had found sticky-coated dark red apple cultivars and mimics thereof to be more attractive than yellow spherical objects. The generally held view that tephritids respond to yellow rectangular panels, an idea stemming from pioneering work by Kennedy *et al.* (1961) and Kring (1967) on other herbivorous insects, was also influential. In a landmark paper, Prokopy (1968b) described the responses of male and female apple maggot flies to rectangular panels of different colours, different shapes and different sphere sizes in the field. Dissection of captured females revealed that sexually immature females responded to yellow panels, while sexually mature individuals of both sexes were most sensitive to dark spherical objects contrasting against the background. Based on this pioneering research, many studies on the visual responses of other *Rhagoletis* species and tephritids in other genera followed (e.g. Prokopy *et al.*, 1975; Greany *et al.*, 1978; Nakagawa *et al.*, 1978; Cytrynowicz *et al.*, 1982; Hill and Hooper, 1985; Sivinski, 1990; Alviña *et al.*, 1993; Mayer *et al.*, 2000; Drew *et al.*, 2003; Piñero *et al.*, 2006).

Interception using traps

Fruit flies in different genera are attracted to orchards when fruit begins to ripen (Aluja and Prokopy, 1992; Aluja *et al.*, 1996; Rull and Prokopy, 2000; Meats *et al.*, 2002) and, as a consequence, traps placed in the periphery of orchards tend to capture greater proportions of foraging flies (Aluja *et al.*, 1986, 1996, 1997a; Papadopoulous *et al.*, 2001). Taking advantage of this fact, if a trapping device is attractive enough, flies can be intercepted in the periphery of an orchard. For example, flies in the genus *Rhagoletis* meet and mate on host fruit (Prokopy *et al.*, 1971) and, as a result, both sexes are attracted strongly to fruit volatiles (reviewed above). Attraction to such volatiles is strong enough to allow for their use in mass trapping and perimeter trapping (Prokopy and Mason, 1996), a method that can ensure more than 95% protection from fruit flies in small orchards (Prokopy *et al.*, 1996). Reynolds *et al.* (1998), Prokopy *et al.* (2003a, 2005) and Rull and Prokopy (2005) refined trap deployment methods to maximize trap efficiency. Currently, a strategy relying on attracticidal spheres (see Green and Wright, Chapter 9, this volume) achieves similar control as pesticide applications in small orchards.

Selective application of insecticides in border rows or in components of the native vegetation surrounding orchards/groves

This approach is not new as it has already been proposed by Nishida and Bess (1950) and Kazi (1976), to lower populations of *B. cucurbitae*, and by Aluja (1993) and Aluja *et al.* (1997a), to control various pestiferous *Anastrepha* species and *T. curvicauda*, respectively. The concept rests on the fact that newly eclosed adult fruit flies are attracted to sources of protein on which they necessarily need to feed to reach sexual maturity. McPhail (1939) first developed the idea of mixing such attractants with insecticides to kill fruit flies. Such an approach also has been applied extensively for male suppression of flies in the genus *Bactrocera* by replacing protein baits with methyl eugenol (Steiner *et al.*, 1965), a compound on which males feed voraciously (to the point where many die), and which plays a role in pheromone production (Nishida *et al.*, 1988) and mating success (Shelly and Dewire, 1994). Attracting foraging flies to certain sectors of a grove/orchard allows the possibility of foregoing cover sprays and concentrating pesticide applications in areas where adults are more likely to be (e.g. borders), therefore saving money and reducing environmental impact in various forms. A recent application of this concept was reported by Prokopy *et al.* (2003b), working with the melon fly (*B. cucurbitae*) in Hawaii. These authors were able to reduce melon-fly attack in experimental cucurbit plots significantly by spraying adjacent sorghum and sugarcane plantations with a bait spray.

Push–pull strategy

The 'stimulo-deterrent diversion' concept was reviewed by Miller and Cowles (1990) and, more recently, by Cook *et al.* (2007) and Hassanali *et al.* (2007). Push–pull strategies use a combination of behaviour-modifying stimuli to manipulate the distribution and abundance of pest and/or beneficial insects for pest management. Pests are repelled or deterred away from this resource (push) by using stimuli that mask host apparency or are repellent/deterrent. Pests are diverted away from the crop (pull) simultaneously, by using highly apparent and attractive stimuli such as traps (e.g. yellow panels) or trap crops, facilitating their elimination. Success has been achieved in the case of onions attacked by the onion fly (*Delia antiqua* (Meigen) (Diptera: Anthomyidae)) in the USA and with cereal stem borers in sub-Saharan Africa, by intercropping cereals with repellent and trap crops that can be used later as forage for livestock (Khan *et al.*, 2000).

In the case of fruit flies, this practice was tested formally by Aluja and Boller (1992) in commercial cherry groves in Switzerland. A significant reduction in fruit infestation was achieved by treating one half of tree canopies (top to bottom) with a synthetic HMP. Repelled females were then trapped with visual traps placed on the other half of the canopy. This strategy may also be developed to manage *Anastrepha* pest species (Aluja *et al.*, 2003b). Aluja and Díaz-Fleischer (2006) tested female response to faeces extracts of three species of pestiferous *Anastrepha*. When flies encountered fruit treated with either 1 or 100 mg/ml faeces extract, there were drastic and statistically significant reductions in tree residence time, mean time spent on fruit and in the number of oviposition attempts or actual ovipositions when compared to the control treatment (clean fruit). These responses were almost identical, irrespective of extract origin (i.e. fly species), indicating complete interspecific HMP cross-recognition by all three *Anastrepha* species tested. The strategy also shows promise for flies in the genus *Ceratitis*, such as the Medfly (Arredondo and Díaz-Fleischer, 2006).

Effect of fruiting phenology on fruit-fly attack and the possibility of manipulating it

Yokoyama and Miller (1993) proposed that a pest-free period for walnut husk fly could be implemented in California based on the fruiting phenology of some stone fruit exported to New Zealand. The rationale of their proposal hinged on the fact that some cultivars matured before flies started to emerge and therefore escaped attack. Aluja and Birke (1993), working with *A. obliqua* (a notorious pest of mangos throughout Latin America), reported that females of this species preferred to lay eggs in their native host, *S. purpurea*, when offered an option between the exotic mango and the native *S. purpurea*. Since *S. purpurea* ripens before most commercially grown mango cultivars, such trees could be used as population sinks if large numbers of parasitoids are released at the peak of the fruiting period. Considering the remarkable attractiveness of *S. purpurea* fruit to *A. obliqua* females, pilot tests were carried out in Apazapan, Veracruz, Mexico, to determine if the highly ephemeral/short fruiting period could be expanded/delayed for 2–3 weeks by treating trees with gibberellic acid (GA_3), as such a delay would keep *A. obliqua* females away from the mango trees (cultivar 'Manila'). Although the results with GA_3 were negative (M. Aluja, unpublished information), other alternatives will be explored, as the potential for such a manipulation in terms of managing *A. obliqua* is highly promising. Finally, Rull and Prokopy (2004a), working with apple maggot flies, concluded that fast-ripening apple cultivars were highly attractive to females early in the season and that natural host odour and synthetic lures did not compete in attracting apple maggot flies but rather had a synergistic effect. Based on this, they suggested that trap deployment strategies in apple orchards should take into account fruiting phenology to maximize fly captures (Rull and Prokopy, 2005).

Manipulation of fruit resistance to fruit-fly attack (e.g. gibberellic acid)
The last environmental strategy we would like to discuss, applicable at the patch, single-tree or single-fruit spatial scales, is the manipulation of fruit resistance. Greany *et al.* (1983) explored the biochemical resistance of citrus fruit to attack by the Caribbean fruit fly, discovering that natural resistance varied according to fruit senescence and cultivar. It was later proposed that natural resistance could be enhanced by applying a plant-growth regulator (gibberellic acid, GA_3) to delay senescence (Greany and Shapiro, 1993). Applying 10 ppm of GA_3 plus surfactant to grapefruit reduced susceptibility to Caribbean fruit fly attack significantly when compared to a control (Greany *et al.*, 1991). Rossler and Greany (1990) reduced the susceptibility of oranges and grapefruit to Medfly attack in the laboratory by applying different doses of GA_3 to the fruit before colour break. Kaur and Rup (2003) assessed the effect of four plant-growth regulators: coumarin, kinetin, GA_3 and indole-3-acetic acid (IAA), on the development of the melon fruit fly, *B. cucurbitae*. The four compounds exerted growth- and development-inhibitory effects on flies. Coumarin was most potent, followed by kinetin, GA_3 and IAA. At concentrations of 125, 625 and 3125 μg/ml, coumarin, kinetin and GA_3 caused 100% mortality in first-instar larvae. By contrast, Birke *et al.* (2006) found that treatment of grapefruit with GA_3 in the field failed to enhance the resistance of fruit to *A. ludens* attack, due to the fact that females of these species were capable of bypassing toxic oils in the peel of treated fruit by depositing eggs deep in the albedo region with their long aculeus. Furthermore, these authors found that if an *A. ludens* female encountered a harder, GA_3-treated fruit, it would lay larger clutches. Expanding on this finding, Díaz-Fleischer and Aluja (2003) found that while ovipositing, *A. ludens* females exhibited a high degree of behavioural plasticity and that 'tricking' them therefore was not an easy endeavour.

Question of scale and type of grower

As alluded to in the beginning of this chapter, the success/feasibility of environmental manipulation schemes for fruit-fly management depends in many ways on the type of grower and the size of the orchard. There are various types of agroecosystems that are ideally suited for such schemes: organic groves, backyard gardens and small-scale (0.5–5 ha) semi-commercial orchards owned by growers interested in keeping a wide array of fruit trees, as opposed to a monoculture. In contrast, mid- and large-scale commercial monocultures (10–50 and 50–500 ha, respectively) are less suited or unsuited.

A good example of an ideal situation for environmental manipulation can be found in the backyard or small-scale semi-commercial orchards typical of rural areas in Latin America, Africa and some Asian countries. Owners (often, very poor) are interested in having access to a wide array of fruit over the entire year and, to that end, they plant as many tree species as their land can sustain and, in many cases, also plant several cultivars of their preferred fruit (Fig 7.4). For highly polyphagous flies such as *C. capitata*, *B. dorsalis* or *A. ludens*, such an environment presents ideal conditions for population build-up and maintenance. Applying insecticides is frequently not an option as growers lack the resources or because people live, literally, under the tree canopies (Fig 7.4). For these growers, adding as many mortality agents as possible (e.g. parasitoids, ants, wasps) without having to pay for them is often the only option. In the case of monitoring tools, Piñero *et al.* (2003) and Aluja and Piñero (2004) showed that diluted human urine placed in a recycled plastic soda bottle was a feasible alternative to costly baits and traps.

Future Research Needs

Many of the environmental manipulation schemes proposed/discussed here sound appealing but their effectiveness still needs to be tested formally and quantified (e.g. the use of colonial spider webs). We would therefore hope that the scenarios outlined by us will appeal to students and researchers throughout the world as potential topics for their theses/dissertations or research projects.

Research topics of particular interest for future investigation are: finding ways of increasing resistance in fruit to the attack by fruit flies through genetic engineering; the identification of repellent plants/compounds; identification and experimentation with predators in the field (particularly finding ways of keeping them in the orchards); finding ways of enhancing the efficacy of native parasitoids; large-scale testing of the efficacy of synthetic HMPs; and the ideal design of mixed orchards.

Acknowledgements

We thank six anonymous reviewers, who acted as official referees, for their valuable comments and suggestions for improvement. We would also like to acknowledge the important contribution of Alberto Anzures-Dadda (Instituto de Ecología, AC) throughout the process of writing this chapter. Without his help, we never would have been able to complete the task. We thank Aline Schünemann (Universidad Nacional Autónoma de México), who provided much appreciated, last minute suggestions for improvement. Parts of the research reported here was financed through grants to Martín Aluja from the Mexican Campaña Nacional Contra Moscas de la Fruta (Convenios SAGARPA-IICA/IICA-INECOL), the Mexican Consejo Nacional de Ciencia y Tecnología (CONACyT Grants D111-903537, 0436P-N9506 and SEP-2004-CO1-46846-Q), CONACyT – Texas A&M Project ('The natural enemies of *Rhagoletis* spp. (Diptera: Tephritidae) in Mexico, with emphasis on the apple maggot, *Rhagoletis pomonella*'), the International Foundation for Science (Grant Number C/1741-1), the US Department of Agriculture (ARS funds managed by John Sivinski) and the Instituto de Ecología, AC. Martín Aluja also acknowledges support from CONACyT through a Sabbatical Year Fellowship (Ref. 79449) and thanks Benno Graf and Jörg Samietz (Forschungsanstalt Agroscope Changins-Wädenswil ACW) for providing ideal working conditions to finish writing this chapter.

References

Addicott, J.F., Aho, J.M., Antolin, M.F., Padilla, D.K., Richardson, J.S. and Saluk, D.A. (1987) Ecological neighbourhoods: scaling environmental patterns. *Oikos* 49, 340–346.

Agnarsson, I. (2004) Morphological phylogeny of cobweb spiders and their relatives (Araneae, Araneoidea, Theridiidae). *Zoological Journal of the Linnean Society* 141, 447–626.

Agnarsson, I. (2006) A revision of the New World *eximius* lineage of *Anelosimus* (Araneae, Theridiidae) and a phylogenetic analysis using worldwide exemplars. *Zoological Journal of the Linnean Society* 146, 453–593.

Agnarsson, I. and Kuntner, M. (2005) Madagascar: an unexpected hotspot of social *Anelosimus* spider diversity (Araneae: Theridiidae). *Systematic Entomology* 30, 575–592.

Altieri, M.A., Martin, P.B. and Lewis, W.J. (1983) A quest for ecologically based pest management systems. *Journal of Environmental Management* 7, 91–99.

Aluja, M. (1993) *Manejo Integrado de la Mosca de la Fruta*. Trillas, México.

Aluja, M. (1996) Future trends in fruit fly management. In: McPheron, B.A. and Steck, G.J. (eds) *Fruit Fly Pests: A World Assessment of their Biology and Management*. St. Lucie Press, Delray Beach, Florida, pp. 309–320.

Aluja, M. (1999) Fruit fly (Diptera: Tephritidae) research in Latin America: myths, realities and dreams. *Anais da Sociedade Entomologica do Brasil* 28, 565–594.

Aluja, M. and Birke, A. (1993) Habitat use by adults of *Anastrepha obliqua* (Diptera: Tephritidae) in a mixed mango and tropical plum orchard. *Annals of the Entomological Society of America* 86, 799–812.

Aluja, M. and Boller, E.F. (1992) Host marking pheromone of *Rhagoletis cerasi*: field deployment of synthetic pheromone as a novel cherry fruit fly management strategy. *Entomologia Experimentalis et Applicata* 65, 141–147.

Aluja, M. and Díaz-Fleischer, F. (2006) Foraging behavior of *Anastrepha ludens*, *A. obliqua* and *A. serpentina* in response to feces extracts containing host marking pheromone. *Journal of Chemical Ecology* 32, 367–389.

Aluja, M. and Liedo, P. (1986) Future perspectives on integrated management of fruit flies in México. In: Mangel, M., Carey, J.R. and Plant, R. (eds) *Pest Control: Operation and Systems Analysis in Fruit Fly Management*. Proceedings of the NATO Advanced Workshop. Springer-Verlag, New York, pp. 12–48.

Aluja, M. and Mangan, R. (2008) Fruit fly (Diptera: Tephritidae) host status determination: critical conceptual, methodological, and regulatory considerations. *Annual Review of Entomology* 53, 473–502.

Aluja, M. and Piñero, J. (2004) Testing human urine as a low-tech bait for *Anastrepha* spp. (Diptera: Tephritidae) in small guava, mango, sapodilla and grapefruit orchards. *Florida Entomologist* 87, 41–50.

Aluja, M. and Prokopy, R.J. (1992) Host search behaviour by *Rhagoletis pomonella* flies: inter-tree movement patterns in response to wind-borne fruit volatiles under field conditions. *Physiological Entomology* 17, 1–8.

Aluja, M. and Prokopy, R.J. (1993) Host odor and visual stimulus interaction during intratree host finding behavior of *Rhagoletis pomonella* flies. *Journal of Chemical Ecology* 19, 2671–2696.

Aluja, M., Celedonio-Hurtado, H., Liedo, P. and Guillen, J. (1986) Some results of general interest for control of *Anastrepha* spp. (Diptera: Tephritidae). In: Cavalloro, R. (ed.) *Fruit Flies of Economic Importance 84*. A.A. Balkema, Rotterdam, The Netherlands, pp. 209–216.

Aluja, M., Prokopy, R.J., Buonaccorsi, J. and Cardé, R.T. (1993) Wind tunnel assays of olfactory responses of female *Rhagoletis pomonella* flies to apple volatiles: effect of wind speed and odour release rate. *Entomologia Experimentalis et Applicata* 68, 99–108.

Aluja, M., Celedonio-Hurtado, H., Liedo, P., Cabrera, M., Castillo, F., Guillén, J. and Rios, E. (1996) Seasonal population fluctuations and ecological implications for management of *Anastrepha* fruit flies

(Diptera: Tephritidae) in commercial mango orchards in Southern Mexico. *Journal of Economic Entomology* 89, 654–667.

Aluja, M., Jimenez, A., Camino, M., Piñero, J., Aldana, L., Castrejon, V. and Valdez, M.E. (1997a) Habitat manipulation to reduce papaya fruit fly (Diptera: Tephritidae) damage: orchard design, use of trap crops, and border trapping. *Journal of Economic Entomology* 90, 1567–1576.

Aluja, M., Jiménez, A., Piñero, J., Camino, M., Aldana, L., Valdés, M.E., Castrejón, V., Jácome, I., Dávila, A. and Figueroa, R. (1997b) Daily activity patterns and within-field distribution of papaya fruit flies (Diptera: Tephritidae) in Morelos and Veracruz, Mexico. *Annals of the Entomological Society of America* 90, 505–520.

Aluja, M., López, M. and Sivinski, J. (1998) Ecological evidence for diapause in four native and one exotic species of larval-pupal fruit fly (Diptera: Tephritidae) parasitoids in tropical environments. *Annals of the Entomological Society of America* 91, 821–833.

Aluja, M., Piñero, J., Jácome, I., Díaz-Fleisher, F. and Sivinski, J. (2000) Behavior of flies in the genus *Anastrepha*. In: Aluja, M. and Norrbom, A.L. (eds) *Fruit Flies (Diptera: Tephritidae): Phylogeny and Evolution of Behavior*. CRC Press, Boca Raton, Florida, pp. 375–406.

Aluja, M., Díaz-Fleischer, F., Papaj, D.R., Lagunes, G. and Sivinski, J. (2001a) Effects of age, diet, female density and the host resource on egg load in *Anastrepha ludens* and *Anastrepha obliqua* (Diptera: Tephritidae). *Journal of Insect Physiology* 47, 975–988.

Aluja, M., Jácome, I. and Macías-Ordóñez, R. (2001b) Effect of adult nutrition on male sexual performance in four Neotropical fruit fly species of the genus *Anastrepha* (Diptera: Tephritidae). *Journal of Insect Behavior* 14, 759–775.

Aluja, M., Díaz-Fleischer, F., Edmunds, A.J.F. and Hagmann, L. (2003a) Isolation, structural determination, synthesis, biological activity and application as control agent of the host marking pheromone (and derivatives thereof) of the fruit flies of the type *Anastrepha* (Diptera: Tephritidae). Patent Number US 6,555,120 B1, 29 April 2003.

Aluja, M., Rull, J., Sivinski, J., Norrbom, A.L., Wharton, R.A., Macías-Ordóñez, R., Díaz-Fleischer, F. and López, M. (2003b) Fruit flies of the genus *Anastrepha* (Diptera: Tephritidae) and associated native parasitoids (Hymenoptera) in the tropical rainforest biosphere reserve of Montes Azules, Chiapas, Mexico. *Environmental Entomology* 32, 1377–1385.

Aluja, M., Sivinski, J., Rull, J. and Hodgson, P.J. (2005) Behavior and predation of fruit fly larvae (*Anastrepha* spp.) (Diptera: Tephritidae) after exiting the fruit in four types of habitats in tropical Veracruz, Mexico. *Environmental Entomology* 34, 1507–1516.

Alviña, A., Frías, D., González, Ch.R. and Guerra, A.H. (1993) Distribución geográfica y respuesta visual de *Rhagoletis tomatis* Foote (Diptera: Tephritidae) a trampas esféricas y rectángulos de diferentes colores. *Acta Entomológica Chilena* 18, 185–194.

Alyokhin, A.V., Mille, C., Messing, R. and Duan, J.J. (2001) Selection of pupation habitats by Oriental fruit fly larvae in the laboratory. *Journal of Insect Behavior* 14, 57–67.

Andrén, H. (1994) Effects of habitat fragmentation on bird and mammals in landscapes with different proportions of suitable habitat: a review. *Oikos* 71, 355–366.

Anton, Ch., Zeisset, I., Musche, M., Durka, W., Boomsma, J.J. and Settele, J. (2007) Population structure of a large blue butterfly and its specialist parasitoid in a fragmented landscape. *Molecular Ecology* 16, 3828–3838.

Arredondo, J. and Díaz-Fleischer, F. (2006) Oviposition deterrents for the Mediterranean fruit fly, *Ceratitis capitata* (Diptera: Tephritidae), from fly faeces extracts. *Bulletin of Entomological Research* 6, 35–42.

Asquith, D. and Colburn, R. (1971) Integrated pest management in Pennsylvania apple orchards. *Bulletin of the Entomological Society of America* 17, 89–91.

Averill, A.L. and Prokopy, R.J. (1987) Intraspecific competition in the Tephritid fruit fly *Rhagoletis pomonella*. *Ecology* 68, 878–886.

Back, E.A. and Pemberton, C.E. (1918a) The Mediterranean fruit fly. *United States*

Department of Agriculture Bulletin 640, 1–44.

Back, E.A. and Pemberton, C.E. (1918b) The melon fly. *United States Department of Agriculture Bulletin* 643, 1–32.

Baker, E.W., Stone, W.E., Plummer, C.C. and McPhail, M. (1944) A review of studies on the Mexican fruit fly and related species. *United States Department of Agriculture Miscellaneous Publications* 531.

Balagawi, S., Vijaysegaran, S., Drew, R.A.I. and Raghu, S. (2005) Influence of fruit traits on oviposition preference and offspring performance of *Bactrocera tryoni* (Froggatt) (Diptera: Tephritidae) on three tomato (*Lycopersicon lycopersicum*) cultivars. *Australian Journal of Entomology* 44, 97–103.

Banks, J.E. and Ekbom, B. (1999) Modelling herbivore movement and colonization: pest management potential of intercropping and trap cropping. *Agricultural and Forest Entomology* 1, 165–170.

Barbosa, P. (ed.) (1998) *Conservation Biological Control*. Academic Press, San Diego, California.

Barry, J.D., Blessinger, T. and Morse, J.G. (2004) Recapture of sterile Mediterranean fruit flies (Diptera: Tephritidae) in California's preventative release program. *Journal of Economic Entomology* 97, 1554–1562.

Bateman, M.A. (1977) Dispersal and species interaction as factors in the establishment and success of tropical fruit flies in new areas. *Proceedings of the Ecological Society of Australia* 10, 106–112.

Beier, P. and Noss, R.F. (1998) Do habitat corridors provide connectivity? *Conservation Biology* 12, 1241–1252.

Bhar, R. and Fahrig, L. (1998) Local vs. landscape effects of woody field borders as barriers to crop pest movement. *Conservation Ecology* [online] 2, 3 www.consecol.org/vol2/iss2/art3/

Bianchi, F.J.J.A., Booij, C.J.H. and Tscharntke, T. (2006) Sustainable pest regulation in agricultural landscapes: a review on landscape composition, biodiversity and natural pest control. *Proceedings of the Royal Society B* 273, 1715–1725.

Birke, A., Aluja, M., Greany, P., Bigurra, E., Pérez-Staples, D. and McDonald, R. (2006) Long aculeus and behavior of *Anastrepha ludens* renders gibberellic acid ineffective as an agent to reduce 'Ruby Red' grapefruit susceptibility to the attack of this pestiferous fruit fly in commercial citrus orchards. *Journal of Economic Entomology* 99, 1184–1193.

Blay, S. and Yuval, B. (1997) Nutritional correlates of reproductive success of male Mediterranean fruit flies (Diptera: Tephritidae). *Animal Behaviour* 54, 59–66.

Boller, E.F. and Aluja, M. (1992) Oviposition deterring pheromone of *Rhagoletis cerasi* L.: biological activity of 4 synthetic isomers and HMP discrimination of two host races as measured by an improved laboratory bioassay. *Journal of Applied Entomology* 113, 113–119.

Boucher, T.J., Ashley, R.A., Adams, R.G. Jr and Morris, T.F. (2001) Effect of trap position, habitat, and height on the capture of pepper maggot flies (Diptera: Tephritidae). *Journal of Economic Entomology* 94, 455–461.

Boucher, T.J., Ashley, R., Durgy, R., Sciabarrasi, M. and Calderwood, W. (2003) Managing the pepper maggot (Diptera: Tephritidae) using perimeter trap cropping. *Journal of Economic Entomology* 96, 420–432.

Bressan-Nascimento, S. (2001) Emergence and pupal mortality factors of *Anastrepha obliqua* (Macq.) (Diptera: Tephritidae) along the fruiting season of the host *Spondias dulcis* L. *Neotropical Entomology* 30, 207–215.

Bruce, T.J.A., Wadhams, L.J. and Woodcock, C.M. (2005) Insect host location: a volatile situation. *Trends in Plant Science* 10, 269–274.

Burgess, J.W. (1978) Social behavior in group-living spider species. *Symposia of the Zoological Society of London* 42, 69–78.

Burns, R.E., Harris, D.L., Moreno, D.S. and Eger, J.E. (2001) Efficacy of spinosad bait sprays to control Mediterranean and Caribbean fruit flies (Diptera: Tephritidae) in commercial citrus in Florida. *Florida Entomologist* 84, 672–678.

Buskirk, R.E. (1981) Sociality in the Arachnida. In: Herman, H.R. (ed.) *Social Insects*. Academic Press, London, pp. 282–367.

Cardé, R.T. and Minks, A.K. (1995) Control of moth pests by mating disruption: successes and constraints. *Annual Review of Entomology* 40, 559–585.

Carle, S.A., Averill, A.L., Rule, G.S., Reissig, H.W. and Roelofs, W.L. (1987) Variation of host fruit volatiles attractive to the apple maggot fly *Rhagoletis pomonella* (Diptera: Tephritidae). *Journal of Chemical Ecology* 13, 795–805.

Celedonio-Hurtado, H., Aluja, M. and Liedo, P. (1995) Adult population fluctuations of *Anastrepha* species (Diptera: Tephritidae) in tropical orchard habitats of Chiapas, Mexico. *Environmental Entomology* 24, 861–869.

Chen, C.C., Dong, Y.J., Li, C.T. and Liu, K.Y. (2006) Movement of the oriental fruit fly, *Bactrocera dorsalis* (Hendel) (Diptera: Tephritidae), in a guava orchard with special reference to its population changes. *Formosan Entomology* 26, 143–159.

Chetkiewicz, C.L., St Clair, C.C. and Boyce, M.S. (2006) Corridors for conservation: integrating pattern and process. *Annual Review of Ecology, Evolution and Systematics* 37, 317–342.

Chust, G., Garbin, L. and Pujade-Villar, J. (2007) Gall wasps and their parasitoids in cork oak fragmented forests. *Ecological Entomology* 32, 82–91.

Cirio, U. (1971) Reperti sul mecanismo stimolorisposta nell'ovidepozitione del dacus oleae Gmelin (Diptera: Tephritidae). *Redia* 52, 577–600.

Clarke, A.R., Allwood, A., Chinajariyawong, A., Drew, R.A.I., Hengsawad, C., Jirasurat, M., Kong Krong, C., Kritsaneepaiboon, S. and Vijaysegaran, S. (2001) Seasonal abundance and host use patterns of seven *Bactrocera* Macquart species (Diptera: Tephritidae) in Thailand and Peninsular Malaysia. *The Raffles Bulletin of Zoology* 49, 207–220.

Cook, S.M., Khan, Z.R. and Pickett, J.A. (2007) The use of push–pull strategies in integrated pest management. *Annual Review of Entomology* 52, 375–400.

Copeland, R.S., Wharton, R.A., Luke, Q. and De Meyer, M. (2002) Indigenous hosts of *Ceratitis capitata* (Diptera: Tephritidae) in Kenya. *Annals of the Entomological Society of America* 95, 672–694.

Cornelius, M.L., Nergel, L., Duan, J.J. and Messing, R.H. (2000) Responses of female oriental fruit flies (Diptera: Tephritidae) to protein and host fruit odors in field cage and open field tests. *Environmental Entomology* 29, 14–29.

Craig, T.P., Abrahamson, W.G., Itami, J.K. and Horner, J.D. (1999) Oviposition preference and offspring performance of *Eurosta solidaginis* on genotypes of *Solidago altissima*. *Oikos* 86, 119–128.

Croft, B.A. (1978) Potential for research and implementation of integrated pest management of deciduous tree fruits. In: Smith, E.H. and Pimentel, D. (eds) *Pest Control Strategies*. Academic Press, New York, pp. 101–115.

Cronin, J.T. and Reeve, J.D. (2005) Host-parasitoid spatial ecology: a plea for a landscape synthesis. *Proceedings of the Royal Society B* 272, 2225–2235.

Cronin, J.T., Hyland, K. and Abrahamson, W.G. (2001). The pattern, rate, and range of within-patch movement of a stem-galling fly. *Ecological Entomology* 26, 16–24.

Cunningham, R.T., Nakagawa, S., Suda, D.Y. and Urago, T. (1978) Tephritid fruit fly trapping: liquid food baits in high and low rainfall climates. *Journal of Economic Entomology* 71, 762–763.

Cytrynowicz, M., Morgante, J.S. and de Souza, H.M.L. (1982) Visual responses of South American fruit flies, *Anastrepha fraterculus*, and Mediterranean fruit flies, *Ceratitis capitata*, to colored rectangles and spheres. *Environmental Entomology* 11, 1202–1210.

Dalby-Ball, G. and Meats, A. (2000) Effects of fruit abundance within a tree canopy on the behaviour of wild and cultured Queensland fruit flies, *Bactrocera tryoni* (Froggatt) (Diptera: Tephritidae). *Australian Journal of Entomology* 39, 201–207.

Dalby-Ball, G. and Meats, A. (2002) The role of foliage in differential landing of the Queensland fruit fly, *Bactrocera tryoni* (Froggatt) (Diptera; Tephritidae) on odiferous and odourless fruit models. *General and Applied Entomology* 31, 29–34.

Dean, R.W. and Chapman, P.J. (1973) Bionomics of the apple maggot in Eastern New York. *Research Agriculture* 3, 1–62.

Debouzie, D. (1989) Biotic mortality factors in tephritid populations. In: Robinson, A.S.

and Hooper, G. (eds) *World Crop Pests. Fruit Flies: Their Biology, Natural Enemies and Control, 3B*. Elsevier, Amsterdam, pp. 221–227.

Díaz-Fleischer, F. and Aluja, M. (2000) Behavior of Tephritid flies: a historical perspective. In: Aluja, M. and Norrbom, A.L. (eds) *Fruit Flies (Diptera: Tephritidae): Phylogeny and Evolution of Behavior*. CRC Press, Boca Raton, Florida, pp. 39–69.

Díaz-Fleischer, F. and Aluja, M. (2003) Clutch size in frugivorous insects as a function of host firmness: the case of the tephritid fly *Anastrepha ludens*. *Ecological Entomology* 28, 268–277.

Díaz-Fleischer, F., Papaj, D.R., Prokopy, R.J., Norrbom, A.L. and Aluja, M. (2000) Evolution of fruit fly oviposition behavior. In: Aluja, M. and Norrbom, A.L. (eds) *Fruit Flies (Tephritidae): Phylogeny and Evolution of Behavior*. CRC Press, Boca Raton, Florida, pp. 811–841.

Dimou, I., Koutsikopoulos, C., Economopoulos, A.P. and Lykakis, J. (2003) Depth of pupation of the wild olive fly *Dacus oleae* (Gmel.) (Diptera: Tephritidae) as affected by soil abiotic factors. *Journal of Applied Entomology* 127, 12–17.

Dirks, C.O. (1935) Larval production and adult emergence of the apple fruit fly in relation to apple varieties. *Journal of Economic Entomology* 28, 198–203.

Doutt, R.L. (1959) The biology of parasitic Hymenoptera. *Annual Review of Entomology* 4, 161–182.

Drew, R.A.I. and Yuval, B. (2000) The evolution of fruit fly feeding behaviour. In: Aluja, M. and Norrbom, A.L. (eds) *Fruit Flies (Diptera: Tephritidae): Phylogeny and Evolution of Behavior*. CRC Press, Boca Raton, Florida, pp. 731–750.

Drew, R.A.I., Prokopy, R.J. and Roming, M.C. (2003) Attraction of fruit flies of the genus *Bactrocera* to colored mimics of host fruit. *Entomologia Experimentalis et Applicata* 107, 39–45.

Duyck, P.F., David, P. and Quilici, S. (2004) A review of relationships between interspecific competition and invasions in fruit flies (Diptera: Tephritidae). *Ecological Entomology* 29, 511–520.

Ebeling, W.T., Nishida, T. and Bess, H.A. (1953) Field experiments on the control of the melon fly, *Dacus cucurbitae*. *Hilgardia* 21, 563–591.

Eilenberg, J., Hajek, A. and Lomer, C. (2001) Suggestions for unifying the terminology in biological control. *BioControl* 46, 387–400.

Elzinga, J.A., van Nouhuys, S., van Leeuwen, D.K. and Biere, A. (2007) Distribution and colonization ability of three parasitoids and their herbivorous host in a fragmented landscape. *Basic and Applied Ecology* 8, 75–88.

Epsky, N.D. and Heath, R.R. (1998) Exploiting the interactions of chemical and visual cues in behavioral control measures for pest tephritid fruit flies. *Florida Entomologist* 81, 273–282.

Epsky, N.D., Deuben, B.D., Heath, R.R., Lauzon, C.R. and Prokopy, R.J. (1997) Attraction of *Anastrepha suspensa* (Diptera: Tephritidae) to volatiles from avian fecal material. *Florida Entomologist* 80, 270–277.

Estevez, B., Domon, G. and Lucas, E. (2000) Use of landscape ecology in agroecosystem diversification towards phytoprotection. *Phytoprotection* 81, 1–14.

Fahrig, L. (1998) Effect of habitat patch characteristics on abundance and diversity of insects in an agricultural landscape. *Ecosystems* 1, 197–205.

Fahrig, L. (2003) Effects of habitat fragmentation on biodiversity. *Annual Review of Ecology, Evolution and Systematics* 34, 487–515.

Falcy, M.R. and Estades, C.F. (2007) Effectiveness of corridors relative to enlargement of habitat patches. *Conservation Biology* 5, 1341–1346.

Feder, J.L. and Forbes, A.A. (2007) Habitat avoidance and speciation for phytophagous insect specialists. *Functional Ecology* 21, 585–597.

Feder, J.L., Hunt, T.A. and Bush, G.L. (1993) The effects of climate, host plant phenology, and host fidelity on the genetics of hawthorn and apple infesting races of *Rhagoletis pomonella*. *Entomologia Experimentalis et Applicata* 69, 117–135.

Feder, J.L., Stolz, U., Lewis, K.M., Perry, W., Roethele, J.B. and Rogers, A. (1997) The effects of winter length on the genetics of apple and hawthorn races of *Rhagoletis pomonella* (Diptera: Tephritidae). *Evolution* 51, 1862–1876.

Fein, B.L., Reissig, W.H. and Roelofs, W.L. (1982) Identification of apple volatiles attractive to the apple maggot. *Journal of Chemical Ecology* 8, 1473–1487.

Fernandes, E.C.M. and Nair, P.K.R. (1986) An evaluation of the structure and function of tropical home gardens. *Agricultural Systems* 21, 279–310.

Filchak, K.E., Roethele, J.B. and Feder, J.L. (2000) Natural selection and sympatric divergence in the apple maggot *Rhagoletis pomonella*. *Nature* 407, 739–742.

Fischer, J. and Lindenmayer, D.B. (2007) Landscape modification and habitat fragmentation: a synthesis. *Global Ecology and Biogeography* 16, 265–280.

Fitt, G.P. (1990) The role of interspecific interactions in the dynamics of Tephritid populations. In: Robinson, A. and Hooper, G.H.S. (eds) *World Crop Pests. Fruit Flies: Their Biology, Natural Enemies and Control, 3B*. Elsevier, Amsterdam, pp. 281–300.

Forman, R.T.T. (1995) *Land Mosaics: The Ecology of Landscapes and Regions*. Cambridge University Press, Cambridge, UK.

Foster, S.P. and Harris, M.O. (1997) Behavioral manipulation methods for insect pest management. *Annual Review of Entomology* 42, 123–146.

Geier, P.W. and Clark, L.R. (1961) *An Ecological Approach to Pest Control*. Proceedings of the 8th Technical Meeting of the International Union for the Conservation of the Nature and Natural Resources, 1960, Warsaw. E.J. Brill, Leiden, The Netherlands, pp. 10–18.

Gould, F. and Stinner, R.E. (1984) Insects in heterogeneous habitats. In: Huffaker, B. and Rabb, R.L. (eds) *Ecological Entomology*. Wiley, New York, pp. 427–449.

Greany, P.D. and Shapiro, J.P. (1993) Manipulating and enhancing citrus fruit resistance to the Caribbean fruit fly (Diptera: Tephritidae). *Florida Entomologist* 76, 258–263.

Greany, P., Burditt, D. Jr, Agee, H.R. and Chambers, D.L. (1978) Increasing effectiveness of visual traps for the Caribbean fruit fly, *Anastrepha suspensa* (Diptera: Tephritidae), by use of fluorescent colors. *Entomologia Experimentalis et Applicata* 23, 20–25.

Greany, P.D., Styer, S.C., Davis, P.L., Shaw, P.E. and Chambers, D.L. (1983) Biochemical resistance of citrus to fruit flies. Demonstration and elucidation of resistance to the Caribbean fruit fly, *Anastrepha suspensa*. *Entomologia Experimentalis et Applicata* 34, 40–50.

Greany, P.D., McDonald, R.E., Schroeder, W.J. and Shaw, P.E. (1991) Improvement in efficacy of gibberellic acid treatments in reducing susceptibility of grapefruit to attack by Caribbean fruit fly. *Florida Entomologist* 74, 570–580.

Green, T.A., Prokopy, R.J. and Hosmer, D.W. (1994) Distance of response to host tree models by female apple maggot flies, *Rhagoletis pomonella* (Walsh) (Diptera: Tephritidae): interaction of visual and olfactory stimuli. *Journal of Chemical Ecology* 20, 2393–2413.

Greenblatt, J.A. and Lewis, W.J. (1983) Chemical environment manipulation for pest insect control. *Environmental Management* 7, 35–41.

Guillén, L., Aluja, M., Equihua, M. and Sivinski, J. (2002) Performance of two fruit fly (Diptera: Tephritidae) pupal parasitoids (*Coptera haywardi* [Hymenoptera: Diapriidae] and *Pachycrepoideus vindemiae* [Hymenoptera: Pteromalidae]) under different environmental soil conditions. *Biological Control* 23, 219–227.

Gurr, G.M., Wratten, S.D. and Altieri, M.A. (eds) (2004) *Ecological Engineering for Pest Management: Advances in Habitat Manipulation for Arthropods*. CSIRO Publishing, Collingwood, Australia.

Haddad, N.M., Bowne, D.R., Cunningham, A., Danielson, B.J., Levey, D.J., Srgent, S. and Spira, T. (2003) Corridor use by diverse taxa. *Ecology* 84, 609–615.

Hajek, A. (2004) *Natural Enemies: An Introduction to Biological Control*. Cambridge University Press, Cambridge, UK.

Hancock, D.L., Osborne, R., Broughton, S. and Glesson, P. (2000) Eradication of *Bactrocera papayae* (Diptera: Tephritidae) by male annihilation and protein baiting in Queensland, Australia. In: Tan, K.H. (ed.) *Area-Wide Control of Fruit Flies and Other Insect Pests*. Penerbit Universiti Sains

Malaysia, Pulau Pinung, Malaysia, pp. 381–388.

Harris, E.J. and Lee, C.Y.L. (1986) Seasonal and annual occurrence of Mediterranean fruit flies (Diptera: Tephritidae) in Makaha and Waianae valleys, Oahu, Hawaii. *Environmental Entomology* 15, 507–512.

Harris, E.J. and Lee, C.Y.L. (1987) Seasonal and annual distribution of the Mediterranean fruit fly (Diptera: Tephritidae) in Honolulu and suburban areas of Oahu, Hawaii. *Environmental Entomology* 16, 1273–1282.

Harris, E.J., Vargas, R.I. and Gilmore, J.E. (1993) Seasonality in occurrence and distribution of Mediterranean fruit fly (Diptera: Tephritidae) in upland and lowland areas on Kauai, Hawaii. *Environmental Entomology* 22, 404–410.

Hassanali, A., Herren, H., Khan, Z.R., Pickett, J.A. and Woodcock, C.M. (2007) Integrated pest management: the push–pull approach for controlling insect pests and weeds of cereals, and its potential for other agricultural systems including animal husbandry. *Philosophical Transactions of the Royal Society B: Biological Sciences* 363, 611–621.

Hassel, M.P. and Southwood, T.R.E. (1978) Foraging strategies of insects. *Annual Review of Ecology and Systematics* 9, 75–98.

Heath, R.R., Landolt, P.J., Robacker, D.C., Dueben, B.O. and Epsky N.D. (2000) Sexual pheromones of tephritid flies: clues to unravel phylogeny and behavior. In: Aluja, M. and Norrbom, A.L. (eds) *Fruit Flies (Diptera: Tephritidae): Phylogeny and Evolution of Behavior*. CRC Press, Boca Raton, Florida, pp. 793–809.

Hendrichs, J. (1990) Social spiders as living traps of *Anastrepha* fruit flies. Grant Proposal. Department of Entomology, University of Massachusetts, Amherst, Massachusetts.

Hendrichs, J. and Hendrichs, M.A. (1990) Mediterranean fruit fly (Diptera: Tephritidae) in nature: location and diel pattern of feeding and other activities on fruiting and non-fruiting hosts and non-hosts. *Annals of the Entomological Society of America* 83, 632–641.

Hendrichs, M.A. and Hendrichs, J. (1998) Perfumed to be killed: interception of Mediterranean fruit fly (Diptera: Tephritidae) sexual signaling by predatory foraging wasps (Hymenoptera: Vespidae). *Annals of the Entomological Society of America* 91, 228–234.

Hendrichs, J. and Prokopy, R.J. (1990) Where do apple maggot flies find food in nature? *Fruit Notes* Summer, 1–3.

Hendrichs, J., Katsoyannos, B.I., Papaj, D.R. and Prokopy, R.J. (1991) Sex-differences in movement between natural feeding and mating sites and tradeoffs between food consumption, mating success and predator evasion in Mediterranean fruit flies (Diptera: Tephritidae). *Oecologia* 86, 223–231.

Hendrichs, J., Katsoyannos, B.I., Wornoayporn, V. and Hendrichs, M.A. (1994) Odour-mediated foraging by yellowjacket wasps (Hymenoptera: Vespidae): predation on leks of pheromone-calling Mediterranean fruit fly males (Diptera: Tephritidae). *Oecologia* 99, 88–94.

Hendrichs, J., Robinson, A.S., Cayol, J.P. and Enkerlin, W. (2002) Medfly areawide sterile insect technique programmes for prevention, supression or eradication: the importance of mating behavior studies. *Florida Entomologist* 85, 1–13.

Hill, A.R. and Hooper, G.S.H. (1985) Attractiveness of various colours to Australian tephritid fruit flies in the field. *Entomologia Experimentalis et Aplicata* 35, 119–128.

Hodgson, P.J., Sivinski, J., Quintero, G. and Aluja, M. (1998) Depth of pupation and survival of fruit fly (*Anastrepha* spp.: Tephritidae) pupae in a range of agricultural habitats. *Environmental Entomology* 27, 1310–1314.

Hoyt, S.C. (1969) Integrated chemical control of insects and biological control of mites on apple in Washington. *Journal of Economic Entomology* 62, 74–86.

Hulthen, A.D. and Clarke, A.R. (2006). The influence of soil type and moisture on pupal survival of *Bactrocera tryoni* (Froggatt) (Diptera: Tephritidae). *Australian Journal of Entomology* 45, 16–19.

Inayatullah, C., Khan, L., Manzoor-Ul, H. and Ata-Ul, M. (1991) Weather-based models to predict the population densities of melon fruit fly, *Dacus cucurbitae* Coq. *Tropical Pest Management* 39, 211–215.

Israely, N.B., Yuval, B., Kitron, U. and Nestel, D. (1997) Population fluctuations of adult Mediterranean fruit flies (Diptera: Tephritidae) in a Mediterranean heterogeneous agricultural region. *Environmental Entomology* 26, 1263–1269.

Jang, E.B. and Light, D.M. (1996) Olfactory semiochemicals of tephritids. In: McPherson, B.A. and Steck, G.J. (eds) *Fruit Fly Pests: A World Assessment of their Biology and Management*. St Lucie Press, Delray Beach, Florida, 73–90.

Jia, H.J., Araki, A. and Okamoto, G. (2005) Influence of fruit bagging on aroma volatiles and skin coloration of 'Hakuho' peach (*Prunus persica* Batsch). *Postharvest Biology and Technology* 35, 61–68.

Jirón, L.F. and Hedström, I. (1991) Population fluctuations of economic species of *Anastrepha* (Diptera: Tephritidae) related to mango fruiting phenology in Costa Rica. *Florida Entomologist* 74, 98–105.

Kaspi, R. and Yuval, B. (2000) Post-teneral protein feeding improves sexual competitiveness but reduces longevity of mass-reared sterile male Mediterranean fruit flies (Diptera: Tephritidae). *Annals of the Entomological Society of America* 93, 949–955.

Katsoyannos, B.I. and Boller, E.F. (1976) First field application of oviposition-deterring marking pheromone of the European cherry fruit fly. *Environmental Entomology* 5, 151–152.

Katsoyannos, B.I. and Boller, E.F. (1980) Second field application of oviposition-deterring pheromone of the European cherry fruit fly, *Rhagoletis cerasi* L. (Diptera: Tephritidae). *Zeitschrift für Angewandte Entomologie* 89, 278–281.

Katsoyannos, B.I., Kouloussis, N.A. and Carey, J.R. (1998) Seasonal and annual occurrence of Mediterranean fruit flies (Diptera: Tephritidae) on Chios Island, Greece: differences between two neighboring citrus orchards. *Annals of the Entomological Society of America* 91, 43–51.

Kaur, R. and Rup, P.J. (2003) Influence of some plant growth regulators (PGR) on biochemical profile in the larvae of melon fruit fly *Bactrocera cucurbitae* (Coquillett) (Diptera: Trypetidae). *Entomon* 28, 89–95.

Kazi, A.S. (1976) Studies on the field habits of adult melon fruit fly *Dacus (Strumeta) cucurbitae*, Coquillet. *Pakistan Journal of Science and Industrial Research* 19, 71–76.

Kennedy, J.S., Booth, C.O. and Kershaw, W.S.J. (1961) Host finding by aphids in the field: visual attraction. *Annals of Applied Biology* 49, 1–21.

Khan, Z.R., Ampong-Nyarko, K., Chliswa, P., Hassanali, A., Kimani, S., Lwande, W., Overholt, W.A., Pickett, J.A., Smart, L.E., Wadhams, L.J. and Woodcock, C.M. (1997). Intercropping increases parasitism of pests. *Nature* 388, 631–632.

Khan, Z.R., Pickett, J.A., Van Den Berg, J., Wadhams, L.J. and Woodcock, C.M. (2000) Exploiting chemical ecology and species diversity: stem borer and striga control for maize and sorghum in Africa. *Pest Management Science* 56, 957–962.

Klassen, W. (2000) Area-wide approaches to insect pest management: history and lessons. In: Tan, K.H. (ed.) *Area-wide Control of Fruit Flies and Other Insect Pests*. Penerbit Universiti Sains Malaysia, Pulau Pinung, Malaysia, pp. 75–78.

Klassen, W. and Curtis, C.F. (2005) History of the sterile insect technique. In: Dyck, V.A., Hendrichs, J. and Robinson, A.S. (eds) *Sterile Insect Technique: Principles and Practices in Area-Wide Integrated Pest Management*. Springer, Dordrecht, The Netherlands, pp. 1–28.

Klein, A.M., Steffan-Dewenter, I. and Tscharntke, T. (2006) Rain forest promotes trophic interactions and diversity of trap-nesting Hymenoptera in adjacent agroforestry. *Journal of Animal Ecology* 75, 315–323.

Knipling, E.F. (1955) Possibilities of insect control or eradication through the use of sexual sterile males. *Journal of Economic Entomology* 48, 459–462.

Kogan, M. (1988) Integrated pest management theory and practice. *Entomologia Experimentalis et Applicata* 49, 59–70.

Kovaleski, A., Sugayama, R.L. and Malavasi, A. (1999) Movement of *Anastrepha fraterculus* from native breeding sites into apple orchards in Southern Brazil. *Entomologia Experimentalis et Applicata* 91, 459–465.

Kring, J.B. (1967). Alighting of aphids on colored cards in a flight chamber. *Journal of Economic Entomology* 60, 1207–1210.

Lambin, E.F., Helmut, J.G. and Lepers, E. (2003) Dynamics of land-use and land-cover change in tropical regions. *Annual Review of Environmental Resources* 28, 205–241.

Landis, D.A., Wratten, S.D. and Gurr, G.M. (2000) Habitat management to conserve natural enemies of arthropod pests in agriculture. *Annual Review of Entomology* 45, 175–201.

Landis, D.A., Menalled, F.D., Costamagna, A.C. and Wilkinson, T.K. (2005) Manipulating plant resources to enhance beneficial arthropods in agricultural landscapes. *Weed Science* 53, 902–908.

Landolt, P.J. and Phillips, T.W. (1997) Host plant influences on sex pheromone behavior of phytophagous insects. *Annual Review of Entomology* 42, 371–391.

Landolt, P.J. and Sivinski, J. (1992) Effects of time of day, adult food, and host fruit on incidence of calling by male Caribbean fruit flies (Diptera: Tephritidae). *Environmental Entomology* 21, 382–387.

Lewis, W.J., van Lenteren, J.C., Phatak, S.C. and Tumlinson, J.H. (1997) A total system approach to sustainable pest management. *Proceedings of the National Academy of Sciences of the United States of America* 94, 12243–12248.

Light, D.M. and Jang, E.B. (1996) Plant volatiles evoke and modulate tephritid behaviour. In: McPheron, B.A. and Steck, G.J. (eds) *Fruit Fly Pests. A World Assessment of Their Biology and Management.* St Lucie Press, Delray Beach, Florida, pp. 123–133.

Light, D.M, Jang, E.B. and Dickens, J.C. (1988) Electroantennogram responses of the Mediterranean fruit fly, *Ceratitis capitata*, to a spectrum of plant volatiles. *Journal of Chemical Ecology* 14, 159–180.

Lindquist, D.A. (2000) Pest management strategies: Area-wide and conventional. In: Tan, K.H. (ed.) *Area-wide Control of Fruit Flies and Other Insect Pests.* Penerbit Universiti Sains Malaysia, Pulau Pinung, Malaysia, pp. 13–19.

Linn, C. Jr., Nojima, S. and Roelofs, W. (2005) Antagonist effects of non-host fruit volatiles on discrimination of host fruit by *Rhagoletis* flies infesting apple (*Malus pumila*), hawthorn (*Crataegus* spp.), and flowering dogwood (*Cornus florida*). *Entomologia Experimentalis et Applicata* 114, 97–105.

McPhail, M. (1939) Protein lures for fruit flies. *Journal of Economic Entomology* 32, 758–761.

Mangan, R.L. and Moreno, D.S. (2007) Development of bait stations for fruit fly population suppression. *Journal of Economic Entomology* 100, 440–450.

Marino, P.C. and Landis, D.A. (1996) Effect of landscape structure on parasitoid diversity and parasitism in agroecosystems. *Ecological Applications* 6, 276–284.

Marino, P.C., Landis, D.A. and Hawkins, B.A. (2006) Conserving parasitoid assemblages of North American pest Lepidoptera: does biological control by native parasitoids depend on landscape complexity? *Biological Control* 37, 173–185.

Martelli, G. (1909) Alcune note intorno ai costumi ed ai danni della mosca delle arance (*Ceratitis capitata* Wied.). *Bollettino del Laboratorio di Zoologia Generale e Agraria della Reale Scuola Superiore dAgricoltura in Portici* 4, 120–127.

Matteson, P.C. (2000) Insect pest management in tropical Asian irrigated rice. *Annual Review of Entomology* 45, 549–574.

Mayer, D.F., Long, L.E., Smith, T.J., Olsen, J., Riedl, H., Heath, R.R., Leskey, T.C. and Prokopy, R.J. (2000) Attraction of adult *Rhagoletis indifferens* (Diptera: Tephritidae) to unbaited and odor-baited red spheres and yellow rectangles. *Journal of Economic Entomology* 93, 347–351.

Mazerolle, M.J. and Villard, M.A. (1999) Patch characteristics and landscape context as predictors of species presence and abundance: a review. *Ecosciences* 6, 117–124.

Meats, A. (1989) Abiotic mortality factors – temperature. In: Robinson, A.S. and Hooper, G. (eds) *World Crop Pests, Fruit Flies: Their Biology, Natural Enemies, and Control.* 3B. Elsevier, Amsterdam, pp. 229–240.

Meats, A., Clift, A.D. and Perepelicia, N. (2002) Performance of permanent and supplementary traps for Mediterranean and Queensland fruit flies in South Australia 1975–2001: comparison of male lure and food lure traps. *General and Applied Entomology* 32, 53–57.

Messina, F.J. and Jones, V.P. (1990) Relationship between fruit phenology and infestation by the apple maggot (Diptera: Tephritidae) in Utah. *Annals of the Entomological Society of America* 83, 742–752.

Metcalf, R.L. (1990) Chemical ecology of Dacinae fruit flies (Diptera:Tephritidae). *Annals of the Entomological Society of America* 83, 1017–1030.

Michon, G., Bompart, J., Hecketsweiler, P. and Decatillon, C. (1983) Tropical forest architectural analysis as applied to agroforests in humid tropics: the example of traditional village-agroforests in West Java. *Agroforestry Systems* 1, 117–129.

Miller, J.R. and Cowles, R.S. (1990) Stimulo-deterrent diversion: a concept and its possible application to onion maggot control. *Journal of Chemical Ecology* 16, 197–212.

Montiel-Bueno, A. and Mata, M.S. (1985) A sexual confusion technique in the control of *Dacus oleae* (Gmel.). In: Cavalloro R. and Crovetti, A. (eds) *Integrated Pest Control in Olive Groves*. A.A. Balkema, Rotterdam, The Netherlands, pp. 420–427.

Montoya, P., Liedo, P., Benrey, B., Cancino, J., Barrera, J.F., Sivinski, J. and Aluja, M. (2000) Biological control of *Anastrepha* spp. (Diptera: Tephritidae) in mango orchards through augmentative releases of *Diachasmimorpha longicaudata* (Ashmead) (Hymenoptera: Braconidae). *Biological Control* 18, 216–224.

Moreno, D.S. and Mangan, R.L. (2000) Novel insecticide strategies such as phototoxic dyes in adult fruit fly control and suppression programmes. In: Tan, K.H. (ed.) *Area-Wide Control of Fruit Flies and Other Insect Pests*. Penerbit Universiti Sains Malaysia, Pulau Pinang, Malaysia, pp. 421–432.

Mosqueda-Vázquez, R., de los Santos, F., Becerra-Leor, E.N., Cabrera-Mireles, H., Ortega-Zaleta, D.A. and del Ángel-Pérez, A.L. (1996) *Manual para Cultivar Mango en la Planicie Costera del Golfo de México*. Instituto Nacional de Investigaciones Forestales, Agrícolas y Pecuarias (INIFAP), Campo Experimental Cotaxtla, Veracruz, México.

Murphy, B.C., Wilson, L.T. and Dowell, R.V. (1991) Quantifying apple maggot (Diptera: Tephritidae) preference for apples to optimize the distribution of traps among trees. *Environmental Entomology* 20, 981–987.

Nakagawa, S., Prokopy, R.J., Wong, T.T.Y., Ziegler, J.R., Mitchell, S.M., Urago, T. and Harris, E.J. (1978) Visual orientation of *Ceratitis capitata* flies to fruit models. *Entomologia Experimentalis et Applicata* 24, 193–198.

Nault, B.A. and Kennedy, G.G. (2000) Seasonal changes in habitat preference by *Coleomegilla maculata*: implications for Colorado potato beetle management in potato. *Biological Control* 17, 164–173.

Neilson, W.T.A. (1962) Effects of temperature on development of overwintering pupae of the apple maggot, *Rhagoletis pomonella* (Walsh). *Canadian Entomologist* 94, 924–928.

Neilson, W.T.A. (1964) Some effects of relative humidity on development of pupae of the apple maggot *Rhagoletis pomonella* (Walsh). *Canadian Entomologist* 96, 810–811.

Nestel, D., Carvalho, J. and Nemny-Lavy, E. (2004). The spatial dimension in the ecology of insect pests and its relevance to pest management. In: Hoeowitz, A.R. and Ishaaya, I. (eds) *Insect Pest Management*. Springer-Verlag, Heidelberg, Germany, pp. 45–63.

Nishida, T. and Bess, H.A. (1950). Applied ecology in melon fly control. *Journal of Economic Entomology* 43, 877–883.

Nishida, T., Harris, E.J., Vargas, R.I. and Wong, T.T.Y. (1985) Distributional loci and host fruit utilization patterns of the Mediterranean fruit fly, *Ceratitis capitata* (Diptera: Tephritidae), in Hawaii. *Environmental Entomology* 14, 602–606.

Nishida, R., Tan, K.H., Serit, M., Lajis, N.H., Sukari, A.M., Takahashi, S. and Fukami, H. (1988) Accumulation of phenylpropanoids in the rectal glands of males of the Oriental fruit fly, *Dacus dorsalis*. *Cellular and Molecular Life Sciences* 44, 534–553.

Nufio, C.R. and Papaj, D.R. (2001) Host marking behavior in phytophagous insects and parasitoids. *Entomologia Experimentalis et Applicata* 99, 273–293.

Oatman, E.R. (1964) Apple maggot trap and attractant studies. *Journal of Economic Entomology* 57, 529–531.

Opdam, P., Steingröver, E. and van Rooij, S. (2005) Ecological networks: a spatial concept for multi-factor planning of sustainable landscapes. *Landscape and Urban Planning* 75, 322–332.

Ovruski, S., Aluja, M., Sivinski, J. and Wharton, R.A. (2000) Hymenopteran parasitoids on fruit infesting Tephritidae (Diptera) in Latin America and Southern United States: diversity, distribution, taxonomic status and their use in fruit fly biological control. *International Pest Management Reviews* 5, 81–107.

Ovruski, S.M., Wharton, R.A., Schliserman, P. and Aluja, M. (2005) Abundance of *Anastrepha fraterculus* (Diptera: Tephritidae) and its associated native parasitoids (Hymenoptera) in feral guavas growing in the endangered northernmost Yungas forests of Argentina with an update on the taxonomic status of Opiine parasitoids previously reported in this country. *Environmental Entomology* 34, 807–818.

Owens, E.D. and Prokopy, R.J. (1986) Relationship between reflectance spectra of host plant surfaces and visual detection of host fruit by *Rhagoletis pomonella* flies. *Physiological Entomology* 11, 297–307.

Papadopoulos, N.T., Carey, J.R., Katsoyannos, B.I. and Kouloussis, N.A. (1996) Overwintering of the Mediterranean fruit fly (Diptera: Tephritidae) in northern Greece. *Annals of the Entomological Society of America* 89, 526–534.

Papadopoulos, N.T., Katsoyannos, B.I. and Carey, J.R. (1998) Temporal changes in the composition of the overwintering larval population of the Mediterranean fruit fly (Diptera: Tephritidae) in northern Greece. *Annals of the Entomological Society of America* 89, 526–534.

Papadopoulos, N.T., Katsoyannos, B.I., Carey, J.R. and Kouloussis, N.A. (2001) Seasonal and annual occurrence of the Mediterranean fruit fly (Diptera: Tephritidae) in Northern Greece. *Annals of the Entomological Society of America* 94, 41–50.

Papadopoulos, N.T., Katsoyannos, B.I. and Nestel, D. (2003) Spatial autocorrelation analysis of a *Ceratitis capitata* (Diptera: Tephritidae) adult population in a mixed deciduous fruit orchard in Northern Greece. *Environmental Entomology* 32, 319–326.

Papaj, D.R. (2000) Ovarian dynamics and host use. *Annual Review of Entomology* 45, 423–448.

Pearson, S.M. (2002) Interpreting landscape patterns from organism-based perspectives. In: Gergel, S.E. and Turner, M.G. (eds) *Learning Landscape Ecology: A Practical Guide to Concepts and Techniques*. Springer-Verlag, New York, pp. 187–198.

Peng, R. and Christian, K. (2005) Integrated pest management in mango orchards in the Northern Territory Australia, using the weaver ant *Oecophylla smaragdina* (Hymenoptera: Formicidae) as a key element. *International Journal of Pest Management* 51, 149–155.

Peng, R. and Christian, K. (2006) Effective control of Jarvis's fruit fly, *Bactrocera jarvisi* (Diptera: Tephritidae), by the weaver ant, *Oecophylla smaragdina* (Hymenoptera: Formicidae), in mango orchards in the Northern Territory of Australia. *International Journal of Pest Management* 52, 275–282.

Pérez-Staples, D. and Aluja, M. (2006) Sperm allocation and cost of mating in a tropical tephritid fruit fly. *Journal of Insect Physiology* 52, 839–845.

Pérez-Staples, D., Aluja, M., Macías-Ordóñez, R. and Sivinski, J. (2008) Reproductive trade-offs from mating with a successful male: the case of the tephritid fly *Anastrepha obliqua*. *Behavioral Ecology and Sociobiology* 62, 1333–1340.

Phipps, C.R. and Dirks, C.O. (1933) Notes on the biology of the apple maggot fly. *Journal of Economic Entomology* 26, 349–358.

Pickett, A.D., Patterson, N.A., Schultz, H.T. and Lord, F.J. (1946) The influence of spray programs on the fauna of apple orchards in Nova Scotia, I. An appraisal of the problem and a method of approach. *Scientia Agricola* 26, 590–600.

Pickett, C.H. and Bugg, R.L. (eds) (1998) *Enhancing Biological Control: Habitat Management to Promote Natural Enemies of*

Agricultural Pests. University of California Press, Berkeley, California.

Piedra, E., Zuñiga, A. and Aluja, M. (1993) New host plant and parasitoid record in Mexico for *Anastrepha alveata* Stone (Diptera: Tephritidae). *Proceedings of the Entomological Society of Washington* 95, 127.

Piñero, J., Aluja, M., Vázquez, A., Equihua, M. and Varón, J. (2003) Human urine and chicken feces as fruit fly (Diptera: Tephritidae) attractants for resource-poor fruit growers. *Journal of Economic Entomology* 96, 334–340.

Piñero, J.C., Jácome, I., Vargas, R. and Prokopy, R.J. (2006) Response of female melon fly, *Bactrocera cucurbitae*, to host-associated visual and olfactory stimuli. *Entomologia Experimentalis et Applicata* 121, 261–269.

Potting, R.P.J., Perry, J.N. and Powell, W. (2005) Insect behavioral ecology and other factors affecting the control efficacy of agro-ecosystem diversification strategies. *Ecological Modelling* 182, 199–216.

Prokopy, R.J. (1968a) The influence of photoperiod, temperature and food on the initiation of diapause in the apple maggot. *Canadian Entomologist* 100, 318–329.

Prokopy, R.J. (1968b) Visual responses of apple maggot flies, *Rhagoletis pomonella* (Diptera: Tephritidae): orchard studies. *Entomologia Experimentalis et Applicata* 11, 403–422.

Prokopy, R.J. (1972) Evidence for a pheromone deterring repeated oviposition in apple maggot flies. *Environmental Entomology* 1, 326–332.

Prokopy, R.J. (1981) Epideitic pheromones that influence spacing patterns of phytophagous insects. In: Nordlund, D.A., Jones, R.L. and Lewis, W.J. (eds) *Semiochemicals: Their Role in Pest Control*. Wiley Press, New York, pp. 181–213.

Prokopy, R.J. (1994) Integration in orchard pest and habitat management: a review. *Agriculture Ecosystems and Environment* 50, 1–10.

Prokopy, R.J. and Mason, J. (1996) Behavioral control of apple maggot flies. In: McPheron, B.A. and Steck, G.J. (eds) *Fruit Fly Pests: A World Assessment of Their Biology and Management*. St Lucie Press, Delray Beach, Florida, pp. 555–559.

Prokopy, R.J. and Owens, E.D. (1983) Visual detection of plants by herbivorous insects. *Annual Review of Entomology* 28, 337–364.

Prokopy, R.J. and Vargas, R.I. (1993) Attraction of *Ceratitis capitata* (Diptera: Tephritidae) flies to odor of coffee fruit. *Journal of Chemical Ecology* 22, 807–820.

Prokopy, R.J., Bennett, E.W. and Bush, G.L. (1971) Mating behavior in *Rhagoletis pomonella* (Diptera: Tephritidae). I. Site of assembly. *Canadian Entomologist* 103, 1405–1409.

Prokopy, R.J., Moericke, V. and Bush G.L. (1973) Attraction of apple maggot flies to odour of apples. *Environmental Entomology* 2, 743–749.

Prokopy, R.J., Economopoulos, A.P. and McFadden, M.W. (1975) Attraction of wild and laboratory-cultured *Dacus oleae* flies to small rectangles of different hues, shades, and tints. *Entomologia Experimentalis et Applicata* 18, 141–152.

Prokopy, R.J., Reissig, W.H. and Moericke, V. (1976) Marking pheromones deterring repeated oviposition in *Rhagoletis* flies. *Entomologia Experimentalis et Applicata* 20, 170–178.

Prokopy, R.J., Ziegler, J.R. and Wong, T.T.Y. (1978) Deterrence of repeated oviposition by fruit marking pheromone in *Ceratitis capitata* (Diptera: Tephritidae). *Journal of Chemical Ecology* 4, 55–63.

Prokopy, R.J., Cooley, S.S., Prokopy, J., Quan, Q. and Buonaccorsi, J.P. (1984) Interactive effects of resource abundance and state of adults on residence of apple maggot (Diptera: Tephritidae) flies in host tree patches. *Environmental Entomology* 23, 304–315.

Prokopy, R.J., Drew, R.A.I., Sabine, B.N.E., Lloyd, A.C. and Hamacek, E. (1991) Effect of physiological and experiential state of *Bactrocera tryoni* flies on intra-tree foraging behavior for food (bacteria) and host fruit. *Oecologia* 87, 394–400.

Prokopy, R.J., Papaj, D.R., Hendrichs, J. and Wong, T.T.Y. (1992) Behavioral responses of *Ceratitis capitata* flies to bait spray droplets and natural food. *Entomologia Experimentalis et Applicata* 64, 247–257.

Prokopy, R.J., Chiou, L.H. and Vargas, R.I. (1993a) Effect of source and condition of animal excrement on attractiveness to adults of *Ceratitis capitata* (Diptera: Tephritidae). *Environmental Entomology* 22, 453–458.

Prokopy, R.J., Cooley, S.S., Galarza, L., Bergergweiler, C. and Lauzon, C.R. (1993b) Bird droppings compete with bait sprays for *Rhagoletis pomonella* (Walsh) flies (Diptera: Tephritidae). *Canadian Entomologist* 125, 413–422.

Prokopy, R.J., Cooley, S.S., Prokopy, J.J., Quan, Q. and Buonaccorsi, J.P. (1994a) Interactive effects of resource abundance and state of adults on residence of apple maggot (Diptera: Tephritidae) flies in host tree patches. *Environmental Entomology* 23, 304–315.

Prokopy, R.J., Roitberg, B.D. and Vargas, R.I. (1994b) Effects of egg load on finding and acceptance of host fruit in *Ceratitis capitata* flies. *Physiological Entomology* 19, 124–132.

Prokopy, R.J., Mason, J.L., Christie, M. and Wright, S.E. (1996) Arthropod pest and natural enemy abundance under second-level versus first-level integrated pest management practices in apple orchards: a 4-year study. *Agriculture, Ecosystems and Environment* 57, 35–47.

Prokopy, R.J., Mittenthal, R.E. and Wright, S.E. (2003a) Evaluation of trap deployment patterns for behavioural control of apple maggot flies (Dipt., Tephritidae). *Journal of Applied Entomology* 127, 276–281.

Prokopy, R.J., Miller, N.W., Piñero, J.C., Barry, J.D., Tran, L.C., Oride, L. and Vargas, R.I. (2003b) Effectiveness of GF-120 fruit fly bait spray applied to border area plants for control of melon flies (Diptera: Tephritidae). *Journal of Economic Entomology* 96, 1485–1493.

Prokopy, R.J., Jácome, I. and Bigurra, E. (2005) An index for assigning distances between odor-baited spheres on perimeter trees of orchards for control of apple maggot flies. *Entomologia Experimentalis et Applicata* 115, 371–377.

Raghu, S., Drew, R.A.I. and Clarke, A.R. (2004) Influence of host plant structure and microclimate on the abundance and behaviour of a tephritid fly. *Journal of Insect Behavior* 17, 179–190.

Rauch, G. and Weisser, W.W. (2007) Local and spatial dynamics of a host-parasitoid system in a field experiment. *Basic and Applied Ecology* 8, 89–95.

Reissig, W.H., Barnard, J., Weires, R.W., Glass, E.H. and Dean, R.W. (1979) Prediction of apple maggot fly emergence from thermal unit accumulation. *Environmental Entomology* 8, 51–54.

Reynolds, A.H., Kaknes, A.M. and Prokopy, R.J. (1998) Evaluation of two trap deployment methods to manage the apple maggot fly (Dipt., Tephritidae). *Journal of Applied Entomology* 122, 255–258.

Richter, M.R. (2000) Social wasp (Hymenoptera: Vespidae) foraging behavior. *Annual Review of Entomology* 45, 121–150.

Robacker, D.C. (2007) Attractiveness to *Anastrepha ludens* (Diptera: Tephritidae) of plant essential oils and a synthetic food-odor lure. *Journal of Applied Entomology* 131, 202–208.

Robacker, D.C. and Heath, R. (1996) Attraction of Mexican fruit flies (Diptera: Tephritidae) to lures emitting host-fruit volatiles in a citrus orchard. *Florida Entomologist* 79, 600–602.

Robacker, D.C. and Lauzon, C.R. (2002) Purine metabolizing capability of *Enterobacter agglomerans* affects volatile production and attractiveness to Mexican fruit fly. *Journal of Chemical Ecology* 28, 1549–1563.

Robacker, D.C., Martinez, A.J., Garcia, J.A. and Bartelt, R.J. (1998) Volatiles attractive to the Mexican fruit fly (Diptera: Tephritidae) from eleven bacteria taxa. *Florida Entomologist* 81, 497–508.

Robacker, D.C., Lauzon, C.R. and Xiaodun, H.E. (2004) Volatile production and attractiveness to the Mexican fruit fly of *Enterobacter agglomerans* isolated from apple maggot and Mexican fruit flies. *Journal of Chemical Ecology* 30, 1329–1347.

Roitberg, B.D. and Mangel, M. (1997) Individuals on the landscape: behavior can mitigate landscape differences among habitats. *Oikos* 80, 234–240.

Roitberg, B.D. and Prokopy, R.J. (1984) Host visitation sequence as a determinant of search persistence in fruit parasitic tephritid flies. *Oecologia* 62, 7–12.

Roitberg, B.D. and Prokopy, R.J. (1987) Insects that mark host plants. *Bioscience* 37, 400–406.

Roitberg, B.D., Cairl, R.S. and Prokopy, R.J. (1984) Oviposition deterring pheromone influences dispersal distance in tephritid fruit flies. *Entomologia Experimentalis et Applicata* 35, 217–220.

Ronchi-Teles, B. and da Silva, N. (2005) Flutuação populacional de espécies de *Anastrepha* Schiner (Diptera: Tephritidae) na região de Manaus, AM. *Neotropical Entomology* 34, 733–741.

Rossler, Y. and Greany, P.D. (1990) Enhancement of citrus resistance to the Mediterranean fruit fly. *Entomologia Experimentalis et Applicata* 54, 89–96.

Rull, J. and Prokopy, R.J. (2000) Attraction of apple maggot flies, *Rhagoletis pomonella* (Diptera: Tephritidae) of different physiological states to odour-baited traps in the presence and absence of food. *Bulletin of Entomological Research* 90, 77–88.

Rull, J. and Prokopy, R.J. (2001) Effect of apple-orchard structure on interception of *Rhagoletis pomonella* (Diptera: Tephritidae) flies by odor-baited traps. *Canadian Entomologist* 133, 355–363.

Rull, J. and Prokopy, R.J. (2003) Trap position and fruit presence affect visual responses of apple maggot flies (Dipt., Tephritidae) to different trap types. *Journal of Applied Entomology* 127, 85–90.

Rull, J. and Prokopy, R.J. (2004a) Host-finding and ovipositional boring responses of apple maggot flies (Diptera: Tephritidae) to different apple genotypes. *Environmental Entomology* 33, 1695–1702.

Rull, J. and Prokopy, R.J. (2004b) Revisiting within-tree trap positioning for apple maggot fly (Dipt., Tephritidae) behavioural control. *Journal of Applied Entomology* 128, 195–199.

Rull, J. and Prokopy, R.J. (2005) Interaction between natural and synthetic fruit odor influences response of Apple maggot flies (Diptera: Tephritidae) to visual traps. *Entomologia Experimentalis et Applicata* 114, 79–86.

Saunders, D.A., Hobbs, R.J. and Margules, C.R. (1991) Biological consequences of ecosystem fragmentation: a review. *Conservation Biology* 5, 18–32.

Shea, K., Possingham, H.P., Murdoch, W.W. and Roush, R. (2002) Active adaptive management in insect pest and weed control: intervention with a plan for learning. *Ecological Applications* 12, 927–936.

Shelly, T.E. and Dewire, A.M. (1994). Chemically mediated mating success in male oriental fruit flies (Diptera: Tephritidae). *Annals of the Entomological Society of America* 87, 375–382.

Shelly, T.E. and Kennelly, S.S. (2007) Settlement patterns of Mediterranean fruit flies in the tree canopy: an experimental analysis. *Journal of Insect Behavior* 20, 453–472.

Shelton, A.M. and Badenes-Pérez, F.R. (2006) Concepts and applications of trap cropping in pest management. *Annual Review of Entomology* 51, 285–308.

Sidehurst, M.S. and Jang, E.B. (2006) Female-biased attraction of oriental fruit fly, *Bactrocera dorsalis* (Hendel), to a blend of host fruit volatiles from *Terminalia catappa* L. *Journal of Chemical Ecology* 32, 2513–2524.

Silva, J.G. and Malavasi, A. (1993) The status of honeydew melon as a host of *Anastrepha grandis* (Diptera: Tephritidae). *Florida Entomologist* 76, 516–519.

Sivinski, J. (1990) Colored spherical traps for capture of Caribbean fruit fly *Anastrepha suspensa*. *Florida Entomologist* 73, 123–128.

Sivinski, J.M. (1996) The past and potential of biological control of fruit flies. In: McPherson, B.A. and Steck, G.J. (eds) *Fruit Fly Pests. A World Assessment of their Biology and Management.* St Lucie Press, Delray Beach, Florida, pp. 365–375.

Sivinski, J., Calkins, C.O., Baranowski, R., Harris, D., Brambila, J., Diaz, J., Burns, R., Holler, T. and Dobson, G. (1996) Suppression of a Caribbean fruit fly *Anastrepha suspensa* (Loew) (Diptera: Tephritidae) population through augmentative releases of the parasitoid *Diachasmimorpha longicaudata* (Ashmead) (Hymenoptera: Braconidae). *Biological Control* 6, 177–185.

Sivinski, J., Aluja, M. and López, M. (1997) Spatial and temporal distributions of parasitoids of Mexican *Anastrepha* species (Diptera: Tephritidae) within canopies of fruit trees. *Annals of the Entomological Society of America* 90, 604–618.

Sivinski, J., Piñero, J. and Aluja, M. (2000a) The distributions of parasitoids (Hymenoptera) of *Anastrepha* fruit flies (Diptera: Tephritidae) along an altitudinal gradient in Veracruz, Mexico. *Biological Control* 18, 258–269.

Sivinski, J., Aluja, M., Dodson, G.N., Friedberg, A., Headrick, D.H., Kaneshiro, K.Y. and Landolt, P. (2000b) Topics in the evolution of sexual behaviour in the Tephritidae. In: Aluja, M. and Norrbom, A.L. (eds) *Fruit Flies (Diptera: Tephritidae): Phylogeny and Evolution of Behavior*. CRC Press, Boca Raton, Florida, pp. 751–792.

Sivinski, J., Aluja, M., Piñero, J. and Ojeda, M. (2004) Novel analysis of spatial and temporal patterns of resource use in a group of tephritid flies of the genus *Anastrepha*. *Annals of the Entomological Society of America* 97, 504–512.

Sivinski, J., Holler, T., Pereira, R. and Romero, M. (2007) The thermal environment of immature Caribbean fruit flies, *Anastrepha suspensa* (Diptera: Tephritidae). *Florida Entomologist* 90, 347–357.

Solomon, M.G., Cross, J.V., Fitzgerald, J.D., Campbell, C.A.M., Jolly, R.L., Olszak, R.W., Niemczyk, E. and Vogt, H. (2000) Biocontrol of pests of apples and pears in Northern and Central Europe – 3. Predators. *Biocontrol Science and Technology* 10, 91–128.

Southwood, T.R.E. and Way, M.J. (1970) Ecological background to pest management. In: Rabb, R.L. and Guthrie, F.E. (eds) *Concepts of Pest Management*. North Carolina State University, Raleigh, North Carolina, pp. 6–29.

Stark, J.D., Vargas, R. and Miller, N. (2004) Toxicity of spinosad in protein bait to three economically important tephritid fruit fly species (Diptera: Tephritidae) and their parasitoids (Hymenoptera: Braconidae). *Journal of Economic Entomology* 97, 911–915.

Steffens, R.J. (1983) Ecology and approach to integrated control of *Dacus frontalis* on the Cape Verde Islands. In: Cavalloro, R. (ed.) *Fruit Flies of Economic Importance*. Bakelma, Rotterdam, The Netherlands, pp. 632–638.

Steiner, L.F., Mitchell, W.C., Harris, E.J., Kozuma, T.T. and Fujimoto, M.S. (1965) Oriental fruit fly eradication by male annihilation. *Journal of Economic Entomology* 585, 961–964.

Stern, V.M. (1973) Economic thresholds. *Annual Review of Entomology* 18, 259–280.

Suckling, D.M., Jang, E.B., Carvalho, L.A., Nagata, J.T., Schneider, E.L. and El-Sayed, A.M. (2007) Can ménage-à-trois be used for controlling insects? *Journal of Chemical Ecology* 33, 1494–1504.

Teixeira, L.A.F. and Polavarapu, S. (2001) Post-diapause development and prediction of emergence of female blueberry maggot (Diptera: Tephritidae). *Environmental Entomology* 30, 925–931.

Teixeira, L.A.F. and Polavarapu, S. (2004) Evidence of a heat-induced quiescence during pupal development in *Rhagoletis mendax* (Diptera: Tephritidae). *Environmental Entomology* 34, 292–297.

Thies, C. and Tscharntke, T. (1999) Landscape structure and biological control in agroecosystems. *Science* 285, 893–895.

Thies, C., Roschewitz, I. and Tscharntke, T. (2005) The landscape context of cereal aphid-parasitoid interactions. *Proceedings of the Royal Society* 272, 203–210.

Thomas, D.B. (1993) Survivorship of the pupal stages of the Mexican fruit fly *Anastrepha ludens* (Loew) (Diptera: Tephritidae) in an agricultural and nonagricultural situation. *Journal of Entomological Science B* 28, 350–362.

Thomas, D.B. (1995) Predation on the soil inhabiting stages of the Mexican fruit fly. *Southwestern Entomologist* 20, 61–71.

Tscharntke, T., Bommarco, R., Clough, Y., Crist, T.O., Kleijn, D., Rand, T.A., Tylianakis, J.M., van Nouhuys, S. and Vidal, S. (2007) Conservation biological control and enemy diversity on a landscape scale. *Biological Control* 43, 294–309.

Turner, M. (1989) Landscape ecology: the effect of pattern on process. *Annual Review of Ecology and Systematics* 20, 171–197.

Turner, M. (2005) Landscape ecology: what is the state of the science? *Annual Review of Ecology, Evolution and Systematics* 36, 319–344.

USDA (1999) Rules and Regulations: 7 CFR Part 319 [Docket No. 97–107–3] Importation of Fruits and Vegetables. *Federal Register* 64, 2993 (http://bulk.resource.org/gpo.gov/register/1999/1999_2993.pdf, accessed 27 November 2008).

Van den Berg, J., Rebe, M., De Bruyn, J. and Van Hamburg, H. (2001) Developing habitat management systems for gramineous stemborers in South Africa. *Insect Science and its Application* 21, 381–388.

van Mele, O. and Cuc, N.T.T. (2000) Evolution and status of *Oecopylla smaragdina* (Fabricius) as a pest control agent in citrus in the Mekong Delta, Vietnam. *International Journal of Pest Management* 46, 295–301.

van Mele, O. and Cuc, N.T.T. (eds) (2003) *Ants as Friends: Improving Your Tree Crops with Weaver Ants*. CAB International, Egham, UK.

van Mele, P. and Chien, H.V. (2004) Farmers, biodiversity and plant protection: developing a learning environment for sustainable tree cropping systems. *International Journal of Sustainable Agriculture* 2, 67–76.

van Mele, P., Cuc, N.T.T. and van Huis, A. (2001) Farmers' knowledge, perceptions and practices in mango pest management in the Mekong Delta, Vietnam. *International Journal of Pest Management* 47, 7–16.

van Mele, P., Vayssières, J.-F., van Tellingen, E. and Vrolijks, J. (2007) Effects of an African weaver ant, *Oecophylla longinoda*, in controlling mango fruit flies (Diptera: Tephritidae) in Benin. *Journal of Economic Entomology* 100, 695–701.

Verkerk, R.H.J., Leather, S.R. and Wright, D.J. (1998) The potential for manipulating crop–pest–natural enemy interactions for improved insect pest management. *Bulletin of Entomological Research* 88, 493–501.

Vreysen, M.J.B., Gerardo-Abaya, J. and Cayol, J.P. (2007) Lessons from area-wide integrated pest management (AW-IPM) programmes with an SIT component: an FAO//IAEA perspective. In: Vreysen, M.J.B., Robinson, A.S. and Hendrichs, J. (eds) *Area-Wide Control of Insect Pests: From Research to Field Implementation*. Springer, Dordrecht, Germany, pp. 723–744.

Weibull, A.C., Östman, Ö. and Granqvist, A. (2003) Species richness in agroecosystems: the effect of landscape, habitat and farm management. *Biodiversity and Conservation* 12, 1335–1355.

Wharton, R.A. (1989) Classical biological control of fruit-infesting Tephritidae. In: Robinson, A.S. and Hooper, G. (eds) *World Crop Pests. Fruit Flies: Their Biology, Natural Enemies and Control, 3B*. Elsevier, Amsterdam, pp. 303–313.

Yokoyama, V.Y. and Miller, G.T. (1993) Pest free period for walnut husk fly (Diptera: Tephritidae) and host status of stone fruits for export to New Zealand. *Journal of Economic Entomology* 86, 1766–1772.

Yokoyama, V.Y. and Miller, G.T. (1994) Walnut husk fly (Diptera: Tephritidae) pest free and preovipositional periods and adult emergence for stone fruits exported to New Zealand. *Journal of Economic Entomology* 87, 747–751.

Zhang, A., Linn, Ch., Wright, S., Prokopy, R.J., Wright, S., Prokopy, R., Reissig, W. and Roelofs, W. (1999) Identification of a new blend of apple volatiles attractive to the apple maggot *Rhagoletis pomonella*. *Journal of Chemical Ecology* 25, 1221–1232.

Zwölfer, H. (1983) Life systems and strategies of resource exploitation in tephritids. In: Cavalloro, R. (ed.) *Fruit Flies of Economic Importance*. A.A. Balkema, Rotterdam, The Netherlands, pp. 16–30.

8 Biorational Approaches to Disease Management in Apples

DANIEL R. COOLEY

Department of Plant, Soil and Insect Sciences, Fernald Hall, University of Massachusetts, Amherst, Massachusetts, USA

Introduction

The late R.J. Prokopy characterized his 20 years of experience developing a small, ecologically based apple orchard on his farm in Conway, Massachusetts, as being a 'bottom up' approach to integrated pest management (IPM), in that he planted it new, using disease-resistant cultivars, on a site with no immediate history of apple production (Prokopy, 2003). This *de novo* method contrasts with 'top down' approaches, the most common forms of IPM implementation, which introduce IPM tactics in established commercial orchards. Prokopy was among the first to attempt to use disease-resistant cultivars as the foundation of an IPM system that might eliminate, or at least cut dramatically, the need for fungicides. In the early years of production, diseases were not an issue in the Conway orchard but, as it matured, the summer blemish diseases, sooty blotch and flyspeck, built to unacceptable levels. From then on, Prokopy used two fungicide applications per year, in addition to managing reservoir hosts in borders, to keep the summer blemish diseases under control.

The Conway orchard illustrates the complexity involved in developing bio-intensive disease management in apple agroecosystems. Initially, Prokopy assumed that resistance would eliminate the need for fungicides because apple scab management so dominated disease priorities in New England orchards that it obscured the importance of other diseases. But, he discovered that this single tactic, genetic resistance to scab, did not solve disease issues in the orchard and he came to appreciate the danger of relying on a single tactic, even one as seemingly robust as genetic resistance, to solve a pest problem. As he later pointed out, 'the essence of IPM may be summarized as a decision-based process involving coordinated use of *multiple tactics* for optimizing the control of all classes of pests (insects, pathogens, weeds, vertebrates) in an ecologically and economically sound manner' (Prokopy, 2003). In apples specifically, this means tactics to control apple scab, for example, should be integrated with tactics that target

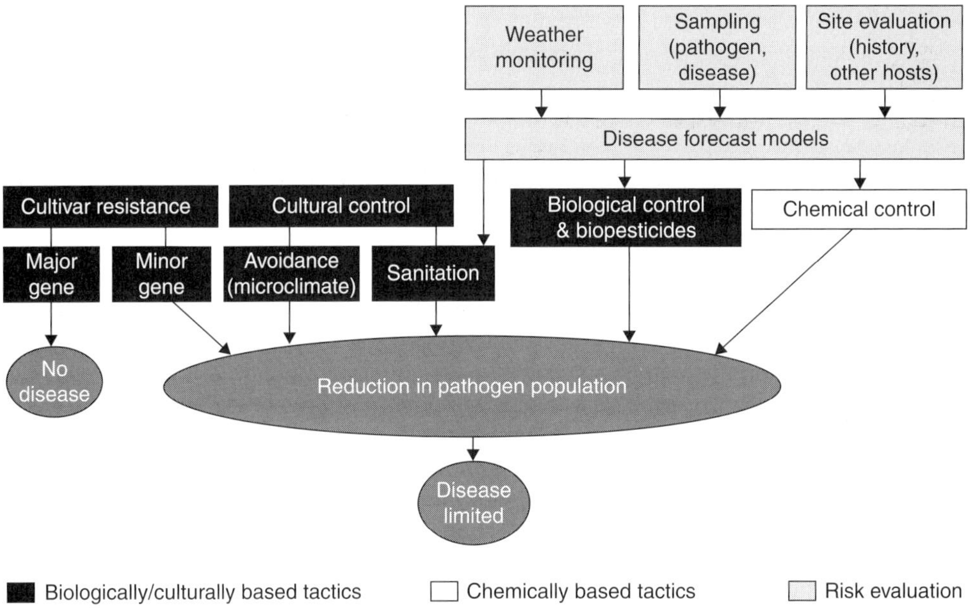

Fig. 8.1. The relationship of general tactics and disease management in apples (MacHardy, 2000b).

other diseases, with tactics for managing insects and with horticultural management such as tree training. Given the complexity of apple production systems, integrating tactics can complicate orchard management.

Disease management alone has several types of tactics with varying degrees of efficacy. Plant pathologists have classified disease management tactics into four general categories: chemical, biological, cultural and genetic management tools (Fig. 8.1) (Agrios, 2005). Disease forecasting models have also been developed in an attempt to use these tools more effectively. The multiple tactic dictum in the context of IPM has often been interpreted to mean that in IPM programmes, chemical control should be de-emphasized relative to other tactics, under the assumption that agricultural chemicals are inevitably more harmful to human health and the environment than other tactics. This idea has been formalized in the concept of biointensive IPM (Benbrook, 1996; Dufour, 2001). In the case of plant diseases, biointensive tactics include biological controls, host genetics and cultural practices (Fig. 8.1). Some forms of chemical control also may be included in biointensive management systems, if the chemicals are judged safe in terms of human health and the environment. Other chemicals, including the most commonly used apple fungicides, captan and mancozeb, have provided an effective and economic means to manage apple diseases for many years, but are often regarded as human health risks (Ragsdale, 2000). Apple IPM programmes have reduced their use, but finding adequate biointensive substitutes for them has been challenging.

Even when biointensive tactics are discovered and developed, they may founder on the rocks of economic necessity. Biointensive IPM approaches tend to be more expensive, more difficult to use and present more risk of failure than conventional chemical approaches (Penrose, 1995; Carisse and Dewdney, 2002). The pesticide use reductions that accrue from biointensive IPM can benefit society and consumers, but the expense, extra management and risk of financial loss are the growers'. Biointensive IPM programmes come with real economic ramifications, and forms of compensation need to be in place to offset risk and increased expense (Penrose, 1995; Rosenberger, 2003).

Ron Prokopy's orchard was an early prototype in the development of biointensive IPM, a proof-of-concept, 20-year experiment. Prokopy realized that IPM develops through increasingly complex levels (Prokopy, 1993) and recognized that truly sustainable orchards ultimately must be supported by policy makers and consumers. It was Prokopy's hope that as biointensive tactics were developed and adopted by producers, consumers and policy makers would become more knowledgeable regarding the value and constraints of sustainable production methods, encouraging the development of markets and regulations that would make more sustainable apple orchards possible.

Relative to other parts of the world, the north-eastern region of North America is a difficult place to grow apples without using chemicals for disease management. The relatively humid, temperate climate is ideal for several fungal pathogens of apple that present fewer problems in drier growing regions, such as the western USA, Chile and Argentina. In eastern Canada and the north-eastern USA, growers typically make from 8 to 12 fungicide applications per year, primarily targeting apple scab, and apple scab fungicides are among the most heavily used pesticides in Quebec (Chouinard *et al.*, 1998). In comparison, growers in Washington state apply 3–5 fungicides per year primarily targeting powdery mildew (Brunner *et al.*, 2003).

An impressive number of biointensive management options may be used against apple diseases in north-eastern North America. While reliability is an issue for some methods, the primary challenge lies in developing a coordinated programme of multiple tactics for commercial orchards. To date, attempts to integrate multiple biointensive disease management tactics complicates management and increases risks. This makes it difficult to meet the requirement that biointensive apple IPM be economically sound. To address this barrier, researchers are developing IPM programmes that combine the most commercially acceptable biointensive tactics with innovative marketing in order to move IPM towards the goal of economic and ecological sustainability.

Key Diseases in Apples

The widely used *APS Compendium of Apple and Pear Diseases*, a practical guide for the diagnosis of pome fruit disorders, describes over 100 diseases (Jones and Aldwinckle, 1990). Fortunately, the number of disease problems requiring regular treatment in apple agroecosystems is more tractable. The New England management guide for tree fruit discusses only about 12 apple diseases and most pathologists working in apples in north-eastern North America list from 10 to 15 economically important diseases (Sutton, 1996; Cooley and Agnello,

2007). In practice, commercial apple growers in the region focus on diseases that have a significant impact on the crop every year or on diseases that kill trees (e.g. fire blight). These may be individual diseases or a group of diseases that can be managed collectively as a single problem, such as rusts, summer blemishes or storage rots.

From a disease management perspective, the apple-growing season in north-eastern North America can be divided into two periods, determined by the key diseases requiring control in each of them. During the first period, from green tip through to fruit set, the primary disease problem is apple scab caused by the fungus *Venturia inaequalis*. Other diseases, including rusts (primarily cedar apple rust, caused by *Gymnosporangium juniperi-virginiana*) and powdery mildew, caused by *Podosphaeria leucotricha*, also must be managed, but this is accomplished generally by selecting fungicides that control rust and mildew, as well as scab. Application timing is determined largely by constraints imposed by scab management. Successful control of primary apple scab infections eliminates the need to manage secondary infections, allowing scab management to end at about the time new fruit have been set. For the rest of the season, the major disease problems include the sooty blotch and flyspeck (SBFS) blemish complex in relatively cooler climates and SBFS plus fruit rots in warmer climates. The number and identification of pathogens involved in the SBFS complex has been in a state of flux in recent years (see the discussion below) and, while other fungi are probably involved, it is agreed generally that flyspeck can be caused by *Schizothyrium pomi* and sooty blotch can be caused by *Peltaster fructicola*, *Leptodontidium elatius* and *Geastrumia polystigmatis*. The important fruit rots include bitter rot (*Colletotrichum gloeosporioides* and *C. acutatum*), black rot (*Botryosphaeria obtusa*) and white rot (*B. dothidea*). In the south-eastern USA, Alternaria leaf blotch (*Alternaria mali*) and Brooks spot (*Mycosphaerella mali*) can cause serious damage (Sutton, 1996). In addition, fire blight (caused by the bacterium, *Erwinia amylovora*) must be monitored in all areas through bloom and treated with streptomycin if necessary. While the growing season can be divided into two periods, early season and summer, demarcated by fruit set, growers also need to manage postharvest diseases. Long-term cold and controlled atmosphere storage of apples can lead to economically important storage rots caused by fungal pathogens, particularly blue mould (*Penicillium expansum*) and grey mould (*Botrytis cinerea*).

Host Resistance

Host resistance is probably the oldest and most effective disease management tool available to agriculture. Historically, people have tended to rely on quantitative and qualitative resistance as the sole tactic to manage important diseases, though, as will be demonstrated with apples, in the context of large-scale monoculture production such reliance can destroy resistance quickly. Resistance is best used as one of multiple disease management tactics and, given that most resistant apple cultivars are resistant to only one disease, scab, other tactics will be needed to control other disease problems.

The apple scab pathosystem has become a model for genomic research in a woody perennial and the extensive literature on the genetics of resistance to *V. inaequalis* in *Malus* × *domestica* has been well reviewed (MacHardy, 1996; Carisse and Dewdney, 2002; Gessler *et al.*, 2006). In North America, there have been three major apple-breeding programmes focused on resistance, including the cooperative programme between Purdue, Rutgers and Illinois Universities (Crosby *et al.*, 1992; Janick, 2006), Cornell University (Brown and Terry, 1997; Brown *et al.*, 2004) and at Agriculture and Agri-Food Canada research stations (presently Kentville, Nova Scotia, Saint-Jean-sur-Richelieu, Quebec and Summerland, British Columbia). The Cornell programme has emphasized multiple disease resistance, including resistance to cedar apple rust, powdery mildew and fire blight, as well as resistance to apple scab. Since the late 1980s, in addition to conventional breeding, recombinant DNA technology has been applied to transforming apples to make them disease resistant (Belfanti *et al.*, 2004; Gessler and Patocchi, 2007). Because apple seeds produce highly variable progeny, it is virtually impossible for conventional breeding methods using backcrosses to reproduce progeny that are both resistant and have all the desirable horticultural characteristics of a clonally propagated parent. Recombinant DNA technology allows the insertion of resistance into existing commercial cultivars.

Malus is a complex genus and the evolution of the domestic apple *M.* × *domestica* Borkh. is still debated (Robinson *et al.*, 2001; Harris *et al.*, 2002). As with cultivated crops in general, apple diseases became more problematic as production systems intensified in the late 19th and early 20th centuries (MacHardy *et al.*, 2001). In the long co-evolutionary history of the apple and its parasites, various resistance genes undoubtedly have arisen in the host, while virulence genes have developed in the parasite populations. While several sources of resistance to *V. inaequalis* have been identified, and eight races of the pathogen classified (Gessler *et al.*, 2006), the majority of named scab-resistant apples, over 70 cultivars, depend on resistance from the V_f gene. This gene confers resistance to five races of *V. inaequalis*, but new races (6 and 7) have overcome V_f resistance and spread in Europe (Parisi *et al.*, 1993; Durel *et al.*, 2003; Calenge *et al.*, 2004; Guerin *et al.*, 2007).

The development of new, virulent *V. inaequalis* races is not surprising given the reliance on V_f. As a rule, pathogens evolve to overcome single qualitative gene resistance when it is distributed widely in a host population. Focusing production on one apple cultivar in a geographic region apparently has selected virulent populations of scab in the past, with Baldwin in the north-eastern USA during the early 20th century and, more recently, with Golden Delicious in Europe being cited as two examples (MacHardy *et al.*, 2001). A strategy to use new resistance genes in a way that will reduce the chance that new scab races will be selected is being developed in Europe, where breeders are pyramiding both qualitative and quantitative resistance genes. To reduce further the chances that a ‘super race’ of scab will develop, it has been suggested that orchards be planted with mixed cultivars, that sanitation be used routinely (MacHardy *et al.*, 2001; Carisse and Dewdney, 2002) and that quarantine and exclusion be used to prevent movement of new scab races (Guerin *et al.*, 2007).

Populations of *V. inaequalis* vary across different parts of the world. This is hardly surprising, as the pathogen exists wherever apples are grown and it can reproduce sexually, while the host population exhibits multiple resistance genes. Using DNA markers, *V. inaequalis* populations from different global regions have been shown to vary, between continents, within regions and even within populations (Tenzer *et al.*, 1999; Xu *et al.*, 2008). *V. inaequalis* populations apparently adapt to host populations. Isolates of *V. inaequalis* from orchards in China were less virulent against the long-established English cv. Cox than were isolates from English orchards (Xu *et al.*, 2008). Where a new race of *V. inaequalis* capable of infecting V_f-resistant cultivars has been established in Europe, the genetic variability of the pathogen population is much smaller than in other *V. inaequalis* populations, suggesting a founder effect caused by a population bottleneck induced by the resistant apple cultivars (Guerin and Le Cam, 2004). A founder effect also may be responsible for the relatively low variability in neutral genetic markers in a Chinese population of *V. inaequalis* from a new area of apple production far from established production areas. Such evidence supports the importance of using multiple genes as a basis for scab resistance and using quarantine, sanitation and planting orchards containing multiple types of resistance in order to improve the sustainability of resistance.

Resistance and virulence with respect to apple powdery mildew is similar to that in the scab pathosystem. While resistance to *P. leucotricha* has been known for some time, and incorporated into commercially acceptable cultivars, strains of mildew have overcome some forms of resistance in Europe (Korban and Dayton, 1983; Urbanietz and Dunemann, 2005; Caffier and Parisi, 2007). Resistance breeding against rust diseases and fire blight also has been carried out to a limited extent. The effects of the deployment of this resistance on pathogen virulence remain unknown (Chen and Korban, 1987; Khan *et al.*, 2006).

Disease-resistant apple cultivars face other hurdles to their widespread commercial acceptance. In the 30 years after Prokopy planted his orchard, commercial growers generally have not accepted those disease-resistant cultivars that have been released, despite significant attempts to develop IPM systems based around them (Merwin *et al.*, 1994). Commercialized disease-resistant apple cultivars are generally rated highly in terms of consumer appeal and acceptance. Problems come largely in the area of marketing. Introduction of any new apple cultivar requires that it be produced in quantity and be recognized by consumers, a process that can take years (Murphy and Willett, 1991).

While disease-resistant apples can reduce the number of fungicides needed (Ellis *et al.*, 1998), the cultivars presently available do not make it possible to grow apples commercially without any fungicides. To date, most disease-resistant apple cultivars are resistant only to scab, though some cultivars are also resistant to mildew, fire blight and rust (Warner, 1990; Crosby *et al.*, 1992; MacHardy, 1996; Gessler *et al.*, 2006; Janick, 2006). While fungicide applications early in the growing season largely target scab, growers generally select materials that also will control rust and powdery mildew. Without these fungicide applications and without resistance to rust and mildew, these diseases become problems. Even in cultivars such as Liberty and Freedom, specifically selected for resistance to cedar apple rust and powdery mildew, these pathogens may cause disease, though symptoms are atypical (Rosenberger *et al*., 1994). Other key diseases for which there appears to be no resistance, notably SBFS, will require treatment. In the absence of fungicides, parasitism by unidentified microbes may cause damage and premature leaf senescence, resulting in fruit drop, decreased fruit size and decreased bud formation (Rosenberger *et al*., 1996).

Dwarfing rootstocks can generate much higher apple yields than seedling or semi-dwarf rootstocks (e.g. M.7) and have become the standard used in northeastern apple orchards (Autio and Southwick, 1986). However, the most common dwarfing rootstocks, M.9 and M.26, are highly susceptible to fire blight, a syndrome called rootstock blight (Vanneste and Eden-Green, 2000). Additionally, some rootstocks (e.g. M.9 EMLA) increase the susceptibility of a given scion to fire blight (Cline *et al*., 2001; Jensen *et al*., 2003). Hence, susceptible rootstocks can and do exacerbate fire blight epidemics in dwarf apple orchards in the eastern USA. *E. amylovora* can migrate asymptomatically through tolerant scions to susceptible rootstocks, where it kills cambium, phloem and xylem tissue, causing significant damage or death (Momol *et al*., 1998b).

Several rootstocks have been developed that are moderately to highly resistant to fire blight (Cline *et al*., 2001; Norelli *et al*., 2003a; Russo *et al*., 2007). The degree and nature of the resistance varies. For example, resistant rootstocks are normally selected in screening using greenhouse leaf or stem inoculations, but the B.9 rootstock expresses phenotypic resistance in that greenhouse inoculations of young plants are susceptible but grafted, older rootstocks in the field are highly resistant (Russo *et al*., 2008). Rootstock breeders have met the challenge of producing clones that grow trees with horticultural characteristics similar to the predominant M.9 and M.26 rootstocks, and several are available commercially (Cummins and Aldwinckle, 1994; Robinson *et al*., 2003a; Russo *et al*., 2007).

Some fire blight resistant rootstocks, including a number of the Geneva clones, are also resistant to root rots caused by *Phytophthora* spp. (Brown and Terry, 1997; Carisse and Khanizadeh, 2006). On wet sites, several *Phytophthora* sp. have been associated with root rot and collar rot of apples (Latorre *et al*., 2001). Where disease pressure is significant, treatment with fungicides such as metalaxyl or fosetyl-Al are warranted. Obviously, selecting a resistant or tolerant rootstock could limit or eliminate the need for such treatments.

Genetic resistance is the logical foundation to a biointensive disease management strategy in apples. To take full advantage of disease resistance, resistant scions should be grafted to resistant rootstocks. This has been done on a very limited basis, for example, in a test using the cultivar 'Liberty' grafted to a set of resistant rootstocks (Robinson *et al*., 2003b), but the study focused on productivity rather than disease. While presently available rootstock clones offer both disease resistance and commercially acceptable productivity, it appears that more effort is required to produce broadly acceptable, disease-resistant commercial varieties. Successful new commercial apple cultivars, while quite marketable, have been notably susceptible to fire blight and other diseases. Perhaps increased understanding of resistance genes and the apple genome in general will enable breeders to develop more easily cultivars that combine disease resistance with marketability, either through classic breeding or genetic engineering. Ultimately, growers will need to cooperate in planting mixed cultivars, with a diversity of resistance genes, in conjunction with other disease management tactics, in order to maintain resistance in the face of ever-adaptable pathogen populations.

Cultural Practices

Cultural practices for disease management in apples include inoculum destruction, pruning or mowing to reduce humidity and improve drying in trees, removing blossoms or otherwise mitigating disease impacts through physical actions on the plant or its environment. These tactics are rarely in themselves enough to manage a key disease, but generally are used in conjunction with other tactics to improve disease management and reduce the need for chemicals.

MacHardy combined and refined different sanitation methods to destroy leaf litter on apple orchard floors, thereby reducing the potential ascospore dose (PAD) in orchards, making it possible to eliminate early-season fungicide applications in some orchards (MacHardy, 2000b; Sutton *et al.*, 2000). Shredding apple leaves and/or spraying them with urea in the autumn makes them more susceptible to microbial degradation and consumption by earthworms, thereby destroying the substrate required by *V. inaequalis* for the formation of ascospores (Holb *et al.*, 2006). The tactic can be used in conjunction with conventional fungicides or with organic methods to reduce scab incidence significantly (Vincent *et al.*, 2004; Holb, 2006; Gomez *et al.*, 2007). Obviously, concentrated sources of scab inoculum outside an orchard, as may occur with abandoned trees, could decrease the usefulness of this tactic, though proponents argue that biologically significant quantities of ascospores do not travel more than 20–30 m (Holb *et al.*, 2004). On one hand, this sanitation practice potentially saves spray applications; on the other, it requires special equipment and additional labour, both of which add expense to scab management. The potential for overwintering conidial inoculum is another complicating factor that must be considered in some regions (Becker *et al.*, 1992; Holb *et al.*, 2005; Holb, 2006). Ultimately, destruction of scab inoculum on the orchard floor is a useful tactic in an overall apple disease management programme, but has not been widely adopted commercially, in part because growers do not see concrete, measurable benefits (Carisse and Dewdney, 2002).

Pruning apple trees improves apple quality and quantity through the management of fruiting branches, increased fruit colour and removal of disease inoculum. Removing inoculum includes cutting out cankers and mummified fruit that provide inoculum for several diseases, primarily bitter rot, black rot, bot rot and fire blight. Careful removal and destruction of these inoculum sources, through flail chopping or burning, allowed a 50% reduction of fungicides in the southeast without increased disease (Horton *et al.*, 1991). In addition to removing inoculum, pruning can change the apple canopy microclimate and pesticide spray deposition, and thereby influence disease (Cooley *et al.*, 1997). Heavy winter pruning reduces apple scab in trees managed using organic and IPM methods (Holb, 2005), apparently largely as a result of the decreased duration of leaf wetness. Sooty blotch and flyspeck are also reduced by winter pruning (Ocamb-Basu *et al.*, 1988) and summer pruning (Cooley *et al.*, 1997), as a result of both decreased relative humidity in the canopy and because spray deposition is improved (Cooley and Lerner, 1994).

Sooty blotch and flyspeck signs can be removed from apples, leaving the fruit unblemished. Hendrix (1991) found that dipping apples for 7 min in 500 ppm chlorine, then brushing and rinsing with fresh water reduced sooty blotch from 100% to 0% and flyspeck from 100% to 27% (Hendrix, 1991). Batzer *et al.* (2002) examined several disinfestation methods, showing that a 7 min dip in 800 ppm chlorine, followed by brushing and rinsing, improved USDA grades from an average of 40% Extra Fancy to 100%, increasing the market value of the fruit by over 20%. Lower rates of chlorine were also effective. The postharvest treatment can be used as a backup and supplement to preventative fungicide sprays, allowing growers to take more risks with SBFS management.

Pre-storage heat treatment has been shown to eliminate rots caused by *P. expansum* and *B. cinerea* in apples (Lurie, 1998). The heat treatments were carried out using hot air (Klein *et al.*, 1997) and hot water (Fallik, 2004). Control appears to come from both direct destruction or inhibition of the pathogen and stimulation of the production of inhibitory compounds in the fruit (Fallik *et al.*, 1996). In some cultivars, not only is disease controlled, but overall quality of the fruit is improved, maintaining firmness and slowing the ripening process (Shao *et al.*, 2007). While heat alone apparently works well against these key postharvest pathogens, it has been suggested that it be used in an integrated approach with other tactics, such as biocontrol agents (Spadaro *et al.*, 2004; Conway *et al.*, 2005).

Research in New York has shown that the vast majority of inoculum for blue mould is carried on storage bins (Rosenberger *et al.*, 2006).Therefore, management should be based around rigorous sanitation of bins, at least between seasons, using high-pressure hot water, quaternary ammonia or a similar antimicrobial. In addition, if a dump water system is used, the water should be chlorinated and pH maintained at levels that will maintain the chlorine's efficacy. Finally, rotten fruit need to be removed from the bins after each use.

Cultural controls generally require labour input and, in the north-eastern USA, farm labour is both difficult to find and expensive. Often, they are not an alternative to chemical controls but a supplement that reduces the amount of pesticide needed. In biointensive IPM, the justification for cultural controls is often that it will reduce disease pressure, allowing the use of less toxic but relatively less effective chemicals or biocontrols. For example, leaf litter destruction enhances scab control in organic orchards, but does not eliminate the need for applications of sulfur or copper, and the combination of leaf litter destruction and the relatively weak fungicides still allows 20% fruit infection at harvest (Gomez *et al.*, 2007). To be incorporated into an IPM programme, cultural controls must be combined with practices that reliably provide tangible benefits in terms of reduced cost, improved control or both.

Biopesticides

Plant pathologists generally define biological control as use by humans of one species of organism to eliminate or control another species of organism. The US

Environmental Protection Agency (EPA) and two industry groups, the Biopesticide Industry Alliance (McFarland, Wisconsin, USA) and the International Biocontrol Manufacturers Association (Mourenx, France), use the term biopesticide, defined as pesticides that have fundamentally different modes of action from conventional pesticides and hence present fewer risks to health and the environment (EPA, 2007). This eclectic group of products includes less broadly toxic antimicrobials and chemicals that stimulate plant physiology or metabolism to make hosts more resistant to pathogens (Rosenberger, 2003). For example, disease management products based around materials such as potassium bicarbonate, neem oil, the bacterial protein harpin or *Bacillus subtilis* preparations would all be biopesticides. The difference between the classic plant pathologists' definition of biological control and the broader definition of biopesticides creates overlap and occasional ambiguity when classifying products.

For example, potassium bicarbonate is fundamentally a chemical control, but because it is a relatively benign chemical with minimal environmental impact, it is also considered a biopesticide and can be termed a biorational chemical. Similarly, induced resistance or systemic acquired resistance (SAR) crosses nomenclature boundaries. Because SAR is ultimately the expression of pre-existing resistance mechanisms in plants, SAR is genetically based; however, the products that stimulate SAR may be chemicals, proteins from microbes or the microbes themselves. As a group, these products can be considered resistance activators. The key difference between activators and classic biological or chemical controls lies in the fact that the latter interact directly with pathogens to achieve disease control, while activators stimulate the plant, which then produces compounds that interact with pathogens.

The industry recognizes several barriers to the widespread adoption of biopesticides, including questions regarding product efficacy, cost, lack of awareness that the products exist, different use-patterns and risk-averse customers (Marrone, 2007). As a group, biopesticides are new products and have yet to be tested widely and incorporated into IPM programmes. In this discussion, biocontrols, activators and biorational chemicals for apple diseases will be presented as subcategories of the more general term, biopesticides.

Biocontrols

Biological control products either destroy pathogens or displace them from hosts and a number have been developed for commercial use on apples. However, biocontrols work by attempting to change the balance of natural systems and this may often mean that some damage results, though there is a significant reduction in damage relative to untreated apples. Unfortunately, the economics of the apple industry do not tolerate even small amounts of damage.

Experiments with biocontrols used against the blossom phase of fire blight illustrate the point. Serenade (a strain of *B. subtilis*), BlightBan A506 (a strain of *Pseudomonas fluorescens*) (Stockwell and Stack, 2007) and formulations of *Pantoea agglomerans/E. herbicola* such as Blossomtime have been examined extensively in university tests in the eastern USA (Momol *et al*., 1998a; Yoder *et al*.,

1999, 2003, 2004, 2007; Jones and Ehret, 2000; Reddy *et al.*, 2000, 2001; Maxson *et al.*, 2001; Aldwinckle *et al.*, 2002; Jones *et al.*, 2003; Penev and Aldwinckle, 2003; Aldwinckle and Penev, 2004; Sundin and Ehret, 2004a,b, 2005; Travis *et al.*, 2004; Sundin *et al.*, 2005, 2006a,b; Werner *et al.*, 2005, 2006a,b; Ehret *et al.*, 2007). Serenade reduced blossom blight significantly in 12 of 17 tests (71%), BlightBan in 4 of 13 tests (31%) and *P. agglomerans* in 6 of 10 tests (60%). Serenade, BlightBan and *P. agglomerans* reduced blossom blight by averages of 38, 19 and 31%, respectively, relative to the untreated controls in these tests. Such performance fell below that of streptomycin, the standard commercial treatment, which reduced blossom blight consistently by 70 to 74% in the same set of tests.

While such performance means that streptomycin remains the primary fire blight treatment, these biocontrol agents perform better when used as part of a sequence of applications that also include streptomycin. Following a streptomycin application, Serenade has been recommended as a second application and, if a third application were required, it would be streptomycin (Cooley and Agnello, 2007). In areas with existing steptomycin-resistance, BlightBan A506 has been shown to improve fire blight control when combined with streptomycin (Stockwell and Stack, 2007).

The performance of biocontrols against other apple diseases has been similar. Where biocontrols reduced disease below that in untreated controls, the reduction did not match that of standard chemical fungicides. For example, some treatments of Serenade suppressed scab, powdery mildew and rust significantly in New York (Rosenberger *et al.*, 2000). The incidence of fruit scab in unsprayed Jerseymac and Ginger Gold was 85 and 28%, respectively, and in the best Serenade treatment, was 30 and 3%, respectively. However, in the best standard fungicide treatment (a sequence of cyprodinil, mancozeb, captan and benomyl) incidence was significantly lower at 0.2 and 0.1%, respectively. Similarly, in another test in North Carolina, Serenade reduced SBFS severity significantly to 28% compared with control severity of 52%, but severity in a captan treatment was 15% less than for Serenade (Sutton *et al.*, 2007).

Theoretically, biological control has more promise against postharvest diseases, largely because the postharvest environment is carefully controlled. However, few of the approved biocontrols are in commercial use. Products aimed at postharvest pathogens have been developed, such as Bio-Save (a strain of *P. syringae*) and Aspire (a strain of the yeast *Candida oleophila*) (Wisniewski *et al.*, 2001; Janisiewicz and Korsten, 2002). Bio-Save has been used to control rot and wound pathogens of apple such as *P. expansum* in storage (ChandGoyal and Spotts, 1997; Janisiewicz and Jeffers, 1997; Stockwell and Stack, 2007).

In eastern North America, where apples commonly are dipped in fungicide and anti-scald agents before storage, the biocontrols are far less cost-effective than standard fungicides. In the western USA, where packing lines spray fungicide on fruit, the economics are more favourable for biocontrols. Still, because they lack the post-infection capabilities and residual protection provided by chemical fungicides, biocontrols have not been adopted widely. This has also led to exploring the options for combining biocontrols with standard postharvest fungicides. Combining biocontrol yeast with thiabendazole improved the activity of both agents and was effective even where postharvest pathogens were resistant to thiabendazole (ChandGoyal and Spotts, 1997). Biocontrols have been combined with newer fungicides, such as cyprodinil, in an integrated approach both to improve disease control and to attempt to manage resistance development (Zhou *et al.*, 2002; Errampalli and Brubacher, 2006).

Another approach to biological control employs microbes to destroy or inhibit the growth of disease inoculum when it is in either a dormant or saprophytic phase. This approach has been investigated extensively as a means of reducing ascospore production by *V. inaequalis*, and the history of this research has been reviewed (Carisse and Dewdney, 2002). Targeting inoculum can be used in conjunction with a PAD analysis to delay the first spray. Several microbes have been found that suppress ascospore release in the field by up to 85% and tests in orchards in Quebec using *Microsphaeropsis ochracea* reduced the number of required fungicide applications by up to 40% (Carisse and Rolland, 2004). The approach is promising, but a commercially available product for the reduction of scab inoculum has yet to be developed.

Activators

Several biorational chemicals can induce resistance in plants by stimulating physiological pathways and generating compounds that are either toxic to the pathogen or isolate the pathogen and disease damage to a limited and tolerable area (Kessmann *et al.*, 1994). This induced resistance, or systemic acquired resistance (SAR), is a phenomenon that has been known for some time, but only recently is it being applied to crops (Sticher *et al.*, 1997; Vallad and Goodman, 2004). SAR is triggered by exposing plants to virulent or avirulent microbes, microbial products or specific chemicals. The SAR response takes some time to develop fully, from several hours to days. Once developed, the plant can become resistant to a broad range of pathogens, though the response generally declines without repeated stimulation. SAR elicitors do not exert direct selective pressure on pathogens, so there is less chance that resistance to them will develop. The elicitors are reportedly less toxic to humans and the environment, making them good candidates for biointensive IPM systems.

Prohexadione-calcium is used primarily for horticultural purposes, specifically as a regulator to slow or stop tree growth. Prohexadione-Ca also can control disease, apparently by activating plant defences, though the reduction in tissue growth itself can increase disease resistance (Norelli and Miller, 2004; Rademacher, 2004). Less vigorous growth generally is believed to make apples more resistant to fire blight. However, the primary mode of action of prohexadione-Ca appears to be the induction of phytoalexin-like compounds, specifically 3-deoxyflavonoids such as luteoforol. The disease management capability of prohexadione-Ca has been exploited primarily to control the shoot blight phase of fire blight (Norelli *et al.*, 2003b), though it may be useful against other diseases. It is uniquely effective in reducing shoot blight infections, but presents some problems because, at the relatively high rates needed to have an effect on the disease, tree growth and long-term yield can be affected negatively.

Other SAR elicitors have been developed and marketed. For example, Messenger, an SAR elicitor based on the harpin protein from bacteria (Peng *et al.*, 2003), and Actigard (acibenzolar-*S*-methyl), a synthetic chemical elicitor (Brisset *et al.*, 2000; Maxson-Stein *et al.*, 2002), have been evaluated in several trials against fire blight in New York and Michigan over multiple years. In these trials, four of five with Messenger and five of seven with Actigard showed significant blossom blight reduction. However, the per cent control relative to untreated trees was 32% for Messenger and 35% for Actigard, compared with averages of 70% and 78% control for streptomycin in the two sets of tests. Tests of elicitors against other key diseases have shown similar results: in some tests, they do not have a significant effect on disease and, when they do, the effect is generally less than that of standard controls.

Biorational chemicals

Chemicals such as potassium bicarbonate and phosphorous acid compounds (phosphites and phosphonates) have been formulated for use as agricultural fungicides because they break down quickly and have very low non-target toxicity. Others, such as some formulations of sulfur and copper, have been approved by the Organic Materials Review Institute or similar organic certification boards and may also be considered biorational, though it is not always clear that they have low non-target toxicity (Eisler, 1998; Bunemann *et al.*, 2006). These chemicals generally are being compared to established chemicals that may be considered more toxic, such as captan or mancozeb. However, commercial growers in North America have not used them widely against apple diseases.

Phosphorous acid, a metabolite of the systemic fungicide aluminium tris-*O*-ethyl phosphonate (fosetyl aluminium; Aliette), is particularly promising. It was used originally to treat crown, collar and root rots caused by *Phytophthora* spp. (Johnson *et al.*, 2004). Other formulations that produce phosphorous acids are now available and, while they have been used generally against oomycete pathogens, they are being examined against other fungal and bacterial pathogens of apple. To date, the combination of a phosphite (ProPhyte) with captan can have a synergistic or added effect when used against SBFS. In North Carolina and New York trials, neither fungicide was effective alone but, applied together, they were highly effective (Sutton *et al.*, 2006, 2007; Rosenberger *et al.*, 2008a,b). Another test has shown that phosphite can enhance the activity of captan or mancozeb against apple scab and, perhaps more intriguing, phosphite sprays applied to the soil under trees at silver tip can control apple scab (Rosenberger *et al.*, 2008c). While it is unlikely that phosphorous acid fungicides by themselves will provide adequate control of key apple diseases, they may be useful in combination with reduced rates of conventional fungicides.

Calcium chloride and similar calcium salts in apples are used primarily as nutritional amendments that improve fruit firmness and reduce storage disorders (Conway *et al.*, 1991), but they have also been shown to have activity against some apple pathogens, particularly fruit rots and blemishes (Biggs, 1999, 2004; Biggs *et al.*, 1993, 2000). In a series of experiments, Biggs showed that calcium foliar sprays in the field can reduce Alternaria rot and bitter rot and have some effect on white rot, scab and cedar apple rust. Used alone, calcium sprays do not reduce SBFS significantly, though calcium does enhance captan activity, allowing captan rates to be reduced (Cooley *et al.*, 2007a). Like phosphorous acid, sprays of calcium salts alone do not control disease at levels that are commercially acceptable, but they do provide some disease suppression and may be used in conjunction with other chemicals or tactics to reduce disease to commercially acceptable levels (Biggs *et al.*, 2000; Cooley *et al.*, 2007a).

Some of the oldest fungicidal chemicals, such as sulfur or copper compounds, were once used widely but generally have been discarded in favour of chemicals that are more effective and less phytotoxic. In recent comparisons, these materials reduce disease incidence, but they also tend to be significantly less effective than conventional fungicides against diseases such as scab and SBFS (Babadoost *et al.*, 2004; Travis *et al.*, 2006a,b). Often, to get effective disease control, these products must be applied more frequently or at rates and times that damage apples.

The performance of biopesticides against apple diseases has been inconsistent to date. As the example of fire blight illustrates, the performance of biocontrols against plant diseases can vary between highly effective and completely ineffective. In an attempt to understand the factors that influence the success or failure of biocontrols in general, Ojiambo and Scherm (2006) conducted a meta-analytical review of 53 biocontrol experiments with a wide range of results. The common ideas that generally biocontrols work better in greenhouses than in the field, or that biocontrols are used more effectively against soilborne pathogens rather than against aerial pathogens, were not supported by the analysis. However, efficacy did vary by type of anatagonist: *r*-selected antagonists (organisms with short generation times and large numbers of short-lived progeny) were better than *k*-selected antagonists (long generation times, relatively fewer, long-lived progeny); and *Bacillus* spp. generally were less effective than other types of biocontrol organisms. It is early in the development of a fundamental understanding of biological controls and biopesticides in plant disease management and it will take several years to identify factors that will improve reliability to the point that they will be of use in commercial biointensive IPM systems.

Chemical Control

It is difficult to overestimate the impact fungicides have had on apple production. Bordeaux mix, sulfur and lime sulfur were used in the early part of the 20th century, but growers recognized these as imperfect solutions and tolerated significant disease damage. Cooperative Extension in New England established an inspection programme to certify that growers had achieved a level of 90% 'clean fruit' (Thies, 1947), and it was not an easily achieved goal. With the introduction of new organic chemicals and airblast sprayers in the 1940s and 1950s, apple disease management was transformed radically. Regular fungicides applied with an airblast sprayer quickly became a central component in apple production systems and, by the early 1960s, the new technology enabled most growers to produce consistently crops with disease damage that was 'negligible' (Lewis and Hickey, 1972).

Today, fungicide use on apples in the USA exceeds that in almost any other food crop (Gianessi and Marcelli, 2000). Fungicides have given growers far greater control over the quality and quantity of their crops, and increased profitability. At the same time, it has increased consumer expectations for visually attractive, blemish-free fruit. Producers generally consider fungicides and other pesticides the only economical means for controlling diseases in fruit. In fact, IPM in apples remains largely integrated *pesticide* management (Ehler, 2006), a matter of using pesticides effectively and efficiently rather than alternatives to them (Gadoury *et al.*, 1989; Penrose, 1995; Sutton, 1996; Biggs *et al.*, 2000; MacHardy, 2000b; Carisse and Dewdney, 2002). IPM programmes in apples have tended to reduce fungicide use initially, primarily by using weather monitoring, forecasting models, evaluations of inoculum density and maturity and selective chemical use. Substitution of other tactics for pesticides has been more difficult. For example, while New England apple IPM programmes reduced fungicide use in apples by up to 40% from 1978 to 1994, average fungicide use actually increased by 26% from 1994 to 2004 (Cooley, unpublished data). Unfortunately, this pattern mirrors that seen for IPM programmes in many other areas and crops (Epstein and Bassein, 2003).

Increased fungicide use in apples in recent years is the result of several factors, primarily pathogen resistance to fungicides and decreased tolerance for disease damage. Pathogen resistance to fungicides and antibiotics has been a significant problem in apple diseases such as apple scab (*V. inaequalis*), fire blight (*E. amylovora*) and blue mould (*P. expansum*). Fungal pathogens have developed resistance both to older fungicides, such as benomyl and dodine (Jones, 1981; Sutton, 1996; Yoshimura *et al.*, 2004), and to newer chemicals, such as the ergosterol biosynthesis inhibitors, strobilurins and anilinopyrimidines (Köller *et al.*, 1997, 2004, 2005b; Holb and Schnabel, 2007). In addition, some older broad-spectrum fungicides (e.g. dichlone, difolatan) are no longer available. Bacterial pathogens in some regions outside of the north-eastern USA have become resistant to the primary antibiotic, streptomycin, and have developed increased tolerance to other chemicals including copper compounds (Loper *et al.*, 1991; Sholberg *et al.*, 2001; McManus *et al.*, 2002). These resistance issues eliminate or constrain use of some of the most effective chemicals applied for apple disease management.

The decrease in fungicide use in apples during the 1980s and early 1990s and the subsequent increase during the past decade is linked to the demethylation-inhibiting (DMI) fungicides. The DMIs were labelled for commercial use on apples in 1987. Because the effective interval for DMI post-infection activity was listed as 96 h, but in reality could extend to twice that long, scab management programmes soon developed that had intervals of 10 days or more between fungicide applications, intervals that significantly exceeded that of any other fungicide (Wilcox *et al.*, 1992). Growers could time four DMI applications to coincide with acaracide and insecticide applications as appropriate, making the overall pest management programme more efficient. In commercial orchards with low scab incidence in the previous year, Gadoury *et al.* showed that the first fungicide applications could be applied as late as the pink bud phenological stage, rather than starting 2–4 weeks earlier at green tip (Gadoury *et al.*, 1989). Forms of this programme were adopted widely, reducing the five to eight early-season fungicide applications typically used against scab to four applications (Cooley and Autio, 1997). Unfortunately, by the mid 1990s, *V. inaequalis* had developed resistance to the DMIs in some commercial orchards (Köller *et al.*, 1997). The problem has grown worse since, and many apple orchards in the north-east have developed significant DMI resistance (Köller *et al.*, 2005a). As a result, growers must use fungicides that must be applied more frequently.

Since the development of significant DMI resistance, new classes of fungicides have been introduced for use on apples, such as the strobilurins or QoIs (quinone outside inhibitors) and anilinopyrimidines; however, these fungicides do not have the long-term post-infection activity of the DMIs and so must be used at shorter intervals. In addition, more recently developed chemicals generally cost significantly more than older fungicides. Rosenberger pointed out that the cost of pest management was one of the few expenses that growers could control and, given the thin margin between profit and loss for apple producers in the USA, growers will opt to use less expensive chemicals (Rosenberger, 2003). Apple growers have tended to use older fungicides such as captan or mancozeb as materials of choice, but these older chemicals face new use restrictions via regulation. This regulation is exemplified by the Food Quality Protection Act in the USA (Ragsdale, 2000). It is estimated that three of the oldest types of apple fungicides, the ethylene bisdithiocarbamates (EBDCs), the dithiocarbamates and captan, are applied to approximately half of the apple acreage in the USA. They represent 58% of the total fungicides applied to apples. While still labelled for use, these fungicides have been classified as B2 carcinogens (EPA, 1999, 2005) and, as such, have raised concerns with consumer groups (Hettenbach and Wiles, 2000). In 1989, on the basis of such concerns, the EPA restricted the amount of the EBDC fungicides that could be used in a season and lengthened the interval between last application and harvest to 77 days. This, in turn, made it more difficult to manage the sooty blotch/flyspeck complex and contributed ultimately to an overall increase in fungicide use in apples (Cooley and Manning, 1995; Williamson and Sutton, 2000).

In general, more recently developed fungicides tend to have fewer toxicity problems and hence are more acceptable in biointensive IPM programmes. The strobilurins, for example, were derived originally from a naturally occurring fungus and were promoted as an environmentally benign group of compounds that had low mammalian toxicity and that broke down rapidly (Ypema and Gold, 1999). Yet, the EPA has rated one strobilurin, kresoxim-methyl, a potential human carcinogen. At the same time, a similar strobilurin, trifloxystrobin, has not been flagged by the EPA as potentially carcinogenic. Both strobilurins are registered for use on apples and both manage the same diseases with similar efficacy (Rosenberger, 2003), yet in one eco-label management programme for apples, trifloxystrobin may be used but kresoxim-methyl may not (Green, 2007).

Forecasting and Models

Plant pathologists have long used models to determine the best time to treat plant diseases. Historically, forecasting models have worked in conjunction with curative pesticides, though appropriate models theoretically may be used to guide any management tactic. Models have been the major tool used to direct chemical applications on an 'as needed' basis. Disease forecasting models for apples typically have reduced the number of chemical applications below levels used when application schedules are based strictly on a schedule or apple phenology, while either improving or maintaining fruit quality. Consequently, these reductions lower the possible negative effects on human health and the environment.

Typically, IPM programmes have monitored the presence and abundance of pests to determine whether there is an economic threat, and hence a need to treat. Microbial pathogens present a monitoring challenge because, unlike mites, insects, weeds or vertebrate pests, they cannot be seen and counted *in situ*. In some situations, direct observation of pathogens in the laboratory has been used to evaluate inoculum density and maturity, but such evaluations require compound microscopes and, sometimes, additional specialized equipment, skill in microscopic techniques and time. Consequently, these techniques have not been adopted widely by most growers and consultants. For practical purposes, growers and crop consultants must either wait until disease is visible, at which point it may be difficult or impossible to slow an epidemic, or use readily available information about the host and environment to infer that treatment is needed. Using models that relate this kind of information to predict the risk of disease is referred to as disease forecasting, and this tactic has been developed for key apple diseases.

Apple scab

One of the oldest forecasting models for any plant disease was developed by Mills (1944) to help apple growers time sulfur applications for scab management. Mills related temperature and the hours of leaf wetness to the risk of infection by *V. inaequalis*. Using Mills's model, the first fungicide application for apple scab was made with the first predicted infection period. Repeated applications were made according to the residual efficacy of fungicide and subsequent infection periods. Mills's periods became more useful as fungicides such as dodine and dichlone, with longer post-infection activity than sulfur, were developed. Rather than assuming every rain event represented a risk of infection, sprays could be applied after infection had started, as long as the fungicide had adequate post-infection activity to arrest it. As researchers have worked with the Mills model and experienced failures to predict infection, adjustments have been made to it, such as accounting for intermittent leaf drying and wetting, light-mediated release of ascospores, or spore development at temperature extremes (Jones *et al.*, 1980; Schwabe, 1980; MacHardy and Gadoury, 1989; Xu *et al.*, 1995; Stensvand *et al.*, 1997).

The maturity of *V. inaequalis* ascospores also has been studied in an attempt to forecast scab risk. *Venturia inaequalis* ascospores mature over a discrete period, coinciding with fruit development and delineating when primary infections can occur. Generally, ascospore release begins at bud break, peaks between the tight cluster stage and petal fall, and ends completely 7–10 days after petal fall (MacHardy, 1996). Year-to-year variability in this pattern relative to tree phenology makes it useful to follow ascospore maturation each season. Several methods have been developed that assess when ascospores start to mature, when the relative number of ascospores is greatest and when all ascospores have been discharged. These include microscopic observations of crushed psuedothecia (Szkolnik, 1969; Agnello *et al.*, 2003), spore discharge in the laboratory (Szkolnik, 1969; Gilpatrick *et al.*, 1972), or using degree-day-based models (Gadoury and MacHardy, 1982; James and Sutton, 1982). The degree-day model developed by Gadoury and MacHardy (1982) has proven to give the best estimate corresponding to the cumulative per cent of trapped ascospores, is a better estimate than direct microscopic observations of pseudothecia or spore discharge tests and is easier than direct observations (Gadoury *et al.*, 2004). The model works very well when wetting occurs frequently, as it does generally in the spring in the eastern USA, but has been adjusted to take into account a hiatus in development when dry periods of a week or more occur (Rossi *et al.*, 1999; Stensvand *et al.*, 2005). It is used in conjunction with inoculum density evaluations to determine when the first scab fungicide application in a growing season should be made and in conjunction with disease incidence evaluations to determine if and when they may stop.

Primary apple scab inoculum density available in an orchard may be estimated using the potential ascospore dose (PAD) model (Gadoury and MacHardy, 1986). PAD is expressed as the number of ascospores available per square metre of orchard floor. Estimates are derived from observations of foliar scab incidence following harvest and the leaf litter density on the orchard floor in the spring. PAD estimates determine the total inoculum available for a season in an orchard and the degree-day ascospore maturity model determines what percentage of that inoculum is available at a given time. Only a small proportion of the total primary inoculum for the year is available at the beginning of ascospore maturation and release, a time coinciding with host bud-break and early leaf development. If the total amount of *V. inaequalis* inoculum available for the year is very low, the proportion of mature ascospores available during the early season will be epidemiologically insignificant and no fungicides will be needed, regardless of infection periods (MacHardy *et al.*, 1993). Specifically, if the calculated PAD is less than 600 ascospores per square metre of orchard, the first fungicide application can be delayed to the pink growth stage or for the first three infection periods, whichever occurs first. This enables growers to eliminate one to three fungicide applications, regardless of what fungicide is used. Originally, PAD determinations required determining scab incidence and intensity in autumn and leaf litter density on orchard floors in spring. Because the procedure was too cumbersome to be adopted widely by growers and consultants, it was simplified to be a count in autumn of the number of scabbed leaves on 600 terminals (MacHardy, 2000a) and simplified further to be a sequential sampling of the presence or absence of scab on terminal leaves (Reardon *et al.*, 2005). Others have used fruit scab incidence of <2% at harvest as a threshold that allows a delay (Wilcox *et al.*, 1992), but harvest fruit scab does not correlate well with PAD and can give erroneous recommendations (Cooley and Autio, 1997). PAD forecasts also can be used to determine whether orchard sanitation procedures that destroy leaf litter and inoculum may be used to lower PAD below the action threshold, thereby allowing the elimination of one or more fungicide applications in early spring (MacHardy, 2000a).

Sooty blotch and flyspeck

The utility of disease forecasting models often depends on the availability of eradicant fungicides that may be applied after an infection has started but before symptoms develop. The summer blemish diseases illustrate this point in apples. Benzimidazoles have been shown to eradicate SBFS, stopping the development of signs after it is applied (Hickey, 1977; Rosenberger *et al.*, 1990). If benzimidazoles are applied prior to the development of signs, then fruit can be kept blemish-free. Researchers in North Carolina developed a model to predict when the first appearance of SBFS signs occurs (Brown and Sutton, 1995). From 1987 to 1994, they collected weather data and determined that SBFS first appeared after 273 accumulated leaf-wetting hours (ALWH), starting at the first rain to occur 10 days after petal fall and counting only wetting that was 4 h or longer. Based on this, they recommended the application of a benzimidazole just prior to the appearance of signs, at a threshold of 200–225 ALWH. After this initial application, they recommended regular fungicide applications at roughly 2-week intervals through the rest of the season. The model generally saved one to two applications per season, because growers could begin summer fungicide applications confidently several weeks later than they would when using a standard calendar schedule.

Using an electronic leaf wetness sensor rather than the mechanical sensor used in the North Carolina work, researchers in Kentucky modified the model threshold to use at 175 ALWH, including all wetting periods regardless of length and starting after the first post-petal fall fungicide application (Hartman, 1995; Smigell and Hartman, 1997). In Kentucky, this approach saved up to four fungicide applications over calendar-based spraying (Williamson and Sutton, 2000) and an average of three applications in the upper Midwest USA (Babadoost *et al.*, 2004). New York researchers have found that signs of SBFS appear in unsprayed trees at 270 ALWH from petal fall. In sprayed orchards, scab sprays applied at petal fall and post-petal fall control flyspeck ascospores, so infections in sprayed orchards are initiated only after 270 ALWH, the time when flyspeck conidia begin blowing into orchards from adjacent wild hosts. Beginning summer treatments at a threshold of 220 ALWH works well under their conditions and allows for a considerable margin of error in measuring wetting periods. However, benzimidazole and strobilurin fungicides were believed to eradicate infections if applied during the first 100 h of leaf wetting following infection, and this post-infection activity could be used to extend the interval between the last scab spray and the first spray for flyspeck. Therefore, the application of a benzimidazole or strobilurin fungicide was recommended at 320 ALWH (Rosenberger *et al.*, 2005; Rosenberger, 2006; Rosenberger and Meyer, 2007). In part, this model is based on findings that the ascospores of *S. pomi* develop over a discrete period corresponding to bloom through early fruit set and the conidial inoculum that causes flyspeck infections on fruit is produced on reservoir hosts surrounding orchards (Cooley *et al.*, 2007b).

The broad variation in action thresholds for SBFS may be related to discoveries that the SBFS complex does not involve just two pathogens, but rather several fungal species, possibly differing by region (Johnson *et al.*, 1997; Batzer *et al.*, 2005). The empirical models have been effective against the complex, but there is probably variability between species in response to ALWH and to fungicides. Variations in the fungal community causing SBFS in different parts of the USA may be responsible for the threshold differences. Alternatively, the type of leaf wetness monitor used and its deployment could also have a significant impact on the measurement of ALWH (see the section on 'Weather monitoring', this chapter). Orchard architecture, distance to inoculum sources and other physical factors in orchards can also affect SBFS development (Ellis *et al.*, 1999; Tuttle *et al.*, 2002). Ellis *et al.* (1999) showed that non-sprayed, fully dwarf apple trees in well-maintained orchards that were further away from reservoir hosts did not develop SBFS signs until over 450 ALWH, while in the same area, larger trees with dense canopies growing near inoculum sources developed SBFS between 225 and 241 ALWH.

An empirical model generally functions best in the region in which it is developed, and it may not function well in other regions. For example, in North Carolina and Kentucky, temperature may be generally in a range that promotes growth of the SBFS fungi indiscriminately, and hence temperature is an unimportant variable in an SBFS model developed in these regions. In climatically cooler regions, temperatures may drop periodically to levels that slow or stop growth of the SBFS fungi, even though wetting conditions favour growth, thereby increasing the number of ALWH necessary for symptom development in cooler climates. Given this, the ALWH model may have to be adjusted based on empirical studies in climatically different regions in order to function well, at least until a better fundamental understanding of SBFS epidemiology is developed.

Fire blight

Presently, there are two widely used fire blight models in apples in North America, MaryBlyt (Steiner, 1990a,b; Steiner and Lightner, 1996) and Cougarblight (Smith, 1993, 1999). Others have been developed and are in use in other parts of the world (Thomson *et al.*, 1982; Berger *et al.*, 1996; Billing, 1999; Shtienberg *et al.*, 1999). These fire blight models are empirical and, while accurate in the region where they were developed, can be less accurate when used in other areas or with different weather monitoring methods (Billing, 2007). For example, neither MaryBlyt nor Cougarblight perform as well in Israel as the model developed in Israel, Fire Blight Control Advisory (FBCA) (Shtienberg *et al.*, 1999, 2003). In northern Italy, FBCA, MaryBlyt and Cougarblight were compared to determine which performed most effectively (Zasso *et al.*, 2006) and, ultimately, aspects of each model were combined in a system adapted to the region, FireFight

(Kuflik *et al.*, 2008). Similarly, comparisons of MaryBlyt and Cougarblight using historical data sets from several regions and sensitivity analysis demonstrated differences in forecasts by region, suggesting that model choice should depend on performance in a given location (Dewdney *et al.*, 2007).

These models generally combine temperature, wetting, phenological information (particularly bloom), the presence and proximity of fire blight, the history of fire blight in the region and any previous antibiotic treatments. Based on this information, the models estimate the risk of fire blight the previous day and over the next 1–3 days. Based on that risk, a recommendation is made regarding the application of a control, usually streptomycin. In many regions, fire blight outbreaks occur sporadically, problematic in one year but absent in others. When fire blight epidemics do erupt, they can cause significant damage. Hence, streptomycin may be applied when it is not needed, contributing to resistance development and increased expense, or it may not be applied when it could have prevented an epidemic. Fire blight models can remove the uncertainty surrounding the need for treatment.

Other models

Models for other apple diseases have been developed but, for several reasons, are used less commonly. If there are no chemicals or other treatments that will stop the progress of an epidemic, then a model has little practical utility. For example, black rot infections can be forecast accurately using a temperature and wetness model, but there are no post-infection fungicides available to treat the disease (Arauz and Sutton, 1989; Sutton, 1996). If adequate management of a disease is achieved with fungicides applied against another, more serious disease, then there is little point in forecasting a need for treatment of the relatively minor disease. A model for the management of cedar apple rust has been developed (Aldwinckle *et al.*, 1980) but, in commercial situations, this disease is controlled by selecting a fungicide to control both apple scab and rusts. Similarly, a model was developed to manage powdery mildew and was shown to decrease fungicide use in Britain (Xu, 1999; Berrie and Xu, 2003), but has not been tested in the eastern USA because fungicides applied to manage scab generally are selected to control powdery mildew also. One model developed to predict powdery mildew risk in the USA (Lalancette and Hickey, 1986) was never validated in the field, probably for the same reason. Finally, if a disease is limited to a specific cultivar or region, a model may be useful, but only on a local basis. A model for Alternaria leaf blotch based on leaf wetness and degree-days from tight cluster has been used to manage the disease effectively (Filajdic, 1994), but Alternaria blotch is limited to the south-eastern USA, largely on Red Delicious, and is not treated in other regions.

Weather monitoring

As can be seen from the discussion above, disease forecast models depend on weather data, particularly information on leaf wetness duration (LWD). In fact, for optimal utility, growers need both recorded data as well as predicted data. If, for example, a model recommends that a fungicide be applied because a wetting threshold has been met but the fungicide cannot be applied because it is raining, the model has limited practical value. A forecast that the threshold will be met and that fungicide application will not be possible when it is met, allows a grower to apply the fungicide earlier. There are several practical issues associated with obtaining both past and forecast weather data (Huber and Gillespie, 1992; Lau *et al.*, 2000; Gleason, 2001; Magarey *et al.*, 2001; Sentelhas *et al.*, 2005; Kim *et al.*, 2006).

Mechanical recording devices based on hemp string sensors have been used extensively in LWD measurement (Zuck and MacHardy, 1981), though such devices have been replaced largely by electronic leaf wetness sensors (Sutton *et al.*, 1984; Lau *et al.*, 2000). There can be differences in LWD as measured by these two types of sensors and between different types of electronic sensors (Magarey *et al.*, 2005). The protocol for the placement of wetness sensors can affect readings significantly (Lau *et al.*, 2000; Batzer *et al.*, 2008). This may account for threshold differences, such as the decrease in threshold made when the SBFS model, which was developed using a deWit hemp string sensor, was adapted to fit data obtained with an electronic sensor (Brown and Sutton, 1995; Hartman *et al.*, 1999). The type of electronic sensor used, its coating material, its placement in the canopy and even the angle of deployment can affect LWD measurements significantly (Lau *et al.*, 2000). Penrose and Nicol (1996) concluded that even with significant microclimate variability in apple canopies, monitoring microclimate and using prediction models still provided a reasonably reliable treatment guide for apple scab (Penrose and Nicol, 1996). From a practical perspective, the issue is not how variable sensors may be, but whether they function within the context of forecast models to give consistently useful disease management information.

As electronic weather monitoring equipment was developed for epidemiological research, microprocessor-based units that recorded and analysed the information for growers in the field were developed. The earliest commercial unit for apples recommended treatments for apple scab and was developed in the early 1980s (Jones *et al.*, 1980, 1984). This unit was used in several parts of the USA and it could save fungicide applications and effectively control scab (Ellis *et al.*, 1984). Electronic units have continued to evolve and growers can now purchase weather stations that will transmit data wirelessly to personal computers and can buy software that incorporates forecasting models for apple diseases.

An alternative to on-site monitoring is subscription to a meteorological service that estimates site-specific data for individual farms via interpolation from a network of ground monitoring sites, such as those run by NOAA's National Weather Service (Gleason *et al.*, 1997; Magarey *et al.*, 2001; Kim *et al.*, 2006). The services provide estimates of past weather, forecasts and disease risk evaluations. To determine risk, the off-site services run established models using data they have generated using their interpolation algorithms. Off-site estimates often differ from on-site measurements and, when used in conjunction with prediction models, the predictions from each data source may differ (Babadoost *et al.*, 2004).

The most frequently seen differences occur in LWD measurements, where off-site predictions tend to overestimate LWD relative to on-site monitors. Such differences should be considered in the context of the wide variability of on-site sensors. Ultimately, the relative performance of two measurement systems should be evaluated in terms of forecast recommendations.

The New York State IPM Programme has developed a hybrid of on-site weather stations and Internet processing of data and disease forecasting, the Network for Environment and Weather Awareness (NEWA), which incorporates apple disease forecasting models. In this system, growers and consultants buy electronic on-site weather stations, which are placed on farms. These stations feed data to a central server via the Internet, where they are evaluated and summarized on a web site accessible to all participants (Carroll, 2007). This allows growers to share weather resources and related disease forecasts.

Biointensive IPM, Standards and the Market

For many years, apple IPM programmes used chemical tactics almost exclusively. Forecast models and integration of the management of different pest classes made the programmes efficient and effective relative to conventional approaches (Gadoury *et al.*, 1989). It has been suggested that as cultural and biological controls, resistant cultivars and forecast models improve, they will replace conventional pesticides in advanced, biointensive IPM systems (Jacobsen, 1997). Yet, data from California suggest that an emphasis on IPM has not led to a significant reduction in chemical use on fruit and vegetable crops (Epstein and Bassein, 2003). Similarly, Penrose showed that in spite of the availability of numerous options that could reduce fungicide use in apples, sociological and financial issues presented significant barriers to their adoption (Penrose, 1995). A decade later, regardless of the impressive list of alternatives described above, the development and adoption of the disease components of biointensive IPM appears to be stalled for the same fundamental reason noted by Penrose: the benefits from reducing pesticide use largely accrue to the public, while the risks are borne by growers.

Growers generally perceive biointensive IPM as risky. The tactics are unfamiliar and growers are uncertain that an IPM alternative will be as effective as a familiar, established control tactic. Since the mid-1990s, apple growers in the north-eastern USA have become increasingly risk averse, as global competition has eroded profit margins, and consequently, they are less willing to try more advanced IPM tactics on even a small part of their farms (see Cooley and Coli, Chapter 6, this volume).

Not only are biointensive tactics perceived as high-risk, they often increase pest management costs. Early in the development of an IPM programme, growers reduced conventional pesticide applications overall and saved money using relatively inexpensive sampling and forecasting methods, but more advanced biointensive techniques were usually more expensive. A California study points out that a decrease in pesticide use that requires an increase in, for example, cultural control practices will generally require more labour, as cultural controls are often labour-intensive (Epstein and Bassein, 2003). Fungicide sprays account for approximately 8% of the total variable costs for an apple orchard in the eastern USA, while labour accounts for 71% (Crassweller, 2008). Because labour is expensive relative to pesticides, producers will be less likely to adopt the cultural control alternative.

Alternative disease management tactics are often more specific than fungicides. A single fungicide application can often manage several diseases. For example, a DMI fungicide will control apple scab, powdery mildew and cedar apple rust. However, using scab-resistant cultivars will not control mildew or rust and grinding apple leaves on the orchard floor will not impact inoculum for the other two diseases. Biointensive management systems that will manage all important apple diseases and integrate with other management practices are more complex than conventional chemical systems.

Growers facing increased risk and expense are very unlikely to adopt biointensive IPM unless they have some compensating motivation, such as increased prices for their products or specifications placed on production practices by the market or government. Prokopy noted that as IPM advanced, it would be part of a broader set of tools used to build sustainable agricultural systems and that inevitably this would require the participation of both the public, which consumes agricultural products, and government, which sets agricultural policy (Prokopy, 1993). Given this, it is not surprising that the areas where biointensive apple IPM programmes have been tried involve niche and regulated markets, represented by eco-label programmes and integrated fruit production (IFP) standards (Hollingsworth, 1995; Cross, 2002). Eco-labels identify a crop to consumers as having been grown using 'ecological' practices and have been developed for a number of crops worldwide since the early 1990s. What qualifies as an 'ecological' production practice remains an open question, depending at least as much on the perceptions of consumers and marketers as on measurable environmental impacts. Consumers generally are willing to pay at least a small premium for eco-labelled apples (Blend and van Ravenswaay, 1999; Loureiro *et al.*, 2002; Moon *et al.*, 2002). From a USA perspective, IFP approaches for apple production allow export to markets such as Europe that have adopted a set of production regulations. IFP programmes have been adopted in the north-eastern USA, for example, in New York state (Carroll and Robinson, 2006) and Massachusetts (Hollingsworth and Coli, 2007).

Overall, both eco-label and IFP approaches provide lists of practices that growers must use, and categorize pesticides in ways that encourage the use of biopesticides and chemicals with relatively low environmental impacts, while limiting or prohibiting the use of more toxic pesticides. These principles have also been extended to various eco-label marketing programmes for apples in the north-eastern USA, most recently the Eco Apple™ programme (Green, 2007). This programme uses several measures of pesticide toxicity and other risk factors (e.g. US EPA classifications, State of California classifications, product labels and International Agency for Research on Cancer) to place chemicals in one of three groups: green – use with justification; yellow – use with justification when green list or other alternatives are not adequate; and red – do not use. In general, pesticides that are useful in the north-eastern USA and do not have hazards, or that have readily mitigated hazards, are placed in the green category. Pesticides with less easily mitigated hazards are placed in the yellow category and their use is limited. Chemicals with risks that cannot be alleviated are placed in the red category and are not used. Biopesticides and chemicals evaluated as having low non-target effects such as *B. subtilis*, some copper compounds (e.g. copper oxychloride sulphate), sulfur, phosphorous acid and cyprodinil are all on the green list. Captan, most EBDCs, thiophanate methyl, most demethylation-inhibiting fungicides, trifloxystrobin, some copper compounds (e.g. copper sulfate) and streptomycin are examples of yellow list materials. Prohibited disease-controlling chemicals include kresoxim-methyl, mefenoxam, thiabendazole, ziram, triadimefon and pyrimethanil.

The Eco Apple programme is just one of many eco-labels, reflecting the lack of standardization in attempts to measure what is, in fact, an ecological production practice. The primary goal of IPM is to optimize pest management using ecologically and economically sound methods. In order to measure progress in IPM, accurate metrics that evaluate the ecological and economic cost and benefits of pest management systems objectively must be used to compare the risks and benefits of different management options accurately (Levitan *et al.*, 1995). In other words, do apples grown using Eco Apple standards have fewer negative environmental and economic impacts than apples grown using other practices?

On its simplest level, the farm-level benefits of a pest management programme may be measured in terms of decreases in disease and other pest damage and costs, including the cost of pesticides, their application and decision support directing treatments. These are relatively easy to estimate (Bowles and Webster, 1995). It is more difficult to measure whether a tactic or programme is ecologically sound, particularly if trying to assign monetary costs (Levitan, 2000; Lewis *et al.*, 2003; Leach and Mumford, 2008). It is quite difficult to measure the external costs and benefits of pesticides, such as changes in health-care costs that may occur if pesticides in food lead to more cancer in the population or, alternatively, if pesticides reduce mycotoxin-related diseases in the population.

One early approach to measuring ecological impacts at the field level was developed by Kovach *et al.* (1992). It involves taking several types of toxicity measures for a given pesticide and combining them in a model to calculate a single numerical estimate of toxicity, the Environmental Impact Quotient (EIQ). This unitless rating does not estimate cost but can give a relative estimate of environmental impacts. In a given agroecosystem, if a pest management approach is ecologically sound, the EIQ should drop relative to existing pest management tactics. As biopesticides and other biorational methods are introduced into a production system, they should lower the EIQ, demonstrating advancement toward an environmentally benign production system.

Kovach compared the EIQ in three different apple pest management systems: a conventional pesticide programme, an IPM system and a typical organic programme. Contrary to what might have been predicted, the organic programme had an EIQ nearly twice that of a typical conventional pesticide programme and over ten times that of an IPM programme (Kovach *et al.*, 1992). This demonstrated that while organic methods might have positive elements with respect to agricultural sustainability, the prohibition of synthetic chemicals in favour of naturally occurring ones did not necessarily decrease toxicity problems.

Yet, consumers generally have come to believe that organic food is 'safer' and 'better for the environment' than conventionally grown products, although, as Kovach demonstrated, this might be based largely on unsupported perceptions (Williams and Hammitt, 2001; Magkos *et al.*, 2006).

It should be noted that different methods of measuring environmental impact might produce very different results. A Washington state study used a proprietary index developed for the Stemilt Growers Responsible Choice eco-label programme to compare the impacts of organic, IPM and conventional apple pest management programmes (Reganold, 2006). Reganold concluded that a conventional apple system had a rating 6.2 times higher (worse) than an organic system, while an IPM system had a rating 4.7 times higher than the organic system. This reversed the earlier EIQ evaluation of Kovach *et al.*, comparing apple production systems (Kovach *et al.*, 1992). Unlike the EIQ, the methods used in the Responsible Choice programme to calculate ecological impacts have not been made public (Consumers Union, 2002), though some pesticide ratings from the programme have been reported (Reed and Nelson, 1995). Without a clear understanding of both models, it is impossible to determine why the two approaches differ so widely in terms of the measurement of relative ecological impacts. When comparing different pest management systems, a rigorous, accurate and transparent environmental impact model will be of the most use in scientific evaluations.

Several models have been developed that attempt to improve the EIQ approach by including more detail in their analyses, such as environmental persistence and long-range transport (Juraske *et al.*, 2007). While the EIQ estimates the ecological impacts of pesticides, it does not attempt an extensive estimate of the external costs or environmental fate, nor does it evaluate environmental impacts in monetary terms. A recent model, the Pesticide Environmental Accounting method, builds on the EIQ and evaluates externalities as costs (Leach and Mumford, 2008). This system may enable better objective comparisons of IPM programmes in environmental terms.

The Future of Biointensive Tactics for Disease Management in Apples

Biointensive apple IPM in north-eastern North America is still in its very early stages of development. The many tactics described here have promise, but have yet to be integrated within the general class of apple disease pests. Farther in the future lies the challenge of integrating them in a comprehensive biointensive IPM system involving arthropod, weed and vertebrate pests, as well as critical horticultural practices such as fruit thinning. The process is at what Prokopy would have referred to as first-level IPM, in which tactics focus on a single pest, in this case individual diseases. Researchers have barely begun to suggest sets of biointensive tactics that will address the whole set of diseases that challenge growers in the region. At the second level of IPM, biointensive tactics for diseases, arthropods and other pests will be integrated. This might, for example, involve using a 'trap tree' approach to either monitor or treat plum curculio (*Conotrachelus nenuphar*) (Piñero and Prokopy, 2003; Piñero *et al.*, 2006) and pesticide-treated, attractant-baited spheres to control apple maggot (*Rhagoletis pomonella*) (Prokopy *et al.*, 2000, 2004). In time, at the third level, multiple tactics would be integrated over the entire orchard production system and could incorporate cultivars and rootstocks that would have resistance or tolerance to most key diseases. But, for the present, in spite of extensive research and development in the individual components of biointensive management of apple diseases, biointensive apple IPM is in its infancy.

Prokopy noted that as IPM advanced, it would be part of a broader set of tools used to build sustainable agricultural systems and that inevitably this would require the participation of both the public, which consumes agricultural products, and government, which sets agricultural policy (Prokopy, 1993). Eco-label programmes and IFP standards show ways that biointensive IPM can be linked with the broader goal of sustainable apple production. It is not yet clear to what extent consumers will willingly pay a premium for fruit grown using biointensive practices and at what level government regulation and subsidization will be used to promote biointensive IPM in the north-eastern USA.

Experience suggests that the adoption of new IPM tactics proceeds most rapidly when growers and other stakeholders are engaged in a participatory research process (see Cooley and Coli, Chapter 6, this volume). Grower involvement in the research and development process will be critical, given grower reticence to try biointensive disease management tactics. Researchers, regulators, producers and marketers will need to work together to develop biointensive IPM systems that present risks, costs and benefits that growers can accept, at least to the extent that they will try these systems on small parts of their orchards. Success or failure of these trials in early-adopter orchards will determine the extent to which biointensive IPM practices expand to whole orchards and regions.

References

Agnello, A.M., Reissig, W.H., Kovach, J. and Nyrop, J.P. (2003) Integrated apple pest management in New York State using predatory mites and selective pesticides. *Agriculture Ecosystems and Environment* 94, 183–195.

Agrios, G.N. (2005) *Plant Pathology.* Elsevier Academic Press, Burlington, Massachusetts.

Aldwinckle, H.S. and Penev, R. (2004) Field evaluation of materials for control of fire blight infection of apple blossoms, 2003. *Fungicide and Nematicide Tests*. Report 59 PF016. DOI:10.1094/FN59.

Aldwinckle, H.S., Pearson, R.C. and Seem, R.C. (1980) Infection periods of *Gymnosporangium juniperi-virginianae* on apple. *Phytopathology* 70, 1070–1073.

Aldwinckle, H.S., Gustafson, H.L., Heidenreich, G., Penev, R. and LoGiudice, N. (2002) Field evaluation of materials for control of fire blight infection of apple blossoms, 2001. *Fungicide and Nematicide Tests*. Report 57 PF01. DOI:10.1094/FN57.

Arauz, L.F. and Sutton, T.B. (1989) Temperature and wetness duration requirements for apple infection by *Botryosphaeria obtusa*. *Phytopathology* 79, 440–444.

Autio, W.R. and Southwick, F. (1986) The effects of rootstock and root-interstem combination on the growth, productivity, and anchorage of a spur and standard strain Delicious apple tree. *Fruit Varieties Journal* 40, 128–133.

Babadoost, M., McManus, P.S., Helland, S.N. and Gleason, M.L. (2004) Evaluating a wetness-based warning system and reduced-risk fungicides to manage sooty blotch and flyspeck of apple. *HortTechnology* 14, 51–57.

Batzer, J.C., Gleason, M.L., Weldon, B., Dixon, P.M. and Nutter, F.W. (2002) Evaluation of postharvest removal of sooty blotch and flyspeck on apples using sodium hypochlorite, hydrogen peroxide with peroxyacetic acid, and soap. *Plant Disease* 86, 1325–1332.

Batzer, J.C., Gleason, M.L., Harrington, T.C. and Tiffany, L.H. (2005) Expansion of the sooty blotch and flyspeck complex on apples based on analysis of ribosomal DNA gene sequences and morphology. *Mycologia* 97, 1268–1286.

Batzer, J.C., Gleason, M.L., Taylor, S.E., Koehler, K.J. and Monteiro, J. (2008) Spatial heterogeneity of leaf wetness duration in apple trees and its influence on performance of a warning system for sooty blotch and flyspeck. *Plant Disease* 92, 164–170.

Becker, C.M., Burr, T.J. and Smith, C.A. (1992) Overwintering of conidia of *Venturia inaequalis* in apple buds in New York orchards. *Plant Disease* 76, 121–126.

Belfanti, E., Silfverberg-Dilworth, E., Tartarini, S., Patocchi, A., Barbieri, M., Zhu, J., Vinatzer, B.A., Gianfranceschi, L., Gessler, C. and Sansavini, S. (2004) The HcrVf2 gene from a wild apple confers scab resistance to a transgenic cultivated variety. *Proceedings of the National Academy of Sciences of the United States of America* 101, 886–890.

Benbrook, C.M. (1996) *Pest Management at the Crossroads*. Consumers Union of United States, Inc, Yonkers, New York.

Berger, F., Zeller, W., Gutsche, V. and Rossberg, D. (1996) A new fire blight forecasting system with first results in southwest Germany. *Acta Horticulturae* 411, 155–161.

Berrie, A.M. and Xu, X.M. (2003) Managing apple scab (*Venturia inaequalis*) and powdery mildew (*Podosphaera leucotricha*) using ADEM™. *International Journal of Pest Management* 49, 243–249.

Biggs, A.R. (1999) Effects of calcium salts on apple bitter rot caused by two *Colletotrichum* spp. *Plant Disease* 83, 1001–1005.

Biggs, A.R. (2004) Effect of inoculum concentration and calcium salts on infection of apple fruit by *Botryosphaeria dothidea*. *Plant Disease* 88, 147–151.

Biggs, A.R., Ingle, M. and Solihati, W.D. (1993) Control of *Alternaria* infection of fruit of apple cultivar Nittany with calcium chloride and fungicides. *Plant Disease* 77, 976–980.

Biggs, A.R., Hogmire, H.W. and Collins, A.R. (2000) Assessment of an alternative IPM program for the production of apples for processing. *Plant Disease* 84, 1140–1146.

Billing, E. (1999) Fire blight assessment: Billing's integrated system (BIS) and its evaluation. *Acta Horticulturae* 489, 399–405.

Billing, E. (2007) Challenges in adaptation of plant disease warning systems to new locations: re-appraisal of Billing's integrated system for predicting fire blight in a warm dry environment. *Phytopathology* 97, 1036–1039.

Blend, J.R. and van Ravenswaay, E.O. (1999) Measuring consumer demand for ecolabeled apples. *American Journal of Agricultural Economics* 81, 1072–1077.

Bowles, R.G. and Webster, J.P.G. (1995) Some problems associated with the analysis of the costs and benefits of pesticides. *Crop Protection* 14, 593–600.

Brisset, M.N., Cesbron, S., Thomson, S.V. and Paulin, J.P. (2000) Acibenzolar-*S*-methyl induces the accumulation of defense-related enzymes in apple and protects from fire blight. *European Journal of Plant Pathology* 106, 529–536.

Brown, E.M. and Sutton, T.B. (1995) An empirical model for predicting the first symptoms of sooty blotch and flyspeck of apples. *Plant Disease* 79, 1165–1168.

Brown, S.K. and Terry, D.E. (1997) The Cornell University apple breeding program: past, present, and future. *Fruit Varieties Journal* 51, 199–204.

Brown, S.K., Maloney, K.E., Hemmat, M. and Aldwinckle, H.S. (2004) Apple breeding at Cornell: genetic studies of fruit quality, scab resistance and plant architecture. *Acta Horticulturae* 663, 693–697.

Brunner, J., Jones, W., Beers, E., Tangren, G., Dunley, J., Xiao, C. and Grove, G. (2003) A decade of pesticide use and IPM practices in Washington's apple orchards. *Agrichemical and Environmental News*, 1–15 May. Washington State University.

Bunemann, E.K., Schwenke, G.D. and Van Zwieten, L. (2006) Impact of agricultural inputs on soil organisms – a review. *Australian Journal of Soil Research* 44, 379–406.

Caffier, V. and Parisi, L. (2007) Development of apple powdery mildew on sources of resistance to *Podosphaera leucotricha*, exposed to an inoculum virulent against the major resistance gene Pl-2. *Plant Breeding* 126, 319–322.

Calenge, F., Faure, A., Goerre, M., Gebhardt, C., Van de Weg, W.E., Parisi, L. and Durel, C.E. (2004) Quantitative trait loci (QTL) analysis reveals both broad-spectrum and isolate-specific QTL for scab resistance in an apple progeny challenged with eight isolates of *Venturia inaequalis*. *Phytopathology* 94, 370–379.

Carisse, O. and Dewdney, M. (2002) A review of non-fungicidal approaches for the control of apple scab. *Phytoprotection* 83, 1–29.

Carisse, O. and Khanizadeh, S. (2006) Relative resistance of newly released apple rootstocks to *Phytophthora cactorum*. *Canadian Journal of Plant Science* 86, 199–204.

Carisse, O. and Rolland, D. (2004) Effect of timing of application of the biological control agent *Microsphaeropsis ochracea* on the production and ejection pattern of ascospores by *Venturia inaequalis*. *Phytopathology* 94, 1305–1314.

Carroll, J. (2007) New York State IPM, Network for Environment and Weather Awareness (http://www.nysipm.cornell.edu/newa/default.asp, accessed August 2007).

Carroll, J.E. and Robinson, T.L. (2006) *New York Integrated Fruit Production Protocol for Apples*. New York State Agricultural Experiment Station, Geneva, New York.

ChandGoyal, T. and Spotts, R.A. (1997) Biological control of postharvest diseases of apple and pear under semi-commercial conditions using three saprophytic yeasts. *Biological Control* 10, 199–206.

Chen, H. and Korban, S.S. (1987) Genetic variability and the inheritance of resistance to cedar-apple rust in apple. *Plant Pathology* 36, 168–174.

Chouinard, G., Vincent, C., Laplante, G. and Morin, Y. (1998) Apple IPM and research at the farm level in Quebec: the Quebec Apple Network. In: Vincent, C. and Smith, R. (eds) *Orchard Pest Management in Canada*. Agriculture and Agri-Food Canada, Saint-Jean-sur-Richelieu, Quebec, Canada, pp. 33–52.

Cline, J.A., Hunter, D.M., Bonn, W.G. and Bijl, M. (2001) Resistance of the Vineland series of apple rootstocks to fire blight caused by *Erwinia amylovora*. *Journal of the American Pomological Society* 55, 218–221.

Consumers Union (2002) Stemilt (http://www.eco-labels.org/labelPrint.cfm?LabelID=99, accessed September 2007).

Conway, W.S., Sams, C.E., Abbott, J.A. and Bruton, B.D. (1991) Postharvest calcium treatment of apple fruit to provide broad-spectrum protection against postharvest pathogens. *Plant Disease* 75, 620–622.

Conway, W.S., Leverentz, B., Janisiewicz, W.J., Saftner, R.A. and Camp, M.J. (2005) Improving biocontrol using antagonist mixtures with heat and/or sodium bicarbonate to control postharvest decay of apple fruit. *Postharvest Biology and Technology* 36, 235–244.

Cooley, D.R. and Agnello, A.M. (eds) (2007) *2007 New England Tree-Fruit Pest Management Guide*. Cornell Cooperative Extension Geneva, New York.

Cooley, D.R. and Autio, W.R. (1997) Disease-management components of advanced integrated pest management in apple orchards. *Agriculture Ecosystems and Environment* 66, 31–40.

Cooley, D.R. and Lerner, S.M. (1994) Summer pruning increases pesticide coverage in apple canopies. *Phytopathology* 84, 1370.

Cooley, D.R. and Manning, W.J. (1995) Estimating the risks and benefits of pesticides – considering the agroecosystem and integrated pest management in the use of EBDC fungicides on apples. *Environmental Pollution* 88, 315–320.

Cooley, D.R., Gamble, J.W. and Autio, W.R. (1997) Summer pruning as a method for reducing flyspeck disease on apple fruit. *Plant Disease* 81, 1123–1126.

Cooley, D.R., Autio, W.R. and Krupa, J. (2007a) Calcium salts and other alternatives to conventional fungicides for management of the summer disease complex of apple, 2006. *Plant Disease Management Reports*. Rept. 1: PF051 DOI:10.1094/PDMR01.

Cooley, D.R., Lerner, S.M. and Tuttle, A.F. (2007b) Maturation of thyriothecia of *Schizothyrium pomi* on the reservoir host *Rubus allegheniensis*. *Plant Disease* 91, 136–141.

Crassweller, R. (ed.) (2008) *Tree Fruit Production Guide*. The Pennsylvania State University, State College, Pennsylvania.

Crosby, J.A., Janick, J., Pecknold, P.C., Korban, S.S., O'Connor, P.A., Ries, S.M., Goffreda, J. and Voordeckers, A. (1992) Breeding apples for scab resistance – 1945–1990. *Fruit Varieties Journal* 46, 145–166.

Cross, J.V. (2002) Guidelines for integrated production of pome fruits in Europe, 3rd edition. *IOBC WPRS Bulletin* 25, 1–8.

Cummins, J.N. and Aldwinckle, H.S. (1994) Challenges to Malling 9, M.26 and M.7 – new rootstocks from Geneva. *Compact Fruit Tree* 27, 25–27.

Dewdney, M.M., Biggs, A.R. and Turechek, W.W. (2007) A statistical comparison of the blossom blight forecasts of MARYBLYT and Cougarblight with receiver operating characteristic curve analysis. *Phytopathology* 97, 1164–1176.

Dufour, R. (2001) Biointensive integrated pest management (IPM). National Center for Appropriate Technology/Appropriate Technology Transfer for Rural Areas (http://www.attra.ncat.org/attra-pub/ipm.html, accessed 10 March 2008).

Durel, C.E., Parisi, L., Laurens, F., Van de Weg, W.E., Liebhard, R. and Jourjon, M.F. (2003) Genetic dissection of partial resistance to race 6 of *Venturia inaequalis* in apple. *Genome* 46, 224–234.

Ehler, L.E. (2006) Integrated pest management (IPM): definition, historical development and implementation, and the other IPM. *Pest Management Science* 62, 787–789.

Ehret, G.R., Sundin, G.W. and McGhee, G.C. (2007) Efficacy of bacterial antagonists in controlling fire blight on Jonathan apple in 2006. *Plant Disease Management Reports*. Rept. 1: PF038. DOI:10.1094/PDMR01.

Eisler, R. (1998) Copper hazards to fish, wildlife and invertebrates: a synoptic review. *Contaminant Hazard Reviews*. Patuxent Wildlife Research Center. US Geological Survey, Laurel, Maryland.

Ellis, M.A., Madden, L.V. and Wilson, L.L. (1984) Evaluation of an electronic apple scab predictor for scheduling fungicides with curative activity. *Plant Disease* 68, 1055–1057.

Ellis, M.A., Ferree, D.C., Funt, R.C. and Madden, L.V. (1998) Effects of an apple scab-resistant cultivar on use patterns of inorganic and organic fungicides and economics of disease control. *Plant Disease* 82, 428–433.

Ellis, M.A., Madden, L.V. and Wilson, L. (1999) Evaluation of an empirical model for predicting sooty blotch and flyspeck of apples in Ohio. *Ohio State University Research Bulletin – Research Circular* 299-99, 1–6.

EPA (1999) Reregistration eligibility decision (RED): Captan, EPA-738-R-99-015. Environmental Protection Agency Office of Pesticide Programs, US Government Printing Office, Washington, DC.

EPA (2005) Reregistration eligibility decision for mancozeb: EPA 738-R-04-012. Environmental Protection Agency Office of Pesticide Programs, US Government Printing Office, Washington, DC.

EPA (2007) What are biopesticides? US Environmental Protection Agency Office of Pesticide Programs (http://www.epa.gov/pesticides/biopesticides/whatarebiopesticides.htm, accessed 16 August 2007).

Epstein, L. and Bassein, S. (2003) Patterns of pesticide use in California and the implications for strategies for reduction of pesticides. *Annual Review of Phytopathology* 41, 351–375.

Errampalli, D. and Brubacher, N.R. (2006) Biological and integrated control of postharvest blue mold (*Penicillium expansum*) of apples by *Pseudomonas syringae* and cyprodinil. *Biological Control* 36, 49–56.

Fallik, E. (2004) Prestorage hot water treatments (immersion, rinsing and brushing). *Postharvest Biology and Technology* 32, 125–134.

Fallik, E., Grinberg, S., Gambourg, M., Klein, J.D. and Lurie, S. (1996) Prestorage heat treatment reduces pathogenicity of *Penicillium expansum* in apple fruit. *Plant Pathology* 45, 92–97.

Filajdic, N. (1994) Management of Alternaria blotch of apples. PhD thesis, North Carolina State University, Raleigh, North Carolina.

Gadoury, D.M. and MacHardy, W.E. (1982) A model to estimate the maturity of ascospores of *Venturia inaequalis*. *Phytopathology* 72, 901–904.

Gadoury, D.M. and MacHardy, W.E. (1986) Forecasting ascospore dose of *Venturia inaequalis* in commercial apple orchards. *Phytopathology* 76, 112–118.

Gadoury, D.M., Machardy, W.E. and Rosenberger, D.A. (1989) Integration of pesticide application schedules for disease and insect control in apple orchards of the northeastern United States. *Plant Disease* 73, 98–105.

Gadoury, D.M., Seem, R.C., MacHardy, W.E., Wilcox, W.F., Rosenberger, D.A. and Stensvand, A. (2004) A comparison of methods used to estimate the maturity and release of ascospores of *Venturia inaequalis*. *Plant Disease* 88, 869–874.

Gessler, C. and Patocchi, A. (2007) Recombinant DNA technology in apple. *Advances in Biochemical Engineering/Biotechnology* 107, 113–132.

Gessler, C., Patocchi, A., Sansavini, S., Tartarini, S. and Gianfranceschi, L. (2006) *Venturia inaequalis* resistance in apple. *Critical Reviews in Plant Sciences* 25, 473–503.

Gianessi, L.P. and Marcelli, M.B. (2000) *Pesticide Use in US Crop Production: 1997.* National Center for Food and Agricultural Policy, Washington, DC.

Gilpatrick, J.D., Smith, C.A. and Blowers, D.R. (1972) A method of collecting ascospores of *Venturia inaequalis* for spore germination studies. *Plant Disease Reporter* 56, 39–42.

Gleason, M.L. (2001) Disease-warning systems. In: Maloy, O.C. and Murray, T.D. (eds) *Encyclopedia of Plant Pathology.* John Wiley and Sons, New York, pp. 367–370.

Gleason, M.L., Parker, S.K., Pitblado, R.E., Latin, R.X., Speranzini, D., Hazzard, R.V., Maletta, M.J., Cowgill, W.P. and Biederstedt, D.L. (1997) Validation of a commercial system for remote estimation of wetness duration. *Plant Disease* 81, 825–829.

Gomez, C., Brun, L., Chauffour, D. and Le Vallee, D.D. (2007) Effect of leaf litter management on scab development in an organic apple orchard. *Agriculture Ecosystems and Environment* 118, 249–255.

Green, T. (ed.) (2007) *Eco-Apple Protocol and Grower Self-Assessment.* IPM Institute of North America, Inc, Madison, Wisconsin.

Guerin, F. and Le Cam, B. (2004) Breakdown of the scab resistance gene Vf in apple leads to a founder effect in populations of the fungal pathogen *Venturia inaequalis*. *Phytopathology* 94, 364–369.

Guerin, F., Gladieux, P. and Le Cam, B. (2007) Origin and colonization history of newly virulent strains of the phytopathogenic fungus *Venturia inaequalis*. *Fungal Genetics and Biology* 44, 284–292.

Harris, S.A., Robinson, J.P. and Juniper, B.E. (2002) Genetic clues to the origin of the apple. *Trends in Genetics* 18, 426–430.

Hartman, J.R. (1995) Evaluation of fungicide timing for sooty blotch and flyspeck control, 1994. *Fungicide and Nematicide Tests* 50, 10.

Hartman, J.R., Parisi, L. and Bautrais, P. (1999) Effect of leaf wetness duration, temperature, and conidial inoculum dose on apple scab infections. *Plant Disease* 83, 531–534.

Hendrix, F.F. (1991) Removal of sooty blotch and flyspeck from apple fruit with a chlorine dip. *Plant Disease* 75, 742–743.

Hettenbach, T. and Wiles, R. (2000) A few bad apples...pesticides in your produce. Environmental Working Group, Washington, DC.

Hickey, K.D. (1977) Sooty blotch and flyspeck suppression with two late-season fungicide sprays. *Fungicide and Nematicide Tests* 34, 8–9.

Holb, I.J. (2005) Effect of pruning on apple scab in organic apple production. *Plant Disease* 89, 611–618.

Holb, I.J. (2006) Effect of six sanitation treatments on leaf litter density, ascospore production of *Venturia inaequalis* and scab incidence in integrated and organic apple orchards. *European Journal of Plant Pathology* 115, 293–307.

Holb, I.J. and Schnabel, G. (2007) Differential effect of triazoles on mycelial growth and disease measurements of *Monilinia fructicola* isolates with reduced sensitivity to DMI fungicides. *Crop Protection* 26, 753–759.

Holb, I.J., Heijne, B., Withagen, J.C.M. and Jeger, M.J. (2004) Dispersal of *Venturia inaequalis* ascospores and disease gradients from a defined inoculum source. *Journal of Phytopathology* 152, 639–646.

Holb, I.J., Heijne, B. and Jeger, M.J. (2005) The widespread occurrence of overwintered conidial inoculum of *Venturia inaequalis* on shoots and buds in organic and integrated apple orchards across the Netherlands. *European Journal of Plant Pathology* 111, 157–168.

Holb, I.J., Heijne, B. and Jeger, M.J. (2006) Effects of integrated control measures on earthworms, leaf litter and *Venturia inaequalis* infection in two European apple orchards. *Agriculture Ecosystems and Environment* 114, 287–295.

Hollingsworth, C. (1995) Integrated fruit production (IFP): a status report. *Fruit Notes* 60(4), 14–15.

Hollingsworth, C. and Coli, W.M. (2007) IPM guidelines: apple (http://www.umass.edu/umext/ipm/guidelines/ipm_guidelines_apple.html, accessed 21 March 2008).

Horton, D.L., Pfeiffer, D.G. and Hendrix, F.F. (1991) Southeastern apple integrated pest management. In: Hullar, T.L. and National Research Council (eds) *Sustainable Agriculture Research and Education in the Field: A Proceedings.* National Academy Press, Washington, DC, pp. 165–182.

Huber, L. and Gillespie, T.J. (1992) Modeling leaf wetness in relation to plant disease epidemiology. *Annual Review of Phytopathology* 30, 553–577.

Jacobsen, B.J. (1997) Role of plant pathology in integrated pest management. *Annual Review of Phytopathology* 35, 373–391.

James, J.R. and Sutton, T.B. (1982) A model for predicting ascospore maturation of *Venturia inaequalis*. *Phytopathology* 72, 1081–1085.

Janick, J. (2006) The PRI apple breeding program. *Hortscience* 41, 8–10.

Janisiewicz, W.J. and Jeffers, S.N. (1997) Efficacy of commercial formulation of two biofungicides for control of blue mold and gray mold of apples in cold storage. *Crop Protection* 16, 629–633.

Janisiewicz, W.J. and Korsten, L. (2002) Biological control of postharvest diseases of fruits. *Annual Review of Phytopathology* 40, 411–441.

Jensen, P.J., Rytter, J., Detwiler, E.A., Travis, J.W. and McNeillis, T.W. (2003) Rootstock effects on gene expression patterns in apple tree scions. *Plant Molecular Biology* 53, 493–511.

Johnson, D.A., Inglis, D.A. and Miller, J.S. (2004) Control of potato tuber rots caused by oomycetes with foliar applications of phosphorous acid. *Plant Disease* 88, 1153–1159.

Johnson, E.M., Sutton, T.B. and Hodges, C.S. (1997) Etiology of apple sooty blotch disease in North Carolina. *Phytopathology* 87, 88–95.

Jones, A.L. (1981) Fungicide resistance – past experience with benomyl and dodine and future concerns with sterol inhibitors. *Plant Disease* 65, 990–992.

Jones, A.L. and Aldwinckle, H.S. (1990) *Compendium of Apple and Pear Diseases.* American Phytopathological Society, St Paul, Minnnesota.

Jones, A.L. and Ehret, G.R. (2000) Field evaluation of Apogee for fire blight management in 1999. *Fungicide and Nematicide Tests* 55, 15.

Jones, A.L., Lillevik, S.L., Fisher, P.D. and Stebbins, T.C. (1980) A microcomputer-based instrument to predict primary apple scab infection periods. *Plant Disease* 64, 69–72.

Jones, A.L., Fisher, P.D., Seem, R.C., Kroon, J.C. and Vandemotter, P.J. (1984) Development and commercialization of an in-field microcomputer delivery system for weather-driven predictive models. *Plant Disease* 68, 458–463.

Jones, A.L., Ehret, G.R., Maxson-Stein, K. and Sundin, G.W. (2003) Fire blight control with S0208, 2002. *Fungicide and Nematicide Tests*. Report 58 PF001. DOI:10.1094/FN58.

Juraske, R., Anton, A., Castells, F. and Hui jbregts, M.A.J. (2007) PestScreen: a screening approach for scoring and ranking pesticides by their environmental and toxicological concern. *Environment International* 33, 886–893.

Kessmann, H., Staub, T., Hofmann, C., Maetzke, T., Herzog, J., Ward, E., Uknes, S. and Ryals, J. (1994) Induction of systemic acquired disease resistance in plants by chemicals. *Annual Review of Phytopathology* 32, 439–459.

Khan, M.A., Duffy, B., Gessler, C. and Patocchi, A. (2006) QTL mapping of fire blight resistance in apple. *Molecular Breeding* 17, 299–306.

Kim, K.S., Gleason, M.L. and Taylor, S.E. (2006) Forecasting site-specific leaf wetness duration for input to disease-warning systems. *Plant Disease* 90, 650–656.

Klein, J.D., Conway, W.S., Whitaker, B.D. and Sams, C.E. (1997) *Botrytis cinerea* decay in apples is inhibited by postharvest heat and calcium treatments. *Journal of The American Society for Horticultural Science* 122, 91–94.

Köller, W., Wilcox, W.F., Barnard, J., Jones, A.L. and Braun, P.G. (1997) Detection and quantification of resistance of *Venturia inaequalis* populations to sterol demethylation inhibitors. *Phytopathology* 87, 184–190.

Köller, W., Parker, D.M., Turechek, W.W., Avila-Adame, C. and Cronshaw, K. (2004) A two phase resistance response of *Venturia inaequalis* populations to the QoI fungicides kresoxim-methyl and trifloxystrobin. *Plant Disease* 88, 537–544.

Köller, W., Parker, D.M., Turechek, W.W., Rosenberger, D.A., Wilcox, W.F., Caroll, J., Agnello, A.M. and Reissig, W.H. (2005a) Fungicide resistance of apple scab: status quo and management options. *New York Fruit Quarterly* 13(1), 9–17.

Köller, W., Wilcox, W.F. and Parker, D.M. (2005b) Sensitivity of *Venturia inaequalis* populations to anilinopyrimidine fungicides and their contribution to scab management in New York. *Plant Disease* 89, 357–365.

Korban, S.S. and Dayton, D.F. (1983) Evaluation of *Malus* germplasm for resistance to powdery mildew. *Hortscience* 18, 219–220.

Kovach, J., Petzoldt, C., Degnil, J. and Tette, J. (1992) A method to measure the environmental impact of pesticides. *New York's Food and Life Science's Bulletin No 139.* New York Agricultural Experiment Station, Geneva, New York.

Kuflik, T., Pertot, I., Moskovitch, R., Zasso, R., Pellegrini, E. and Gessler, C. (2008) Optimization of Fire blight scouting with a decision support system based on infection risk. *Computers and Electronics in Agriculture* 62, 118–127.

Lalancette, N. and Hickey, K.D. (1986) An apple powdery mildew model based on plant growth, primary inoculum, and fungicide concentration. *Phytopathology* 76, 1176–1182.

Latorre, B.A., Rioja, M.E. and Wilcox, W.F. (2001) Phytophthora species associated with crown and root rot of apple in Chile. *Plant Disease* 85, 603–606.

Lau, Y.F., Gleason, M.L., Zriba, N., Taylor, S.E. and Hinz, P.N. (2000) Effects of coating, deployment angle, and compass orientation on performance of electronic wetness sensors during dew periods. *Plant Disease* 84, 192–197.

Leach, A.W. and Mumford, J.D. (2008) Pesticide environmental accounting: a method for assessing the external costs of individual pesticide applications. *Environmental Pollution* 151, 139–147.

Levitan, L. (2000) 'How to' and 'why': assessing the enviro-social impacts of pesticides. *Crop Protection* 19, 629–636.

Levitan, L., Merwin, I. and Kovach, J. (1995) Assessing the relative environmental impacts of agricultural pesticides: the quest for a holistic method. *Agriculture Ecosystems and Environment* 55, 153–168.

Lewis, F.H. and Hickey, K.D. (1972) Fungicide usage on deciduous fruit trees. *Annual Review of Phytopathology* 10, 399–428.

Lewis, K.A., Brown, C.D., Hart, A. and Tzilivakis, J. (2003) p-EMA (III): overview and application of a software system designed to assess the environmental risk of agricultural pesticides. *Agronomie* 23, 85–96.

Loper, J.E., Henkels, M.D., Roberts, R.G., Grove, G.G., Willett, M.J. and Smith, T.J. (1991) Evaluation of streptomycin, oxytetracycline, and copper resistance of *Erwinia amylovora* isolated from pear orchards in Washington state. *Plant Disease* 75, 287–290.

Loureiro, M.L., McCluskey, J.J. and Mittelhammer, R.C. (2002) Will consumers pay a premium for eco-labeled apples? *Journal of Consumer Affairs* 36, 203–219.

Lurie, S. (1998) Postharvest heat treatments. *Postharvest Biology and Technology* 14, 257–269.

MacHardy, W.E. (1996) *Apple Scab: Biology, Epidemiology and Management.* APS Press, St Paul, Minnesota.

MacHardy, W.E. (2000a) Action threshold for managing apple scab with fungicides and sanitation. *Acta Horticulturae* 525, 123–131.

MacHardy, W.E. (2000b) Current status of IPM in apple orchards. *Crop Protection* 19, 801–806.

MacHardy, W.E. and Gadoury, D.M. (1989) A revision of Mills criteria for predicting apple scab infection periods. *Phytopathology* 79, 304–310.

MacHardy, W.E., Gadoury, D.M. and Rosenberger, D.A. (1993) Delaying the onset of fungicide programs for control of apple scab in orchards with low potential ascospore dose of *Venturia inaequalis*. *Plant Disease* 77, 372–375.

MacHardy, W.E., Gadoury, D.M. and Gessler, C. (2001) Parasitic and biological fitness of *Venturia inaequalis*: relationship to disease management strategies. *Plant Disease* 85, 1036–1051.

McManus, P.S., Stockwell, V.O., Sundin, G.W. and Jones, A.L. (2002) Antibiotic use in plant agriculture. *Annual Review of Phytopathology* 40, 443–465.

Magarey, R.D., Seem, R.C., Russo, J.M., Zack, J.W., Waight, K.T., Travis, J.W. and Oudemans, P.V. (2001) Site-specific weather information without on-site sensors. *Plant Disease* 85, 1216–1226.

Magarey, R.D., Russo, J.M., Seem, R.C. and Gadoury, D.M. (2005) Surface wetness duration under controlled environmental conditions. *Agricultural and Forest Meteorology* 128, 111–122.

Magkos, F., Arvaniti, F. and Zampelas, A. (2006) Organic food: buying more safety or just peace of mind? A critical review of the literature. *Critical Reviews in Food Science and Nutrition* 46, 23–56.

Marrone, P.G. (2007) Barriers to adoption of biological control agents and biological pesticides. Online *CAB Reviews: Perspectives in Agriculture, Veterinary Science, Nutrition and Natural Resources* 2 (http://www.cababstractsplus.org/cabreviews doi: 10.1079/PAVSNNR20072051).

Maxson, K.L., Ehret, G.R. and Jones, A.L. (2001) Field evaluation of Actigard for control of fire blight in apple, 2000. *Fungicide and Nematicide Tests*. Report 56 PF11. DOI:10.1094/FN56.

Maxson-Stein, K., He, S.Y., Hammerschmidt, R. and Jones, A.L. (2002) Effect of treating apple trees with acibenzolar-S-methyl on fire blight and expression of pathogenesis-related protein genes. *Plant Disease* 86, 785–790.

Merwin, I.A., Brown, S.K., Rosenberger, D.A., Cooley, D.R. and Berkett, L.P. (1994) Scab-resistant apples for the Northeastern United States – new prospects and old problems. *Plant Disease* 78, 4–10.

Mills, W.D. (1944) Efficient use of sulfur dusts and sprays during rain to control apple scab. *Bulletin of the New York Agricultural Experiment Station*, Geneva, New York.

Momol, M.T., Aldwinckle, H.S. and Norelli, J.L. (1998a) Evaluation of several products for control of fire blight on apple blossom, 1997. *Fungicide and Nematicide Tests* 53, 22.

Momol, M.T., Norelli, J.L., Piccioni, D.E., Momol, E.A., Gustafson, H.L., Cummins, J.N. and Aldwinckle, H.S. (1998b) Internal movement of *Erwinia amylovora* through symptomless apple scion tissues into the rootstock. *Plant Disease* 82, 646–650.

Moon, W., Florkowski, W.J., Bruckner, B. and Schonhof, I. (2002) Willingness to pay for environmental practices: implications for eco-labeling. *Land Economics* 78, 88–102.

Murphy, C. and Willett, L.S. (1991) Issues in the development and marketing of reduced chemical agricultural products: a look at disease-resistant apple cultivars. *Agricultural Economics Extension Bulletin 91-34*. Cornell University, Ithaca, New York.

Norelli, J.L. and Miller, S.S. (2004) Effect of prohexadione-calcium dose level on shoot growth and fire blight in young apple trees. *Plant Disease* 88, 1099–1106.

Norelli, J.L., Holleran, H.T., Johnson, W.C., Robinson, T.L. and Aldwinckle, H.S. (2003a)

Resistance of Geneva and other apple rootstocks to *Erwinia amylovora*. *Plant Disease* 87, 26–32.

Norelli, J.L., Jones, A.L. and Aldwinckle, H.S. (2003b) Fire blight management in the twenty-first century – using new technologies that enhance host resistance in apple. *Plant Disease* 87, 756–765.

Ocamb-Basu, C.M., Sutton, T.B. and Nelson, L.A. (1988) The effects of pruning on incidence and severity of *Zygophiala jamaicensis* and *Gloeodes pomigena* infections of apple fruit. *Phytopathology* 78, 1004–1008.

Ojiambo, P.S. and Scherm, H. (2006) Biological and application-oriented factors influencing plant disease suppression by biological control: a meta-analytical review. *Phytopathology* 96, 1168–1174.

Parisi, L., Lespinasse, Y., Guillaumes, J. and Kruger, J. (1993) A new race of *Venturia inaequalis* virulent to apples with resistance due to the V_f gene. *Phytopathology* 83, 533–537.

Penev, R. and Aldwinckle, H.S. (2003) Field evaluation of materials for control of fire blight infection of apple blossoms, 2002. *Fungicide and Nematicide Tests* 58, PF004.

Peng, J.L., Dong, H.S., Dong, H.P., Delaney, T.P., Bonasera, J.M. and Beer, S.V. (2003) Harpin-elicited hypersensitive cell death and pathogen resistance require the NDR1 and EDS1 genes. *Physiological and Molecular Plant Pathology* 62, 317–326.

Penrose, L.J. (1995) Fungicide use reduction in apple production – potentials or pipe dreams. *Agriculture Ecosystems and Environment* 53, 231–242.

Penrose, L.J. and Nicol, H.I. (1996) Aspects of microclimate variation within apple tree canopies and between sites in relation to potential *Venturia inaequalis* infection. *New Zealand Journal of Crop and Horticultural Science* 24, 259–266.

Piñero, J.C. and Prokopy, R.J. (2003) Field evaluation of plant odor and pheromonal combinations for attracting plum curculios. *Journal of Chemical Ecology* 29, 2735–2748.

Piñero, J.C., Tuttle, A.F., Agnello, A.M., Faubert, H., Koehler, G., Morin, G., Leahy, K., Los, L., Cooley, D.R. and Prokopy, R.J. (2006) Using odor-baited trap trees as indicators of need and timing of insecticide spray against plum curculio in the Northeast: validation studies 2004–2005. *Fruit Notes* 72(4), 1–11.

Prokopy, R.J. (1993) Stepwise progress toward IPM and sustainable agriculture. *The IPM Practitioner* 15, 1–4.

Prokopy, R.J. (2003) Two decades of bottom-up, ecologically based pest management in a small commercial apple orchard in Massachusetts. *Agriculture Ecosystems and Environment* 94, 299–309.

Prokopy, R.J., Wright, S.E., Black, J.L., Hu, X.P. and McGuire, M.R. (2000) Attracticidal spheres for controlling apple maggot flies: commercial-orchard trials. *Entomologia Experimentalis et Applicata* 97, 293–299.

Prokopy, R.J., Jácome, I., Bigurra, E., Blanco, M., Chandler, B.W. and Wright, S.E. (2004) Evaluation of pesticide-treated spheres for control of apple maggot flies in 2003. *Fruit Notes* 69(2), 21–25.

Rademacher, W. (2004) Prohexadione-Ca induces resistance to fire blight and other diseases. *EPPO Bulletin* 34, 383–388.

Ragsdale, N.N. (2000) The impact of the food quality protection act on the future of plant disease management. *Annual Review of Phytopathology* 38, 577–596.

Reardon, J.E., Berkett, L.P., Garcia, M.E. and Gottlieb, A. (2005) Field evaluation of a new sequential sampling technique for determining apple scab 'risk'. *Plant Disease* 89, 228–236.

Reddy, M.V.B., Norelli, J.L. and Aldwinckle, H.S. (2000) Control of fire blight infection of apple blossoms, 1999. *Fungicide and Nematicide Tests*. Report 55 22. DOI: 10.1094/FN55.

Reddy, M.V.B., Norelli, J.L. and Aldwinckle, H.S. (2001) Biologicals, SAR inducers, copper compounds and other chemicals for blossom blight control on apple, 2000. *Fungicide and Nematicide Tests*. Report 56 PF34. DOI:10.1094/FN56.

Reed, A.N. and Nelson, P. (1995) Responsible choice – an IFP approach. *Acta Horticulturae* 422, 319–320.

Reganold, J.P. (2006) Sustainability of organic, conventional, and integrated apple orchards.

Crop Management. DOI:10.1094/CM-2006-0921-16-PS.

Robinson, J.P., Harris, S.A. and Juniper, B.E. (2001) Taxonomy of the genus *Malus* Mill. (Rosaceae) with emphasis on the cultivated apple, *Malus domestica* Borkh. *Plant Systematics and Evolution* 226, 35–58.

Robinson, T.L., Aldwinckle, H.S., Fazio, G. and Holleran, H.T. (2003a) The Geneva series of apple rootstocks from Cornell: performance, disease resistance, and commercialization. In: Janick, J. (ed.) *XXVI International Horticultural Congress: Genetics and Breeding of Tree Fruits and Nuts*. ISHS, Toronto, Canada, pp. 513–520.

Robinson, T.L., Anderson, L., Azarenko, A., Barritt, B., Baugher, T., Brown, G., Couvillon, G., Cowgill, W.P., Crassweller, R., Domoto, P., Embree, C., Fennell, A., Garcia, M.E., Gaus, A., Granger, R., Greene, G., Hirst, P., Hoover, E., Johnson, S.A. and Kushad, M. (2003b) Performance of Cornell-Geneva apple rootstocks with 'Liberty' as the scion in NC-140 trials across North America. *Acta Horticulturae* 622, 521–530.

Rosenberger, D.A. (2003) Factors limiting IPM-compatibility of new disease control tactics for apples in eastern United States. *Plant Health Progress*. DOI:10.1094/PHP-2003-0826-01-RV.

Rosenberger, D.A. (2006) Timing sprays for flyspeck and sooty blotch. Online *Scaffolds Fruit Journal* 14 (14) (http://www.nysaes.cornell.edu/ent/scafolds/doi: none).

Rosenberger, D.A. and Meyer, F.W. (2007) Control of flyspeck with phosphorous acid, liquid lime-sulfur, and other fungicides, 2006. *Plant Disease Management Reports*. DOI: 10.1094/PDMR01.

Rosenberger, D.A., Meyer, F.W. and Rist, D.L. (1990) Effect of spray timing on incidence and severity of sooty blotch and flyspeck on Liberty apple. *Fungicide and Nematicide Tests* 46, 31.

Rosenberger, D.A., Meyer, F.W. and Engle, C.A. (1994) Early-season diseases occurring on scab-resistant apple cultivars and advanced selections grown in southeastern New-York state. *Fruit Varieties Journal* 48(1), 52–53.

Rosenberger, D.A., Engle, C.A. and Meyer, F.W. (1996) Effects of management practices and fungicides on sooty blotch and flyspeck diseases and productivity of liberty apples. *Plant Disease* 80, 798–803.

Rosenberger, D.A., Ahlers, C.A. and Meyer, F.W. (2000) Controlling diseases on Jerseymac and Ginger Gold apples with biocontrol and contact fungicides, 1999. *Fungicide and Nematicide Tests*. Report 55 25. DOI:10.1094/FN55.

Rosenberger, D.A., Meyer, F.W. and Christiana, R.W. (2005) Controlling summer diseases on trees left unsprayed for 50 days during June and July, 2004. *Fungicide and Nematicide Tests*. Report 60. DOI:10.1094/FN60.

Rosenberger, D.A., Rugh, A. and Meyer, F.W. (2006) Keeping apples disease-free during storage and shipping. *New York Fruit Quarterly* 14(2), 7–12.

Rosenberger, D.A., Meyer, F.W. and Rugh, A. (2008a) Effectiveness of ProPhyt used alone or with other fungicides to control summer disease of apples, 2007. *Plant Disease Management Reports*. Report 2 PF032. DOI:10.1094/PDMR02.

Rosenberger, D.A., Meyer, F.W. and Rugh, A. (2008b) Interactions among fungicides and spry adjuvants in late summer sprays for flyspeck, 2007. *Plant Disease Management Reports*. Report 2 PF031. DOI:10.1094/PDMR02.

Rosenberger, D.A., Meyer, F.W. and Rugh, A. (2008c) Effectiveness of ProPhyt, Inspire and Inspire Super for controlling early-season apple diseases, 2007. *Plant Disease Management Reports*. Report 2 PF027. DOI:10.1094/PDMR02.

Rossi, V., Ponti, I., Marinelli, M., Giosue, S. and Bugiani, R. (1999) Field evaluation of some models estimating the seasonal pattern of airborne ascospores of *Venturia inaequalis*. *Journal of Phytopathology – Phytopathologische Zeitschrift* 147, 567–575.

Russo, N.L., Robinson, T.L., Fazio, G. and Aldwinckle, H.S. (2007) Field evaluation of 64 apple rootstocks for orchard performance and fire blight resistance. *Hortscience* 42, 1517–1525.

Russo, N.L., Robinson, T.L., l'Azio, G. and Aldwinckle, H.S. (2008) Fire blight resistance of Budagovsky 9 apple rootstock. *Plant Disease* 92, 385–391.

Schwabe, W.F.S. (1980) Wetting and temperature requirements for apple leaf infection by *Venturia inaequalis*. *Phytophylactica* 12, 69–80.

Sentelhas, P.C., Gillespie, T.J., Batzer, J.C., Gleason, M.L., Monteiro, J., Pezzopane, J.R.M. and Pedro, M.J. (2005) Spatial variability of leaf wetness duration in different crop canopies. *International Journal of Biometeorology* 49, 363–370.

Shao, X.F., Tu, K., Zhao, Y.Z., Chen, L., Chen, Y.Y. and Wang, H. (2007) Effects of prestorage heat treatment on fruit ripening and decay development in different apple cultivars. *Journal of Horticultural Science and Biotechnology* 82, 297–303.

Sholberg, P.L., Bedford, K.E., Haag, P. and Randall, P. (2001) Survey of *Erwinia amylovora* isolates from British Columbia for resistance to bactericides and virulence on apple. *Canadian Journal of Plant Pathology* 23, 60–67.

Shtienberg, D., Kritzman, G., Herzog, Z., Oppenheim, D., Zilberstaine, M. and Blatchinsky, D. (1999) Development and evaluation of a decision support system for management of fire blight in pears. *Acta Horticulturae* 489, 385–392.

Shtienberg, D., Shwartz, H., Oppenheim, D., Zilberstaine, M., Herzog, Z., Manulis, S. and Kritzman, G. (2003) Evaluation of local and imported fire blight warning systems in Israel. *Phytopathology* 93, 356–363.

Smigell, C.G. and Hartman, J.R. (1997) Evaluation of fungicide timing for sooty blotch and flyspeck control, 1996. *Fungicide and Nematicide Tests* 52, 31.

Smith, T.J. (1993) A predictive model for forecasting fire blight of pear and apples in Washington State. *Acta Horticulturae* 338, 153–160.

Smith, T.J. (1999) Report on the development and use of Cougarblight 98C – a situation-specific fire blight risk assessment model for apple and pear. *Acta Horticulturae* 489, 429–436.

Spadaro, D., Garibaldi, A. and Gullino, M.L. (2004) Control of *Penicillium expansum* and *Botrytis cinerea* on apple combining a biocontrol agent with hot water dipping and acibenzolar-S-methyl, baking soda, or ethanol application. *Postharvest Biology and Technology* 33, 141–151.

Steiner, P.W. (1990a) Predicting apple blossom infections by *Erwinia amylovora* using the Maryblyt model. *Acta Horticulturae* 273, 139–148.

Steiner, P.W. (1990b) Predicting canker, shoot and trauma blight phases of apple fire blight epidemics using the Maryblyt model. *Acta Horticulturae* 273, 149–158.

Steiner, P.W. and Lightner, G.W. (1996) Maryblyt™ 4.3. A predictive program for forecasting fire blight disease in apples and pears. University of Maryland, College Park, Maryland.

Stensvand, A., Gadoury, D.M., Amundsen, T., Semb, L. and Seem, R.C. (1997) Ascospore release and infection of apple leaves by conidia and ascospores of *Venturia inaequalis* at low temperatures. *Phytopathology* 87, 1046–1053.

Stensvand, A., Eikemo, H., Gadoury, D.M. and Seem, R.C. (2005) Use of a rainfall frequency threshold to adjust a degree-day model of ascospore maturity of *Venturia inaequalis*. *Plant Disease* 89, 198–202.

Sticher, L., MauchMani, B. and Metraux, J.P. (1997) Systemic acquired resistance. *Annual Review of Phytopathology* 35, 235–270.

Stockwell, V.O. and Stack, J.P. (2007) Using *Pseudomonas* spp. for integrated biological control. *Phytopathology* 97, 244–249.

Sundin, G.W. and Ehret, G.R. (2004a) Biological control of fire blight with Serenade in 2003. *Fungicide and Nematicide Tests*. Report 59 PF008. DOI:10.1094/FN59.

Sundin, G.W. and Ehret, G.R. (2004b) Fire blight control on Jonathan apple in 2003. *Fungicide and Nematicide Tests*. Report 59 PF007. DOI:10.1094/FN59.

Sundin, G.W. and Ehret, G.R. (2005) Control of fire blight on Jonathan and Golden Delicious apple in 2004. *Fungicide and Nematicide Tests*. Report 60 PF033. DOI: 10.1094/FN60.

Sundin, G.W., Ehret, G.R. and Shane, W.W. (2005) Evaluation of bactericides for fire blight control in a commercial orchard near Benton Harbor, Michigan, 2004. *Fungicide and Nematicide Tests*. Report 60 PF038. DOI:10.1094/FN60.

Sundin, G.W., Ehret, G.R. and McGhee, G.C. (2006a) Biological control of fire blight on Jonathan and Golden Delicious apple in 2005. *Fungicide and Nematicide Tests*. Report 61 PF009. DOI:10.1094/FN61.

Sundin, G.W., Ehret, G.R. and McGhee, G.C. (2006b) Fire blight control on Gala apple in 2005. *Fungicide and Nematicide Tests*. Report 61 PF008. DOI:10.1094/FN61.

Sutton, D.K., MacHardy, W.E. and Lord, W.G. (2000) Effects of shredding or treating apple leaf litter with urea on ascospore dose of *Venturia inaequalis* and disease build-up. *Plant Disease* 84, 1319–1326.

Sutton, J.C., Gillespie, T.J. and Hildebrand, P.D. (1984) Monitoring weather factors in relation to plant disease. *Plant Disease* 68, 78–84.

Sutton, T.B. (1996) Changing options for the control of deciduous fruit tree diseases. *Annual Review of Phytopathology* 34, 527–547.

Sutton, T.B., Brown, E.M., Anas, O. and Meister, C.W. (2006) Efficacy of phosphorus-containing fungicides on summer diseases of apples, 2005. *Fungicide and Nematicide Tests*. Report 61. DOI:10.1094/FN61.

Sutton, T.B., Brown, E.M. and Anas, O. (2007) Summer disease control with reduced risk fungicides, 2006. *Plant Disease Management Reports*. Report 1 PF007. DOI:10.1094/PDMR01.

Szkolnik, M. (1969) Maturation and discharge of ascospores of *Venturia inaequalis*. *Plant Disease Reporter* 53, 534–537.

Tenzer, I., degli Ivanissevich, S., Morgante, M. and Gessler, C. (1999) Identification of microsatellite markers and their application to population genetics of *Venturia inaequalis*. *Phytopathology* 89, 748–753.

Thies, W.H. (1947) Announcing the 'Red Apple Club'. *Fruit Notes* 12 (4 May), 2–3.

Thomson, S.V., Schroth, M.N. and Reil, W.O. (1982) A forecasting model for fire blight of pear. *Plant Disease* 66, 576–579.

Travis, J.W., Halbrendt, N.O., Anderson, E., Wilson, C., Norelli, J.L. and Burchard, E. (2004) Evaluation of biological agents for control of fire blight on apple, 2003. *Biological and Cultural Tests for Control of Plant Diseases*. Report 19 N001. DOI: 10.1094/BC19.

Travis, J.W., Halbrendt, N.O., Rytter, J., Lehman, B. and Jarjour, B. (2006a) Evaluation of alternative fungicide programs for control of flyspeck and sooty blotch on Golden Delicious and Cortland apples in PA, 2005. *Biological and Cultural Tests for Control of Plant Diseases*. Report 21 1–2. DOI:10.1094/BC21.

Travis, J.W., Halbrendt, N.O., Rytter, J., Lehman, B. and Jarjour, B. (2006b) Evaluation of organic alternatives for control of apple scab, 2005. *Biological and Cultural Tests for Control of Plant Diseases*. Report 21 1–3. DOI:10.1094/BC21.

Tuttle, A.F., Bergweiler, C., Hall, J., Reisner, L., Christle, S., Autio, W.R. and Cooley, D.R. (2002) Development of a model for predicting flyspeck risks in blocks of apple trees. *Fruit Notes* 67(1), 5–8.

Urbanietz, A. and Dunemann, F. (2005) Isolation, identification and molecular characterization of physiological races of apple powdery mildew (*Podosphaera leucotricha*). *Plant Pathology* 54, 125–133.

Vallad, G.E. and Goodman, R.M. (2004) Systemic acquired resistance and induced systemic resistance in conventional agriculture. *Crop Science* 44, 1920–1934.

Vanneste, J.L. and Eden-Green, S. (2000) Migration of *Erwinia amylovora* in host plant tissue. In: Vanneste, J.L. (ed.) *Fire Blight: The Disease and Its Causative Agent,* Erwinia amylovora. CAB International, New York, pp. 73–83.

Vincent, C., Rancourt, B. and Carisse, O. (2004) Apple leaf shredding as a non-chemical tool to manage apple scab and spotted tentiform leafminer. *Agriculture Ecosystems and Environment* 104, 595–604.

Warner, J. (1990) Field susceptibility of scab-resistant apple cultivars and selections to cedar apple rust, quince rust and hawthorn rust. *Fruit Varieties Journal* 44, 216–225.

Werner, N.A., Heidenreich, G. and Aldwinckle, H.S. (2005) Field evaluation of materials for control of fire blight infection of apple, 2004. *Fungicide and Nematicide Tests*. Report 60 PF030. DOI:10.1094/FN60.

Werner, N.A., Heidenreich, G. and Aldwinckle, H.S. (2006a) Field evaluation of biological and chemical control of fire blight of apple,

2005. *Fungicide and Nematicide Tests*. Report 61 PF020. DOI:10.1094/FN61.

Werner, N.A., Heidenreich, G. and Aldwinckle, H.S. (2006b) Field evaluation of antibiotics, SAR inducers and copper for control of fire blight of apple, 2005. *Fungicide and Nematicide Tests*. Report 61 PF021. DOI: 10.1094/FN61.

Wilcox, W.F., Wasson, D.I. and Kovach, J. (1992) Development and evaluation of an integrated, reduced-spray program using sterol demethylation inhibitor fungicides for control of primary apple scab. *Plant Disease* 76, 669–677.

Williams, P.R.D. and Hammitt, J.K. (2001) Perceived risks of conventional and organic produce: pesticides, pathogens, and natural toxins. *Risk Analysis* 21, 319–330.

Williamson, S.M. and Sutton, T.B. (2000) Sooty blotch and flyspeck of apple: etiology, biology, and control. *Plant Disease* 84, 714–724.

Wisniewski, M., Wilson, C., El Ghaouth, A. and Droby, S. (2001) Non-chemical approaches to postharvest disease control. *Acta Horticulturae* 553, 407–412.

Xu, X.M. (1999) Modelling and forecasting epidemics of apple powdery mildew (*Podosphaera leucotricha*). *Plant Pathology* 48, 462–471.

Xu, X.M., Butt, D.J. and VanSanten, G. (1995) A dynamic model simulating infection of apple leaves by *Venturia inaequalis*. *Plant Pathology* 44, 865–876.

Xu, X.M., Yang, J.R., Thakur, V., Roberts, A. and Barbara, D.J. (2008) Population variation of apple scab (*Venturia inaequalis*) isolates from Asia and Europe. *Plant Disease* 92, 247–252.

Yoder, K., Byers, R., Cochran, A.E., DeMarsay, A., Royston, W.S. and Kilmer, S.W. (1999) Evaluation of bloom treatments for fire blight suppression on Golden Delicious and Rome Beauty apples, 2000. *Fungicide and Nematicide Tests* 56, PF36.

Yoder, K., Cochran, A.E., Royston, W.S. and Kilmer, S.W. (2003) Fire blight blossom treatments on Golden Delicious and Rome Beauty apples, 2002. *Fungicide and Nematicide Tests* 58, PF025.

Yoder, K., Cochran, A.E., Royston, W.S. and Kilmer, S.W. (2004) Evaluation of fire blight blossom treatments on Golden Delicious and Rome Beauty apples, 2003. *Fungicide and Nematicide Tests* 59, PF030.

Yoder, K., Cochran, A.E., Royston, W.S. and Kilmer, S.W. (2007) Integrated biological and streptomycin treatments for fire blight control on Idared apples, 2006. *Plant Disease Management Reports* 1, PF028.

Yoshimura, M.A., Luo, Y., Ma, Z.H. and Michailides, T.J. (2004) Sensitivity of *Monilinia fructicola* from stone fruit to thiophanate-methyl, iprodione, and tebuconazole. *Plant Disease* 88, 373–378.

Ypema, H.L. and Gold, R.E. (1999) Modification of a naturally occurring compound to produce a new fungicide. *Plant Disease* 83, 4–19.

Zasso, R., Moskovitch, R., Pertot, I. and Gessler, C. (2006) Integrated forecasting systems for monitoring fire blight in Trentino. *Acta Horticulturae* 704, 191–196.

Zhou, T., Northover, J., Schneider, K.E. and Lu, X.W. (2002) Interactions between *Pseudomonas syringae* MA-4 and cyprodinil in the control of blue mold and gray mold of apples. *Canadian Journal of Plant Pathology* 24, 154–161.

Zuck, M.G. and MacHardy, W.E. (1981) Recent experience in timing sprays for control of apple scab – equipment and test results. *Plant Disease* 65, 995–998.

9 Speciation, Consumers and the Market: Profit with a Conscience

THOMAS A. GREEN[1] AND STARKER E. WRIGHT[2]

[1]*IPM Institute of North America, 4510 Regent St., Madison, USA;* [2]*Pest Management Innovations, PO Box 1363, Harpers Ferry, USA*

> But, after all, this is not as anomalous and inexplicable as the fact that the Apple-maggot Fly (*Trypeta pomonella*, Walsh) which exists both in Illinois, New York, and New England, and the larva of which feeds in Illinois upon the native haws, and has never once been noticed to attack imported apple there, should, within the last few years, have suddenly fallen upon the apple, both in New York and New England, and in many localities there, have become a more grievous foe to the fruit than even the imported Apple-worm (*Carpocapsa pomonella*, Linn).
>
> Benjamin Walsh (Walsh and Riley, 1868)

Introduction

In the middle of the 19th century, apple maggot fly (AMF) was noted as a serious pest of cultivated apple in the Hudson Valley of New York, where it had subsisted previously on native hawthorn (*Crataegus* spp.) (Walsh, 1867). Given the rapid westward and southward expansion of the shift to apples from an initially limited geographic area, this detection probably occurred late in the establishment process on the new host (Brown, 1993). While the shift provided certain presumed advantages to AMF including reduced inter- and intraspecific competition (Feder *et al.*, 1995), escape from predators and parasitoids (Brown *et al.*, 1995) and expansion of spatial and temporal availability of suitable host fruit (Feder and Filchak, 1999), emergence of this host-associated race also led to an unparalleled depth of study in both basic and applied science. By establishing persistent populations on apple, AMF entered a new realm of ecological, academic and economic consequence, unwittingly anointing itself a key pest and extraordinary case study for the finer details of evolutionary theory. In addition, by carrying out its exploitation of a new resource in plain sight, AMF provided a partial set of keys to the examination of host and mate selection, ecological adaptation and mechanisms of speciation. Finally, by impacting commercial apple production with nearly 100% damage if left uncontrolled, this pest demanded substantial pesticide use, helped justify apples as a pilot crop for one of the first federally funded integrated pest management (IPM) programmes in the USA and continues to help drive innovations in IPM technology.

The importance of AMF's shift to a new host was recognized immediately by Walsh (1867); expanded on by Illingsworth (1912), Porter (1928), Dean (1942) and Hodson (1943); sharpened by Bush (1969), Papaj and Prokopy (1989), Aluja and Prokopy (1993), Feder (1998) and Berlocher and Feder (2002; Chapter 2, this volume); and applied to management-related objectives by Prokopy (1968b, 1972), Prokopy and Owens (1983), Prokopy *et al.* (1973, 1982, 1988), Kring (1970), Reissig (1975), Fein *et al.* (1982), Carle *et al.* (1987), Agnello *et al.* (1990), Aliniazee (1990), Linn *et al.* (2003), Prokopy and Roitberg (2007) and others.

In this chapter, we examine environmental and health impacts as an important motivation for R.J. Prokopy's work, the transition from basic science to the development and delivery of biologically based pest management tools to growers and the marketing of the fruits of lesser impact production systems to consumers.

The Need for Improved Pest Management and Reduced Impacts

Prokopy's influence has been felt in all aspects of research on the behaviour, ecology and genetics of AMF over the past four decades. However, perhaps his greatest influence has been on tapping biology and behaviour to extract IPM systems including behavioural control strategies for AMF and more than a dozen other key pests of apple in the north-eastern USA (see Cooley and Coli, Chapter 6, this volume). Nearly all of Prokopy's studies focused on revealing the underpinnings of the relationships between individual insects, insects and host trees, pests and host fruit, growers and crops, and crops and ecologically conscious consumers. As a body of work, these studies bridge the gaps between basic behavioural research, growers and the marketplace, and provide an avenue for shared pursuit of a cooperative relationship between apple producers and consumers.

The history of apple and other tree-fruit production is littered with abandoned control strategies that have proved to be potentially more harmful than helpful. Introduced in Massachusetts in 1892 for the control of gypsy moth, *Lymantria dispar* (Linnaeus), lead arsenate soon became a favourite of orchardists and the most widely used agricultural pesticide in the USA from 1919 to the late 1940s. Lead arsenate continues to be an expensive pollutant to remediate at old orchard sites, with contaminated soil removal and replacement costs of up to US$2.47 million/ha (reviewed in Hood, 2006). The product ultimately was banned in 1988. Similarly, DDT used in apple orchards from 1949 to 1969 continues to be detected in air samples drawn at long-time orchard sites at levels 10-times greater than non-orchard sites (Hermanson *et al.*, 2007). Most recently, in 2008, the US Environmental Protection Agency (EPA) announced a complete phase-out by 2012 of all uses of azinphos methyl, a potent neurotoxin and commonly used organophosphate insecticide in apple orchards.

In April 1985, Prokopy presented 'Towards a World of Less Pesticide' at the University of Massachusetts (UMass) as one of four speakers invited annually as part of the 'Chancellor's Lecture Series' (Prokopy, 1986). He acknowledged the 'ingenuity and determination of humans' in maintaining a food supply apace with a world population increasing then at more than 80 million individuals per year and the 'high price' paid in 'human and environmental well-being' due to side effects including an increase in pesticide use. He cited growth in USA pesticide use from 22.7 million kg per year in 1935 to 544.3 million kg in 1985, equivalent to 2.3 kg per person per year. He anticipated world population growth to 9 billion by 2020, requiring even more intensive agriculture and potentially greater pesticide use. Prokopy catalogued pesticide-related costs including 2500 killed and 150,000 injured in Bhopal in 1984 by the accidental release of methyl isocyanate, a chemical used in the manufacture of the pesticides, Sevin and Temik; 5000 deaths and 500,000 injuries worldwide annually due to pesticide poisonings; and groundwater contamination by pesticides used in tobacco and potato production in the Connecticut Valley, where many in his audience lived and worked.

In later life, ever the realist and optimist, Prokopy acknowledged serious, persistent 'impediments to broader adoption of pesticide alternatives' but also significant positive potential for new technologies, and suggested we look to Western Europe for policy initiatives 'truly speeding widespread transition… to biologically based approaches to pest management' (Prokopy, 1998).

Humankind is making limited progress on several of these fronts. For example, world population growth has slowed to about 40 million per year, half the 1985 rate of increase. At the close of 2007, the world population stood at 6.4 billion, including 303 million in the USA (US Census Bureau, 2007). Projecting from current trends, the world population may grow to 7.6 billion in 2020 and reach 9 billion by 2040, 20 years later than projected in 1985.

Conventional pesticide use is declining. World pesticide use was 2.3 billion kg of active ingredient in 2001 (Kiely *et al.*, 2004), down 454 million kg from 1985 usage. USA pesticide use was nearly 635 million kg in 1982, with 454 million kg used in agriculture. This total declined to 544 million kg in 2001, with 411 million kg used in agriculture, or about 1.9 kg per person overall. More recent sales data show an average 3.0% decline per year in inflation-adjusted conventional pesticide sales from 1999 to 2006 (Crop Life International, 2007). During the same period, biotech seed sales increased from US$2.3 to US$6.1 billion.

The toxicity of pesticides is also declining. Between 1992 and 2000, use of those pesticides identified as 'most hazardous' by the US EPA, while still comprising 40% of total pesticide use, declined by 14% (US GAO, 2001). In 1982, organophosphate insecticides made up 58% of all insecticide use in the USA, at 59.4 million kg. This amount declined to 33.1 million kg in 2001. Organophosphates are broad-spectrum (toxic to a wide variety of both beneficial and pest insects) and neurotoxic (they block acetylcholinesterase, an enzyme necessary for normal nerve function in insects, mammals and other organisms).

California, which has the most comprehensive pesticide use tracking system in the USA, reported a decline in overall pesticide use in agriculture, structural pest control, landscape maintenance, rights of way, public health and other uses,

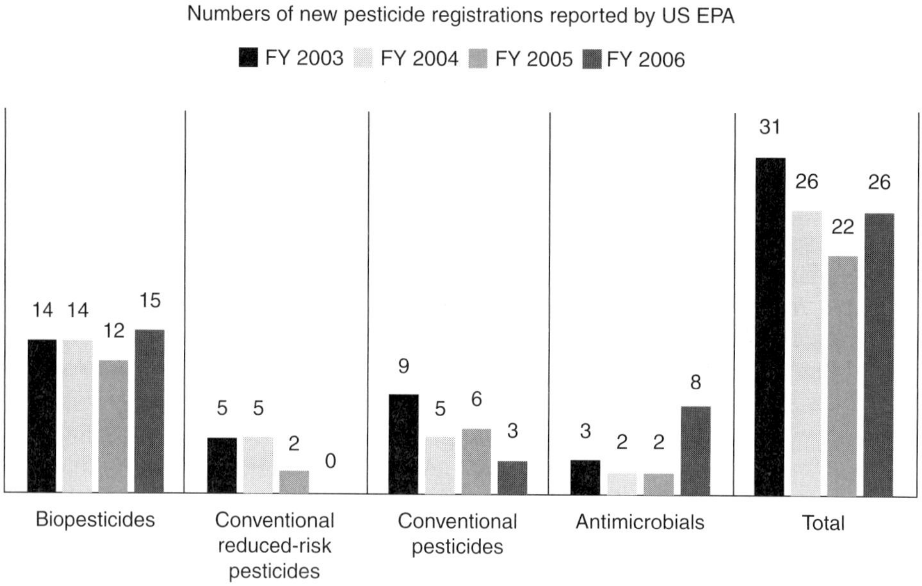

Fig. 9.1. Numbers of new pesticide registrations reported by the US EPA (2007) by classification. The majority of new registrations have met EPA definitions for biopesticides or reduced risk.

from a peak of 98.4 million kg of active ingredient in 1998 to 86.2 million kg in 2006 (California DPR, 2007). California further reported statewide declines from 1996 to 2006 in the use of pesticides containing chemicals known to cause reproductive toxicity (46% reduction) and acetylcholinesterase-inhibiting pesticides (55% decline). Use of pesticides containing active ingredients known to cause cancer has remained steady during the same period.

Both the US EPA and the California Department of Pesticide Regulation have expedited processes for the registration of reduced toxicity pesticides, including biopesticides (naturally occurring substances, microorganisms and pesticidal substances produced by plants containing genetic material introduced specifically to control pests). The pesticide industry has responded in a commendable fashion. Since 2003, the majority of new pesticide registrations have met the EPA's criteria for reduced risk (US EPA, 2007; Fig. 9.1).

Progress also has been made in reducing pesticide impacts. For example, bald eagle nesting pairs in the lower 48 USA states have increased from fewer than 450 in 1963 to now more than 11,000 (Suckling and Hodges, 2007). The bald eagle has been a poster child for the impacts of DDT and other organochlorine pesticides, the effects of which compounded up the food chain and resulted in lethal eggshell thinning in these and other raptors. In a second example, recent restrictions on diazinon and chlorpyrifos, used for structural pests, have led to declining concentrations of these neurotoxins in the umbilical cord blood of infants and increases in fetal growth parameters (Whyatt *et al.*, 2004).

Continued improvements in pest management practices are imperative. A recent review of 51 studies over 10 years found that 96% of fish, 100% of surface water and 33% of major aquifers sampled in the USA from 1992 to 2001 contained one or more pesticides (Gilliom *et al.*, 2007). Of streams sampled in agricultural areas, 10% contained pesticides at concentrations greater than the estimated threshold for impacts on human health. A study of human body contaminants provided a second example, reporting an average of 91 industrial compounds, pollutants and other chemicals in the blood and urine of nine volunteers (Thornton *et al.*, 2002; Houlihan *et al.*, 2003). None of these individuals worked with chemicals occupationally or lived near industrial facilities. Of the total of 167 chemicals found, 17 were pesticides or pesticide breakdown products, 76 were carcinogens, 94 neurotoxins and 79 developmental or reproductive toxins.

Improvements in managing pests are also needed and pesticides will continue to be necessary tools. Malaria continues to kill more than 1 million people annually (WHO, 2006). Asthma incidence is at epidemic levels in the USA, including 6% in children nationally and up to 25% in some urban areas, and growing. Asthma is associated with cockroach allergen sensitivity and exposure (Gruchalla *et al.*, 2005), as well as exposure to pesticides (Salam *et al.*, 2004). Current persistent and emerging pest management challenges include vectored human and animal diseases such as West Nile virus, Eastern equine encephalitis and Lyme disease; plant pests and diseases such as emerald ash borer (*Agrilus planipennis*) and soybean rust (*Phakopsora pachyrhizi*); and more than 170 noxious aquatic, terrestrial or parasitic weeds.

These challenges demand reinvigorated efforts to increase the adoption of basic IPM tactics in more crops (as well as in urban environments), accelerate progress along the IPM continuum towards more biologically based and more fully integrated systems and bridge gaps between producers, pest managers, researchers and educators working in conventional, IPM and organic systems to maximize information sharing and efficiency in bringing about positive change in economics, health and the environment.

Prokopy appreciated the importance of the marketplace in bringing about change and was extremely supportive of Green and Wright in the commercial endeavours described below. He never sought to benefit financially from the commercialization of his developments, freely sharing and publishing the fruits of his and collaborators' work. He deferred to W.M. Coli on matters relating to building consumer awareness and support for IPM, including growers practising IPM, which was a function of the pioneering Massachusetts Partners with Nature programme, described later in this chapter and in Cooley and Coli, Chapter 6, this volume.

Marketplace Power to 'Push' and 'Pull' IPM Forward

The marketplace has tremendous potential to drive improvements in health, environment and economics, goals laid out in the IPM Roadmap (USDA, 2004). The market provides a channel both for researchers to push new developments in IPM technology to potential users and for eco-minded consumers to pull improvements in practices by goods and service providers.

USDA has recognized market-based leverage through a new policy which 'seeks to broaden the use of private sector markets for environmental goods and services through voluntary market-based mechanisms' (USDA, 2006) and has formed a Market-based Environmental Stewardship Council to facilitate the activities necessary to implement the policy.

On the 'pull' side, a recent survey of nearly 3000 USA consumers commissioned by the National Pest Management Association (Harris Interactive, 2007) found that 82% agreed with the statement, 'It is very important to me that the products I use in and around my home are safe for the environment.' Examples of market pull include growing demand for organic food 'produced by farmers who emphasize the use of renewable resources and the conservation of soil and water to enhance environmental quality' (USDA, 2002). Although currently less than 2% of the USA agricultural acreage, USA sales of organic food have grown at rates exceeding 14% per year for the past several years and from US$3.6 billion in 1997 to US$16.7 billion in 2006 (Organic Trade Association, 2007). USA non-food organic product sales grew by 26% between 2005 and 2006, reaching US$938 million. Evidence is increasing that organics may promote consumer health with increased levels of some vitamins and minerals, as well as antioxidants, polyphenols, flavonoids and anthocyanins (Benbrook, 2007). Many additional comparative studies are under way. In addition, children placed on an all-organic diet for just 5 days had lower concentrations of neurotoxic organophosphate pesticides in their urine (Lu *et al*., 2005).

On the 'push' side, a healthy and diverse market for IPM goods and services has been established since the mid-1980s. Evidence includes the expansion of a one-page *IPM Products and Services* (listing 30 pheromones) and a two-page *Suppliers of Beneficial Organisms* (23 biocontrol agents), first published in 1982, into an annual *Directory of Least-Toxic Pest Control Products*, which in 2006 reached 50 pages in length and listed 2500 products and 600 suppliers. The directory is published by the Bio-Integral Resource Center, producer of the monthly *IPM Practitioner*, an independent, non-profit organization devoted to IPM education. Prokopy served on the centre's advisory board until his death.

Products listed in the *Directory* include biocontrol agents, microbial pesticides, traps, pheromones, physical controls (soaps, oils) and monitoring devices. Chemical controls include borates, baits, insect growth regulators and botanicals. Synthetic chemical control products are limited primarily to formulations that have low acute toxicity and reduced potential for non-target exposure, such as insecticidal baits. Suppliers include long-time IPM-speciality monitoring product manufacturers and distributors and biological agent producers, as well as pesticide registrants offering biopesticides alongside conventional pesticides and manufacturers of technologies with multiple uses, including applications in IPM (Table 9.1).

Prokopy was known by his friends in Conway for the 'mystery' boxes of apples that would show up on their doorsteps during harvest time, but nowhere was his generosity more evident or had more impact than in his contribution to pull-side advancements in IPM by sharing freely the fruits of his research, including

Table 9.1. Sample of suppliers of IPM monitoring, data management and non-chemical controls, biocontrol agents and biorational pesticides with year founded, web site and example products listed by primary activity, e.g. manufacturing or distribution. For a more detailed listing of products and suppliers, see BIRC (2007).

Sample manufacturers/developers	Example products and brands
Hercon Environmental (1974), www.herconenviron.com	Disrupt® mating disruptants, Insectape® insecticidal tape, Luretape® pheromone attractants, Scout™ monitoring traps
ISCA Technologies (1996), www.iscatech.com	ISCA monitoring traps and lures, Zumba™ mosquito traps and SkinLure™
Koppert Biological Systems (1968), www.koppertonline.com	Beneficial insects and mites, biological pesticides
Olson Products (1980), www.olsonproducts.com	Insect monitoring traps and trap coating, Stiky™ Fly Traps, pest barriers
Scentry Biologicals (1975 as Albany International), www.scentry.com	NoMate® mating disruptants, Scentry® monitoring traps and lures
Tanglefoot Co (1885), www.tanglefoot.com	Bird repellants, insect traps and trap coating, pest barriers.
Trece (1984), www.trece.com	Cidetrak® mating disruption, Pherocon® and Storgard® monitoring traps and lures
ZedX (1987), www.zedxinc.com	Precision ag services including AgFleet® decision-support system for pest and other ag management applications, Skybit® site-specific weather data and insect and disease forecasting service
Distributors	
Arbico (1979), www.arbico.com	Extensive selection of biological controls and biorational pesticides
GEMPLER'S (1982, purchased Pest Management Supply Co in 1995 which was founded in 1980), www.gemplers.com	Wide variety of products for ag, horticulture and grounds maintenance, including broad selection of IPM products for monitoring and management of insect, disease, rodent and wildlife pests and weeds
Great Lakes IPM (1985), www.greatlakesipm.com	Broad array of products for monitoring and management primarily for insect pests of agricultural crops and stored products
Spectrum Technologies (1987), www.specmeters.com	Wide variety of measuring and monitoring products, including self-developed and manufactured Watchdog weather stations and data loggers and Specware software

the red sphere trap for AMF. As expressed by Marcos Kogan (Deutsch, 2004), 'In a time when there seems to exist a cult of selfishness, Prokopy's generosity and selflessness made him stand out as a model.' Variations on Prokopy's original trap design are now manufactured and sold by several companies and used widely in commercial and hobbyist horticulture to monitor or provide control of AMF. Prokopy developed the original trap in the mid-1960s in Door County,

Wisconsin, and promptly published his results (Prokopy, 1968a) without thought of patents or profits. His trap for AMF served as a model for the development of several additional traps and led to the launch of a new national distribution channel for IPM tools by T.A. Green.

Pest management supply company: from US$1500 to US$1.6 million in annual sales

The birth of New England Insect Traps

In 1980, while Green was enrolled in Prokopy's Integrated Pest Management Course at the University of Massachusetts, Prokopy announced that Ken Spatcher and Patti Powers, sole manufacturers of the red sphere trap, were ready to retire from the trap-making business. In addition to producing a commercial strawberry crop on their farm, the pair were also making Prokopy's white rectangle trap for the tarnished plant bug, *Lygus lineolaris* (Palisot de Beauvois) (Prokopy *et al.*, 1982) and the European apple sawfly, *Hoplocampa testudinea* (Klug) (Prokopy and Owens, 1978). Prokopy previously had persuaded Powers, then working in his laboratory as a technician, to start making the traps so that Massachusetts apple growers would have a source from which to purchase them. Now, strawberry production had become too demanding during the peak trap season for the trap makers to do both.

Green visited the couple at their farm, received instructions on where to obtain raw materials (croquet balls for the red spheres and Tartar Red Dark enamel paint from Sherwin Williams, milk carton stock and non-UV reflecting white paint for the white rectangles) and how to manufacture the traps. Spatcher and Powers were selling less than 1000 traps per year at that point, primarily to growers working with Prokopy in the apple IPM pilot project. Green sold 1500 traps the next year.

Over the next few years, highlights included meeting Harry Ricker, Maine apple grower, at the Massachusetts Fruit Grower Annual meeting. Ricker said something like, 'You know, Tom, your traps saved me US$38,000 last year. I generally spray for apple maggot starting in late June. I didn't catch any flies until August. Saved me six weeks of spraying.' Green's first thought was that at just US$2.50 per trap, he was not charging enough. The second thought was what that conversation might have been like if Ricker had ended up with maggots in his apples!

Other developments included improving the red sphere trap by replacing the small, 1.9 cm screw eye atop the trap with a 6.5 cm screw hook, which also served as a convenient handle when applying and removing Tangle-Trap®, the adhesive used to coat the trap. Metal grommets were added to the holes in the paper, 15 × 20 cm white rectangle traps to keep the wire hangers from pulling through. Eventually, this original milk carton stock was replaced with white high-density polyethylene that did not reflect UV and thus needed no painting or grommets. Unlike the red sphere traps, these rectangles were shipped with the Tangle-Trap pre-applied. The preparation process included keeping the exposed surface of 11.34 kg pails of Tangle-Trap warm with a blow drier and spreading the softened sticky coating by hand with a 10-cm-wide paint scraper. Several hundred traps could be coated per hour in this manner. Finally, a pre-stickied red sphere trap version was produced, with the Tangle-Trap spread in this case with a piece of plastic cut from a milk jug and bent to fit the curve of the sphere. The trap came in two halves, with the pre-stickied halves stacked up inside a cardboard shipping container and joined with a wire hanger when ready to deploy.

Sales growth

The big breakthrough came in 1986, when Roger Swain, science editor of *Horticulture* and host of the 'Victory Garden' television show on the Public Broadcasting Service, featured the trap in a column and on the show. 'The standard against which to judge the effectiveness of any trap', Swain reported, a phrase that helped launch sales of more than 20,000 red sphere traps that year.

Also in 1986, the company operated under a new name, Pest Management Supply Co, reflecting expansion into new product lines and geographic distribution. The business moved out of Green's basement and into a former trucking terminal on a vegetable farm in Hadley, Massachusetts. Additional products included traps and lures, from Scentry, Trece and Hercon, and hygrothermographs adapted for measuring the temperature and duration of leaf wetness for forecasting apple scab infection periods. Also new was a visual monitoring trap for the apple blotch leaf miner, *Phyllonorycter crataegella* (Clemens), developed with Prokopy as part of Green's thesis project in his final year at Hampshire College (Green and Prokopy, 1986). The trap was made of red high-density polyethylene, an unusual colour for a visual insect trap, because most insects cannot discriminate between red and black. The red ball for AMF was red for human aesthetics; there was no significant difference in fly captures on red versus black balls (Prokopy, 1968a).

This new 'species' of visual monitoring trap was developed using the same approach Prokopy used previously, including observing insect behaviour in the field, followed by iterative experiments comparing captures on the sticky surfaces of various sizes, shapes and colours (including clear), with the most successful combination then tested in various positions in the tree canopy. The trap was unique, not only in that red outperformed other colours significantly, but that placement and orientation of the trap could be varied to determine if a pesticide application was needed and, if so, when it should be applied to suppress egg laying on foliage. Optimum placement for early-season detection was vertical on the tree trunk at about knee height, where mating occurred in the morning. A second trap, placed horizontally in the tree canopy, detected initial egg-laying flights, which occurred in the evening once temperatures were suitably warm and wind speeds were light (Prokopy *et al.*, 1997; Green and Prokopy, 1998). This pattern of activity differs from a close species, the spotted tentiform leaf miner, *P. blancardella*, which both mates and oviposits in the evening (Pottinger and LeRoux, 1971).

Observing apple blotch leaf miner activity in the field is remembered with fond affection by Green. This study involved spending several consecutive days in orchards during moth flights, surveying various structures in the orchard (ground cover, upper and lower leaf surfaces in various parts of the canopy, tree trunk and branches) each and every hour, day and night, for the presence and activity of moths. These observations were made over the course of several seasons and during all three moth flights each season. Initially, leks were suspected due to large numbers of male moths converging on a single structure, generally a leaf, during the morning mating hours. However, this soon revealed itself as an artifact of pesticide use. Enormous populations of this insect occurred in managed orchards, having been freed from biological control by the lethal effects of several commonly used insecticides on its natural enemies (Van Driesche and Taub, 1983). Populations built up to the extent that trees appeared 'scorched' due to the many leaf miners present. These severe infestations were not observed in unmanaged trees with intact populations of beneficials. The massing of moths were not true leks, but simply mobs of amorous males responding to a single, calling female.

Over the next several years, print catalogue mailings increased from just over 12,000 pieces in 1991 to *c.*20,000 in 1994. The number of products offered grew from the original 2 to more than 500, including Hercon, Scentry and Trece pheromone lures and monitoring traps; a new apple odour lure for AMF; sweep nets; magnifiers; and weather stations and pest forecasting software. Gross sales grew from US$132,000 in 1990 to over US$500,000 in 1995, when the business was sold to GEMPLER'S, a national mail-order merchant to agriculture and horticulture. Within 2 years at GEMPLER'S, the business (and Green) were firmly established in Wisconsin and sales of the product line grew to US$1.6 million within GEMPLER'S overall *c.*US$10 million business at that time.

Along the way, Green served as an apple IPM field scout under Coli in the Massachusetts IPM Programme (1980–1982), an Extension technician (1983–1984) and a research assistant to Prokopy (1985–1990). He also earned his MS (1986) and PhD (1992) under Prokopy's supervision. He learned not only about insect trap development but also about mail-order industry benchmarks for gross margin and response to direct mail campaigns including catalogues; the science of identifying, segmenting, purchasing and testing mailing lists, including evaluation of response; and square-inch analysis of catalogue product performance whereby each product was assigned a 'rent' or overhead cost based on the cost of production and mailing of the square inches of catalogue space it occupied. Product 'stars' pay for their catalogue space many times over; dogs cannot make the rent and must be reworked or dropped from the catalogue to maintain profitability.

Gross margin is the difference between the wholesale purchase price and the retail price and is critical to meet overhead, sales and marketing expenses and profit targets. Gross margins were slim on many IPM products. For example, margins on monitoring traps and lures were as low as 10% versus the 40–45% ideal, due to both active direct selling by the manufacturers in competition with their distributors and discounted pricing offered by manufacturers to research and Extension users, a substantial segment of the market. In addition, many highly technical and/or crop- and region-specific IPM products do not survive square-inch analysis and require more targeted marketing and sales than the mail order approach provides. Unfortunately, the slim profit margin of the production and marketing of IPM products is reflective of the cost constraints endured by commercial growers who endeavour to implement alternative strategies for pest control.

Concurrent with the growth and development of the IPM supply chain from the AMF trap to a nationally distributed full-IPM-product line, Prokopy and colleagues were pursuing a trap-based approach for the direct control of AMF in commercial orchards. While Prokopy was successful with trap-out as the exclusive method of controlling AMF in his 2-acre orchard (1 acre = 0.404 hectares) (Leskey *et al.*, Chapter 5, this volume), sticky-coated traps were too labour-intensive to be economically viable on a larger scale.

Pursuit of an Economically Viable Trap-out System for AMF

With commercial growers positioned at the interface between pest biology and the marketplace, development of behaviourally based management strategies fell within the broader aspirations of IPM, with an inherent subcontext of improving grower profitability (Prokopy, 2003). Based on the understanding that in the development of a high-level IPM system, economic considerations of the grower are of principal consideration, Wright, Prokopy and collaborators attempted to pursue product development within the contemporary bounds of orchard profitability. Prokopy's research programme focused on developing tools for trap-based control of AMF as a substitute for organophosphate sprays within the broader context of an IPM system, including site selection, wild host removal, plot design, cultivar arrangement, tree architecture and orchard sanitation (Prokopy *et al.*, 1990). Due to the costs of implementing the system, the material and labour costs for the input substitution grossly underestimate the costs of the total management strategy. Likewise, the value of the output is underappreciated, given limited understanding of the long-term ecological benefits of behavioural control and only modest market-based initiatives designed to recapture the value of alternative management strategies (Taylor, 1980). That said, development of a trap-based control strategy to fit within the bounds of modern, conventional orchard management provides the most rigorous standards of effectiveness and efficiency, and offers a comprehensible context for determination of success.

Full success in the development of an alternative pest management strategy cannot be realized until reduction to utility, demonstrated by sustained grower adoption (Luckman and Metcalf, 1994). In the case of behavioural control of AMF, a resultant likelihood of grower adoption lies at the terminus of a stepwise path to commercial utility, broadly defined by a progression from discovery of potential to demonstration of effectiveness, from development of efficiency to establishment of availability and, finally, to promotion of adoption. Success in this process demands study from evolving and overlapping perspectives: first the pest organism, followed by the protected crop, the producer and the consumer. The sequential building blocks of this research project are the foundations of the pursuit of conscientious product development: effective management tools built on a sound biological basis, economically feasible, supported by regulatory priorities, enlightening and satisfying market demand. Details and implications of

Prokopy's studies of the biology of AMF and other key fruit pests are found elsewhere in this volume. In this section, the focus is on the history of the pursuit of efficiency in trap design, along with the current research emphasis and the potential commercial utility of behavioural control of AMF.

Components of the AMF trapping system

For nearly 40 years, Prokopy worked toward the development of a comprehensive trapping system for AMF that could provide a commercially acceptable level of control in the absence of insecticide treatments (Prokopy and Mason, 1996). There are four fundamental components of the trapping system – three physical elements (visual stimulus, odour stimulus and trapping mechanism) and the deployment strategy, the method of trap placement in space and time (Foster and Harris, 1997).

Red or black spheres of 7.5–9.0 cm are supernormal mimics of ripe apple fruit (visual stimulus) and are highly visually attractive to sexually mature AMF. When placed in a fruiting canopy, the resolution of the optimal ranges of trap shape, size and colour allow these mimics to outcompete the visual stimuli of available fruit (Prokopy, 1968a). When coated with a non-drying adhesive such as Tangle-Trap (trapping mechanism) and hung at a density of 2–6 traps per tree (depending on canopy size), visual traps alone can provide 99% control of fruit injury (Prokopy, 1975). Although a legitimate demonstration of the potential for trap-based control, the great number of traps required to achieve this level of crop protection is prohibitively difficult to deploy and maintain (Prokopy *et al.*, 1990). With the identification of attractive mid-range apple odours by researchers at Cornell University (odour stimulus) (Reissig *et al.*, 1982; Averill *et al.*, 1988), a greater flexibility was realized in techniques for trap-based plot protection. Given that the principal threat of AMF damage arises from flies that emerge outside of commercial orchards and immigrate at sexual maturity, odour-baited traps could be placed only on perimeter trees of an apple orchard (deployment strategy). Using this strategy, fruit-foraging flies are attracted to and held within baited canopies, leading to eventual visual contact with sphere traps hung within (Prokopy *et al.*, 1996).

This combined method was demonstrated in 10-acre blocks of commercial apple orchards in Massachusetts from 1991 to 1994, using a single-component odour lure (butyl hexanoate), Tangle-Trap as the trapping mechanism and systematic (5 m) perimeter trap placement as the deployment strategy. In this study, trap density averaged 16 per acre and traps were maintained biweekly. The results of this demonstration were both favourable and encouraging: nearly 150,000 flies were captured in trapped plots, yielding an average of 1.1% injury to fruit in trapped plots compared with an average of 0.7% injury to fruit in neighbouring plots receiving three summer insecticide sprays (Prokopy *et al.*, 1996). Two shortcomings arose from this study: first, preparation and deployment of traps bearing Tangle-Trap as the fly killing agent was far too messy and costly to hold commercial potential; second, to retain trap effectiveness, all spheres needed to be cleaned of all debris biweekly for the entire season, tripling the time cost of deployment. Effectiveness of the combined system was demonstrated but, for potential commercial acceptance, the trapping mechanism demanded refinement.

Development of toxicant-treated spheres

Given the need to develop an effective, inexpensive alternative to Tangle-Trap, Prokopy conceived of unifying behavioural and chemical control in the form of sphere traps coated with a mixture of insecticide and residue extending agent (Duan and Prokopy, 1993). Preliminary studies revealed that most alighting flies could be killed by direct contact alone with a coating of insecticide on a sphere, but the dose required to kill at least 95% of those alighting was far greater than the dose required to kill alighting flies that ingested insecticide along with a feeding stimulant (Duan and Prokopy, 1995). With the objective of using the lowest possible effective dose of toxicant on a trap, Prokopy found that low doses of several organophosphate insecticides could retain their toxicity to flies over a long period of time when mixed with latex paint and applied to spheres. Unfortunately, the residue extending agent (latex paint) was ineffective in preserving sugars; all of the feeding stimulant was washed away when traps received as little as 1 cm of rainfall (Duan and Prokopy, 1993, 1995).

Beginning in 1996, researchers in Prokopy's laboratory sought an alternative to organophosphates for use on toxicant-treated spheres and found that several neonicotinoid insecticides (notably imidacloprid) were safer alternatives and equally effective (Hu and Prokopy, 1998; Prokopy *et al.*, 1998). When mixed with latex paint, residual activity of imidacloprid was preserved to a useful degree – 4% (a.i.) imidacloprid in latex paint applied to spheres retained enough toxicity to flies to kill 80–90% of alighting flies after 12 weeks of field exposure, if traps were treated with feeding stimulant (Wright *et al.*, 1997). The challenge was to develop some method of protecting sugar as a feeding stimulant from the degrading effects of environmental moisture. Three general approaches were taken to ensure a sufficient supply of feeding stimulant on the surface of toxicant-treated traps throughout the 12–15-week period of sphere deployment needed for season-long control of AMF. The first approach involved attempts to protect sucrose physically from the impact of rainfall, either by placing a cone-shaped cap above a treated sphere to deflect rainfall or by encasing the sucrose and insecticide within a perforated hollow sphere, inducing flies to enter the holes and feed on the interior of the trap (Duan and Prokopy, 1992; Reynolds *et al.*, 1996). Neither of these approaches was effective, both apparently violating the visual and structural characteristics by which flies selected and accepted host fruit. The second approach involved a variety of attempts to develop a mechanism for ensuring the slow release of sucrose applied with a binding agent: (i) chemically bonding sucrose to pregelatinized cornflour; (ii) incorporating sucrose into a polymer matrix or mixing with a spreader-sticker agent; and (iii) covering a layer of feeding stimulant with a second coating of linseed oil or shellac to minimize the availability of sugars (Hu and Prokopy, 1996; Hu *et al.*, 1997). No chemical binding or protective approaches proved successful.

The third approach involved entrapping a large amount of sugar in a mixture of starch and glycerin and forming the ingredients into a sphere, creating a large reservoir of feeding stimulant (Hu *et al.*, 1998). On drying, these biodegradable traps were coated with latex paint containing toxicant; under rainfall, the sugar dissolved and bled through the paint layer, providing a continuous supply of feeding stimulant to the sphere surface. These traps were designed to maintain structural integrity throughout the deployment period and degrade during the winter months. Despite encouraging laboratory and semi-field results (Hu *et al.*, 1998; Prokopy *et al.*, 1998), traps of this type were highly prone to premature breakdown, toxicant degradation, fungal growth and widespread consumption by vertebrates (Prokopy *et al.*, 2000; Wright *et al.*, 2000). Although patented, biodegradable traps of this type are no longer produced.

Development of externally renewed toxicant-treated spheres

Given discovery of the potential (1968–1986) and demonstration of the effectiveness (1987–1994), the likelihood of the commercial acceptability of behavioural control for AMF has hinged for 10 years on the development of an efficient trapping mechanism for use with sphere traps. After exhaustive testing of potential improvements, one option remained: a shift to renewing (rather than preserving) the supply of feeding stimulant on reusable traps. Since 1997, efforts have been aimed at the development of a reusable toxicant-treated trap system focusing on the design of a highly concentrated source of feeding stimulant affixed to the top of each sphere, such that, under rainfall, a small amount of feeding stimulant is released on to the sphere surface – as trap-surface sugar is dissolved and runs off under rainfall or heavy dew, it is replaced with sucrose from a source atop the sphere (Wright *et al.*, 1997). Although relatively simple in concept, the mechanism of these traps proved novel and flexible, in that traps of this type used environmental moisture to the benefit of the system. Thus, the trapping system is activated and sustained by the same set of environmental factors (periodic wetting and drying) that activate and sustain the pest population. Therefore, unlike any other control strategy, these traps fit within the dynamic of the threat, i.e. environmental conditions leading to the greatest threat lead to the greatest effectiveness of the control mechanism.

In 1997 and 1998, these caps were formed of nearly pure sucrose, quickly revealing that pure sugar caps were highly prone to breakdown under conditions of high humidity (Wright *et al.*, 1997). In 1999, 4 cm, 25 g, flat-topped caps were press-formed of 85% sucrose bound in 15% paraffin and tested in the field. Although these caps worked for a short time in the field (*c.*3 weeks), they ran out of feeding stimulant well before the period of risk had passed (Prokopy *et al.*, 1999). In 2000, caps consisting of 85% sucrose bound in 15% paraffin were again tested, but cap mass was increased to 50 g and the caps were formed using a hydraulic press mould that stamped eight flutes into the top of each cap, ensuring a uniform pattern of runoff and sucrose distribution. Upon laboratory and field testing, this style of cap was effective through 127 mm of accumulated rainfall, roughly equivalent to 50% of the durability needed (Wright *et al.*, 2000). Like biodegradable traps described previously, these caps were targeted and damaged by rodents; in 2001, an integrated rodent guard was added to the cap design, yielding 90% or greater protection from vertebrate consumption (Wright *et al.*, 2001).

To enhance the longevity of caps in the field, Prokopy and Wright capitalized on the observation that, under rainfall, a small amount of water penetrated and moved through the paraffin/sucrose matrix of each cap. To enhance the availability of sucrose bound within the wax matrix, 2000 field-standard caps (15%

Fig. 9.2. Feeding stimulant cap (a) designed to sit atop sphere traps for AMF. Cap consists of sucrose, paraffin wax and carnauba wax. Caps of this type are formed with eight reservoirs stamped into the cap top to hold and distribute runoff after rainfall events. (b) Depicts an exploded schematic of a 4-piece visually integrated sphere trap. (c) Shows an AMF feeding on rainfall-produced trap-surface residue.

paraffin, 50 g) were reshaped, such that eight shallow reservoirs were pressed into the top of each cap (Fig. 9.2a). As an alternative to fluted caps that channel rainfall off wax/sugar caps, these caps were designed to retain a small amount of water (roughly 5 ml) in reservoirs atop each cap, allowing such retained moisture to percolate through the slightly porous cap body (Wright *et al.*, 2001). This percolation effect yields four advantages over previous cap styles: (i) the slowly developing sucrose-bearing runoff is highly concentrated (nearly saturated) to stimulate fly feeding; (ii) very little sucrose runs off on to fruit and foliage, limiting fungal growth; (iii) the entire mass of sucrose in each cap is used eventually, increasing the endurance of the trap dramatically; and (iv) a very small amount of rainfall or dew is needed to recharge the traps with available feeding stimulant. Field trials in 2001 and 2002 revealed that with modifications of cap size, wax type and proportion and density, reliable season-long control of AMF could be achieved in commercial orchards. In fact, pesticide-treated spheres of this type provided control of injury caused by AMF equal to or greater than control provided by standard whole-plot insecticide sprays (Prokopy *et al.*, 2001). In 2003, the University of Massachusetts applied for patent protection of this feeding stimulant release system (Prokopy *et al.*, 2004a).

Expansion of trap utility

In 2001, Prokopy's research programme arrived at a prototype pesticide-treated sphere system that (in small-plot trials) provided control of AMF injury equal to whole-plot insecticide treatments (Prokopy *et al.*, 2000, 2001, 2003). However, demonstration of effectiveness within the context of orchard management for which the traps were designed (conventional, commercial orchards) led to examination of the potential for utility within other management arenas, notably: (i) certified organic apple production; and (ii) hobbyist/backyard growing. Given that the development of this visual/olfactory trapping system was rooted in an understanding of fundamental fly biology, Prokopy hypothesized that the principles of stimuli and deployment developed for integrated pest management within conventional orchards should translate into effectiveness within other management strategies. However, with this attempt at expanded relevance, two major shortcomings existed in the design of our trap mechanism. First, the toxicant (imidacloprid) was a restricted-use neonicotinoid insecticide, limiting potential implementation to conventionally managed orchards. Second, traps were treated annually with latex paint bearing insecticide, posing both a handling risk and significant labour cost. Therefore, beginning in 2003, the focus of this project shifted to redesigning traps to broaden the potential availability of the system and to reduce the labour associated with annual deployment. First, a new cap formulation and production technique was developed that permitted binding toxicant to the feeding stimulant within the stimulant-release caps, obviating the need for the painting of traps. Second, alternative toxicants were examined; trials revealed that several materials (such as spinosad and avermectin) were highly toxic to AMF when ingested directly with sugars, but did not provide residual activity through the season when applied in latex paint. With the release of a new wettable powder formulation in 2003, spinosad emerged as an excellent candidate toxicant – very lethal to flies at a very low dose, labelled for organic use and chemically compliant to integration within the structural matrix of feeding stimulant caps (Prokopy *et al.*, 2004b).

Development of visually integrated sphere traps

Initially, spheres and caps were patterned after previous models, such that toxicant leached out of attached caps on to the unpainted sphere surface under rainfall or heavy dew. In this system, trap performance was dependent on the residual effectiveness of the feeding stimulant and toxicant that were dripped on to the surface of the trap after exposure to environmental moisture, but laboratory and field results were highly encouraging: trials in 2003 revealed that the performance of spinosad-capped traps (with no toxicant applied to the sphere before deployment) was equal or superior to standard imidacloprid-treated spheres and whole-plot insecticide sprays (Prokopy *et al.*, 2004b). Concurrent trials revealed that the rate of spinosad within a trap could be set at 0.5% (a.i.) (roughly 1 g of toxicant per trap), with little reduction in trap effectiveness (Prokopy *et al.*, 2004b).

In subsequent laboratory trials, exposure to spinosad-capped spheres resulted in 90% or greater mortality to AMF, even after the spheres were exposed in the field for 12 weeks or more. Observations within these laboratory trials revealed that a significant number of flies that were killed by exposure to traps had moved upward on the traps and fed directly on the attached cap (Prokopy *et al.*, 2004b). However, concurrent trials of fly visual responses demonstrated that the presence of a large cap atop the visual stimulus reduced fly response by *c.*30%, presumably by interfering with the flies' visual interpretation of shape and form (Lafleur *et al.*, 2001). Given these findings, the entire trapping system was restructured to: (i) ensure that all alighting flies encountered the feeding stimulant/toxicant matrix directly; and (ii) integrate the feeding stimulant–toxicant matrix into the visual stimulus, such that a reusable trap body and replaceable feeding stimulant–toxicant cap combined in sphere form (Fig. 9.2b).

In 2004, experimental, visually integrated traps were designed and deployed bearing spinosad as the toxicant (Fig. 9.2c), comparing the sustained lethality of traps to that of spheres fitted with previous-model caps, also using spinosad as the toxicant. Again, results were encouraging – visually integrated traps yielded kill of flies on first exposure to traps (mean lethality 92.0%) nearly equal to that provided by previous-model traps (mean lethality 96.5%). Both compared favourably to imidacloprid-treated spheres, which yielded a mean lethality of 65% under identical test procedures (S.E. Wright, 2005, unpublished results). Given that potential alightings on this new trap design were maximized by visual integration and sustained lethality that more than doubled previously tested traps, the protective capacity of these traps was assessed in commercial orchards.

Building efficiency in deployment

In balancing the structural and deployment characteristics of a trapping system for an insect pest, it is important to maintain contact with the process through which the pest is controlled. With toxicant-treated visual traps, there are six distinct steps leading to the control of a single fly: (i) fly enters zone of potential control; (ii) fly detects trap; (iii) fly alights on trap; (iv) fly feeds; (v) fly takes up toxicant; and (vi) fly dies (Duan and Prokopy, 1995). As noted throughout this section, our recent efforts have been focused on steps 4–6 above – tailoring the efficiency of the trap mechanism. However, a perfect trapping mechanism can only control the insects that respond, so the effectiveness of the system is bound to the relative competitiveness of the visual and olfactory stimuli in the framework of the deployment strategy.

While developing an efficient toxicant-treated trap mechanism, Prokopy was also refining the parameters by which a plot-specific deployment strategy could be defined. In 2004, the first field-deployment trials of this complete trapping system were conducted in 28 small (1 acre) demonstration plots across 7 northeastern states. In this trial (a component of a project funded by the USDA-CSREES-Northeast Regional IPM Grants Programme), the protective capacity of perimeter trapping was assessed in comparison with whole-plot insecticide treatments. Traps were deployed at varying densities in accordance with the hypothetical plot-specific potential for AMF injury, determined by evaluation of: (i) tree size; (ii) canopy density; (iii) susceptibility of plot cultivar; and (iv) bordering habitat (Hoffman *et al.*, 2002; Prokopy *et al.*, 2005).

These parameters were hypothesized to constitute the key structural elements challenging the protective capacity of traps, impacting visual trap apparency (tree size and canopy density), likelihood of sustained in-canopy foraging (cultivar susceptibility) and scale of potential immigrating population (bordering habitat). Overall, this project was a successful demonstration of the commercial potential of trap-based control of AMF. However, in 2004, there were substantial control failures in a few of the demonstration plots, attributed to: (i) endemic emergence of AMF, subverting the control potential of perimeter-deployed traps; (ii) poor trap placement; and (iii) unusually susceptible fruit cultivars (S.E. Wright, 2005, unpublished results).

These control failures revealed that while trap-based control of AMF was a suitable management tactic for most plots, the structural and spatial characteristics of deployment could compromise the control potential of the strategy. Furthermore, it was demonstrated in the first year of this study that physical conditions unsupportive of trap-based fly control realistically might be encountered in some commercial orchards. In 2005, the interpretation of risk factors was reassessed, traps were deployed using more conservative density and placement considerations and a majority of control failures were alleviated (S.E. Wright, 2005, unpublished results). Although the overall findings of this study were positive, the practical outcome in terms of conscientious product development was that even the most advanced combined trapping strategy (visually integrated spheres, 5-component odour lure, risk-based perimeter deployment) did not compete well enough with the protected resource to accommodate extremes in structural or spatial variation between plots.

This led to reconsideration of the dynamics of fly response to the basic functional unit of behavioural control of AMF: a single, unbaited visual trap. The intent of this effort is to maximize the efficiency of a single trap's performance within a spatial unit of protected crop, leading to a more competitive trapping system, able to accommodate a greater breadth of structural variation within and across plots. Four decades of Prokopy's behavioural studies with AMF show that trap efficiency within an area of potential control is determined by in-canopy trap apparency coupled with the likelihood of sustained host-seeking by the fly, taken in the context of population size (Prokopy and Roitberg, 2007). From this, the contribution of cultural management is clear – sustainable behavioural control of AMF can be realized only against the backdrop of broadly integrated orchard management, permanently limiting the potential scale of the pest threat (Prokopy and Kogan, 2003). However, the functional components may be improved either by: (i) enhancing the relative apparency of traps within canopies using either improved visual, olfactory or placement factors; or (ii) encouraging foraging flies artificially to continue seeking and rejecting host fruit until arrival on traps using additional olfactory cues (Bernays, 2000).

After 10 years of systematic modifications, Prokopy's research programme yielded a functional trap design, integrating visual and olfactory stimuli with sustained effectiveness and availability of toxicant and feeding stimulant within the bounds of field durability and cost effectiveness. After fulfilling the structural requirements of the trap system, though, the management strategy stands at the precipice of commercial relevance – determination of relative efficiency across the broad range of biotic and abiotic stresses of practical deployment. It remains to be seen whether or not the final technical hurdles will be overcome.

IPM technological advances such as those described here often entail greater risks and higher costs to growers as they are implemented on a commercial scale, often with a crop and pest-specific niche lacking a large market over which to spread development, production, marketing and sales costs. In addition to pursuing government and other third party funding to support these higher costs, a second potential is to press the wholesale and consumer market to purchase preferentially and provide price premiums to growers who adopt these advances.

Integrating Ecology and Fruit Marketing and Sales

Prokopy was very concerned that growers who advance along the IPM continuum remain competitive in the marketplace despite potentially greater risks of crop loss, more damage from some pests and higher costs for newer, more specific and less broadly toxic pest management strategies. He also wanted consumers to know they were getting more value for themselves and the environment by purchasing his own apples, including greatly reduced pesticide use and residues. He was often disappointed by produce managers at local supermarkets who would most often market his very low-spray fruit as conventional, lacking the motivation and/or mechanism to single them out. At the same time, he avoided the debate about whether or not IPM apples should be identified in the marketplace as a new 'brand', perhaps at least in part because this was not a universally well-received idea among growers (Hollingsworth, 1994). Some objected strongly to the concept that growers and crops should be distinguished on the basis of IPM performance, including the amount or toxicity of pesticides applied. Objections were also raised to the Extension's involvement in drawing this distinction as potentially inconsistent with the Extension's mission of providing assistance to all growers.

Eco-labels, IPM and the marketplace

An eco-label is a seal signifying that a product or service has met a set of environmental or social standards. These labels help both wholesale buyers and consumers identify products and support companies providing benefits beyond the primary need satisfied by the product. Eco-labels in agriculture are growing rapidly in number, scope, producers and hectares, including those that require IPM for participation. The US eco-market has been estimated at 23% of adult USA consumers and as large as US$209 billion (French and Rogers, 2005). To be successful, eco-marketers must first meet consumer core requirements for quality, price and convenience. The rapid growth in organic sales since the late 1990s is attributable largely to the improvement in quality, affordability and convenient availability in mainstream stores.

In addition to organic, one of the earliest and most successful programmes, more than 130 eco-labels are now rated online by the watchdog Consumers Union (Consumer Reports, 2007), just like home appliances and other products. A number of eco-labels formally incorporate IPM in their standards, including four (Food Alliance, Forest Stewardship Council, Rainforest Alliance and Protected Harvest) that reported certifying a combined 11.3 million ha in the USA and 94.3 million ha worldwide in 2007 (IPM Institute, 2008a). On the non-agricultural side, the US Green Building Council LEED Standards for existing buildings include IPM as an opportunity to score points towards certification (USGBC, 2007).

Several programmes, including the New England Pest Management Association's IPM Registry (NEPMA, 2007), the Association of Bay Area Government's EcoWise (ABAG, 2007) and the IPM Institute of North America's Green Shield Certified (Green Shield Certified, 2007) now offer IPM certification for structural pest management providers. Links to these and other programmes that require IPM practices for participation are catalogued on the IPM Institute web site (Green, 2008).

Eco Apples, a market success

A fifth programme, Northeast Eco Apples, promotes apples grown according to an IPM protocol as locally produced, tapping regional sentiments for keeping agriculture in the landscape. Sales of the 2008 crop by Red Tomato, a non-profit marketing agent, topped US$1.8 million in 2008, the fourth year of the programme, supplementing grower direct sales. Major buyers include retail grocers, Trader Joes and Whole Foods Market.

Large-scale, fresh apple production in much of the USA is not feasible under organic standards due to pest pressure in wet, humid climates and the limited pesticide options available to organic producers. Working with a team of collaborators, T.A. Green and others developed a protocol designed to minimize impacts on health and the environment. These Eco Apple™ brand apples are marketed to receptive wholesale and retail consumers. On average, growers earn a modest premium of US$1–2 per box, or about 5%, which helps defray the cost of the extra measures they take to reduce resource impacts. The project began in 2005 with 6 growers and 182 ha and now includes 12 growers with more than 283 ha in the programme.

Production

The Eco Apple collaboration includes growers, crop advisors, researchers from Cornell, UMass and Penn State and staff from Extension, USDA-NRCS, USDA-ARS, Red Tomato Inc and the IPM Institute. The collaborators function as a working group and learning community, with ongoing monthly project management conference calls and one annual meeting. The production protocol was developed and is maintained by the IPM Institute, with input from the group. The IPM Institute also certifies that the grower's fruit meets the standards verified by an independent, on-farm, third party inspection. The inspection occurs prior to harvest in the grower's first year in the programme and every third year thereafter. In the intervening years, verification includes a review of grower pesticide application and scouting records, which are submitted to the Institute. Only those apples from blocks meeting the protocol requirements may be marketed under the Eco Apple brand.

The production protocol (IPM Institute, 2008b) includes three main components:

1. Fifty-eight minimum requirements address regulatory requirements, worker safety and pest-specific IPM practices. For example, growers must sample mites and mite predators before applying miticide and may use insecticide for tarnished plant bug only if trap captures are over threshold.

2. Forty-nine optional point-based practices cover pest-specific IPM practices, soil quality, recycling, energy conservation, continuing education, food safety and fruit quality. Growers must implement enough practices to meet a minimum score, e.g. using traps to control AMF without pesticides earns five points. These point-based practices are those that are not as critical to a good IPM programme, or are more challenging or not appropriate in all orchards. The point-based approach allows growers the flexibility to choose those practices that work well for their orchard.

3. Pesticides with the most serious potential health and environmental hazards are not permitted; others may be used with justification. A third category includes pesticides with serious hazards that are permitted only with restrictions that limit the potential for harm. Pesticides are evaluated for placement in the protocol using the following criteria and recognized authorities:

- Acute aquatic toxicity: variety of published sources.
- Acute toxicity: caution, warning or danger label/US EPA.
- Neurotoxin: cholinesterase inhibitor or listed on Toxics Release Inventory maintained by the US EPA.
- Possible, likely, probable carcinogen: US EPA, State of California, International Agency for Research on Cancer.
- Reproductive/developmental toxin: US EPA, State of California.
- Toxic to key natural enemies/secondary pests: Extension recommendations.
- Suspected endocrine disruptor: Illinois EPA, Keith, Colburn or Benbrook lists.
- Broad spectrum: Extension publications.
- Resistance risk: Extension publications.
- Potential or known groundwater contaminant: State of California, variety of published sources.

Marketing

Eco Apples are marketed by growers directly to wholesalers and supermarkets and/or to consumers visiting farm markets. In addition, Red Tomato, a non-profit marketing agent based in Canton, Massachusetts, has developed a nearly year-round programme to supermarket chains. The programme is based on superior firmness and flavour, supporting local growers, keeping farms in the north-east landscape, and production practices that protect health and environment. Marketing and communications support materials include:

- Unique, colourful packaging that communicates the eco-message and provides information about the farm and farmer.
- A distinctive logo and tag lines including 'Trust the Farmer. Know the Orchard. LOVE the Fruit' and 'Born and Raised Here' (Fig. 9.3).
- Posters featuring participating growers and farms for display in stores.
- Point-of-sale brochures explaining the programme in consumer-friendly language.
- Farm tours for wholesale buyers, media, funders and others.
- Information posted at www.redtomato.org and www.ipminstitute.org

eco
apples™
BORN AND RAISED HERE™

Fig. 9.3. Eco Apple logo with one of the tag lines used on packaging, signage and other point-of-purchase materials. (Reproduced with permission from Red Tomato.)

Table 9.2. Changes, from 2004 to 2005, in pesticide use by hazard category (a) as established in the Eco Apple Protocol (IPM Institute, 2008b) and by type (b).

(a) Hazard category	2004 (US$)	2005 (US$)	% change
Use with justification	26,071	24,030	–8
Use with restrictions	93,098	90,878	–2
Do not use	12,330	763	–94
(b) Type	**2004 (US$)**	**2005 (US$)**	**% change**
Fungicide	70,909	69,068	–3
Herbicide	8,230	5,273	–36
Insecticide	36,438	24,389	–33
Miticide	9,372	10,538	12
Other	4,635	4,890	5
Rodenticide	2,571	2,168	–16
Total	134,159	118,331	–12

Results to date

Eco Apple brand apple sales increased from US$400,000 and 18,000 cases in 2005 to US$700,000 and 29,000 cases in 2006. Sales of the 2008 crop by Red Tomato exceeded US$1.8 million, in addition to grower sales to wholesale and retail customers. Overall pesticide cost declined by 12% in the first year (Table 9.2), although a dry season might account for at least part of that reduction. In 2007, all organophosphate insecticide applications to fruit were eliminated. A 2005 post-season grower survey indicated 95% grower satisfaction; access to markets and price and net return were the most important benefits.

Funding

The collaboration has benefited from grants from the USDA NRCS Conservation Innovation Grant programme, US EPA Strategic Ag Initiative, USDA Crops at Risk, USDA CSREES Northeastern IPM Center Working Group Grants and an anonymous foundation. In addition, Red Tomato earns a small commission on each box of apples they sell and smaller licence fees on Eco Apple brand apples sold directly by participating farmers.

Culmination of years of experimentation and effort

The Eco Apple project benefits from the expanding knowledge base contributed to by organic, buy local, IPM, fair trade and other cause-marketing programmes, including several specific to fruit and vegetable production in the north-east USA (Green, 2006). These include Partners with Nature (UMass, 1998), which used crop-specific 'IPM Guidelines' (Hollingsworth *et al.*, 1993; Hollingsworth and Coli, 2000). Growers were awarded points for implementing practices listed in the guidelines and were required to achieve a minimum score to participate. The guidelines have been used to help evaluate growers for incentives offered by the USDA Soil and Water Conservation Service (now USDA-NRCS), to generate sales by advertising seasonally the location and wares of participating growers to potential buyers in the state through advertising in the local media and to evaluate and compare IPM performance between participating and non-participating growers (Hollingsworth and Coli, 1999, unpublished results).

Concurrently, scientists at Cornell University were collaborating with managers at Wegmans Food Markets on a similar approach. Crop and region-specific 'IPM Elements' practice lists were crafted by workgroups including researchers, Extension, consultants, growers and others, again with points assigned to each practice and a required minimum score (Cornell University, 2007). Many of the individual practices within the elements (and guidelines) address the use of traps and other behaviour-based strategies described above.

Wegmans marketed qualifying fresh, frozen and canned fruit and vegetables under an IPM logo and labels (Pool, 1999; Fig. 9.4).

The use of IPM practice lists with assigned point values for evaluating IPM performance was described first by Boutwell and Smith (1981), but both Cornell and UMass Extension scientists involved in these efforts were unaware of the pair's pioneering work until well into the development of their own projects.

In 1992, a new collaboration developed between unlikely partners, Mothers and Others for a Livable Planet, a non-profit environmental advocacy group based in New York City, and north-east apple growers and scientists. This project, called CORE Values, was established shortly after environmental groups including Mothers and Others participated in a campaign to alert parents to the potential health hazard to their children associated with consuming apple juice made from apples treated with daminozide. The compound, a probable human carcinogen according to the US EPA, was contained in the growth regulator, Alar. Mothers and Others and apple growers agreed to work together to launch CORE Values, which loosely followed the European-initiated Integrated Fruit Production (IFP) system (IOBC, 2007) and aimed to reduce the health and environmental impacts of farming.

Partners with Nature, Cornell's IPM Elements and CORE Values have not survived as marketplace forces. Partners with Nature was not welcomed universally by growers and funding was terminated by the Massachusetts Department

Fig. 9.4. Wegmans IPM label on canned vegetables. (Reproduced with permission from Wegmans Food Markets.)

of Food and Agriculture, which indicated grower participation was insufficient (Cooley and Coli, Chapter 6, this volume). In New York, consumer recognition of 'IPM' never exceeded 17% (Pool, 1999), making marketing both challenging and expensive, and the supply of IPM-qualified product could not keep up with Wegman's growth. CORE Values closed its doors just prior to the dissolution of Mothers and Others for a Livable Planet. However, IPM Guidelines and Elements continue to be made available online from multiple states (Green, 2008) and are used by USDA-NRCS for performance assessments and by growers and others for education and self-assessment. An IFP protocol for apples in New York is also available (Carroll and Robinson, 2006), as well as a Good Agricultural Practices (GAP) audit workbook, both including practice lists and scoring mechanisms (Carroll *et al*., 2004). SYSCO, a USA-based food-service distributor with US$36 billion in annual sales, was inspired in part by the Cornell effort to implement an IPM and sustainable agricultural programme that included point-based scoring and annual audits on more than 242,000 ha of fruit and vegetable production. In addition, Food Alliance (www.foodalliance.org), an international eco-label founded in 1999 and requiring IPM practices of participating farmers and ranchers, grew 25% in 2007 to more than 2 million ha.

A number of the current Eco Apple collaborators and growers participated in these related, early and ongoing efforts and their expertise has been a great boon to the programme. Red Tomato's founder and current director is a co-founder of Equal Exchange, the oldest and largest for-profit fair trade company in the USA, with US$28 million in annual sales. The importance of land-grant programmes and resources cannot be overestimated as a contributing factor in our ability to meet the growing marketplace demand and requires our ongoing recognition and support. Eco-labelling has benefited not only from direct participation in programme development and maintenance, but also from the basic and applied research and technology transfer provided by these institutions, programmes and staff which make the practices included in production protocols possible.

Conclusion

AMF, in exploiting apples as a newly introduced host, spawned a new host-associated race, a coterie of scientists devoted to answering both basic and applied questions raised by its evolution, biology, ecology and behaviour, and new links in the supply chain for IPM and reduced-risk pesticide products designed to declare apples a forbidden fruit to its progeny and those of dozens of other apple-loving pests. While still short of Prokopy's personal achievement of a clean, marketable crop with only two insecticide and one or two fungicide applications per year in his own orchard, a successful 'Eco Apple' protocol built on IPM tactics and reduced-risk products has penetrated successfully through to the marketplace.

These accomplishments owe much to Prokopy's tremendous legacy, which will long outlive his physical presence in warm memories, tremendous scientific advances and practical IPM technology and techniques. As Wes Autio expressed, 'Prokopy's boundless energy in support of the apple industry will be sorely missed, and his extensive research contributions will never be forgotten' (Coli, 2004); 'His contribution is something we won't even understand the scope of for many, many years' (Green, 2005).

There is much work to be done in the management arena alone. In late 2007, the Eco Apple working group of growers, scientists, consultants and others identified more than 29 research priorities (Eco Apple Working Group, 2008) and ranked the following among the most urgent:

- Efficacy of the trap-tree approach for plum curculio management.
- Efficacy of insecticide-treated spheres as a control measure for AMF.
- Development of new IPM sampling and monitoring programmes for control of apple pests with reduced-risk insecticides.
- Impacts of eco-labels and eco-marketing, including organic, on region-based marketing of fruits and vegetables.

The 29 outstanding priorities beg questions. Who will do the research required to address them? Who will extend the fruits of research done here and elsewhere to growers, consultants and consumers? The Eco Apple working group is potentially a sustainable learning community for maintaining and improving the protocol going forward, built on the economic incentives provided by a value-added product and support from public and private sector funders. However, Prokopy's research and extension role at UMass remains unfilled, although an effort has been mounted to fund an endowed position. Other university-based scientists in the Eco Apple workgroup and elsewhere fear the same fate for their positions as they reach retirement age. How will the progress in the development of IPM be maintained as new challenges and pests emerge if there are no scientists to take their places? That spectre presents probably the greatest threat to the ongoing improvement and progress towards a greater understanding of current and future apple pests and ongoing reductions in the impacts of production practices.

References

ABAG (2007) Association of Bay Area Governments, Oakland, California, www.abag.ca.gov/ (accessed 23 June 2008).

Agnello, A.M., Spangler, S.M., and Reissig, W.H. (1990) Development and evaluation of a more efficient monitoring system for apple maggot (Diptera: Tephritidae). *Journal of Economic Entomology* 83, 539–546.

Aliniazee, M.T. (1990) Evaluating apple maggot traps in the west. In: Dowell, R.V. and Wilson, L.T. (eds) *Agricultural Science: Apple Maggot in the West: History, Biology, and Control, Publication 3341*. University of California, Oakland, California, pp. 56–66.

Aluja, M. and Prokopy, R.J. (1993) Host odor and visual stimulus interaction during intratree host finding behaviour of *Rhagoletis pomonella* flies. *Journal of Chemical Ecology* 19, 2671–2696.

Averill, A.L., Reissig, W.H. and Roelofs, W.L. (1988) Specificity of olfactory responses in the apple maggot fly. *Entomologia Experimentalis et Applicata* 47, 211–222.

Benbrook, C. (2007) The organics debate; is the premium worth it? Institute of Food Technologies (http://www.organic-center.org/reportfiles/IFT_Webcast%20[Compatibility%20Mode].pdf, accessed 29 November 2007).

Berlocher, S.H. and Feder, J.L. (2002) Sympatric speciation in phytophagous insects: moving beyond controversy? *Annual Review of Entomology* 47, 773–815.

Bernays, E. (2000) *Plant–Insect Interactions: A Synthesis*. In: Gazzoni, D.L. (ed.) Abstract Book I: XXI International Congress of Entomology, Foz de Iguassu, Brazil. Embrapa Soja, Londrina, Brazil, pp. 8–13.

BIRC (2007) *Directory of Least-toxic Pest Control Products*. Bio-Integral Resource Center, Berkeley, California.

Boutwell, J.L. and Smith, R.H. (1981) A new concept in evaluating integrated pest management programs. *Bulletin of the Entomological Society of America* 27, 117–188.

Brown, J.M., Abrahamson, W.G., Packer, R.A. and Way, P.A. (1995) The role of natural enemy escape in a gallmaker host plant shift. *Oecologia* 104, 52–60.

Brown, M.W. (1993) Population dynamics of invading pests: factors governing success. In: Kim, K.C. and McPheron, B.A. (eds) *Evolution of Insect Pests: Patterns of Variation*. Wiley Interscience, New York, pp. 203–218.

Bush, G.L. (1969) Sympatric host race formation and speciation in frugivorous flies of the genus *Rhagoletis* (Diptera, Tephritidae). *Evolution* 23, 237–251.

California DPR (2007) *Summary of Pesticide Use Report Data 2006: Indexed by Commodity*. Department of Pesticide Regulation, Sacramento, California.

Carle, S.A., Averill, A.L., Rule, G.S., Reissig, W.H. and Roelofs, W.L. (1987) Variation of host fruit volatiles attractive to the apple maggot fly *Rhagoletis pomonella*

(Diptera: Tephritidae). *Journal of Chemical Ecology* 13, 795–805.

Carroll, J.E. and Robinson, T.L. (2006) *New York Integrated Fruit Production Protocol for Apples*. New York State Agricultural Experiment Station, Geneva, New York.

Carroll, J.E., Breth, D.I. and Fargione, M.J. (2004) *EUREPGAP Audit Workbook for New York State Tree Fruit Growers*. Cornell Cooperative Extension, Orleans County, Albion, New York.

Coli, W.M. (2004) Remembering Ron. *Fruit Notes* 69(2), 5–9.

Consumer Reports (2007) *Greener Choices. Eco-label Center*. Consumer Reports, Yonkers, New York, www.greenerchoices.org (accessed 23 June 2008).

Cornell University (2007) NYS Elements of IPM, www.nysipm.cornell.edu/elements/ (accessed 23 June 2008).

Crop Life International (2007) Annual report (http://www.croplifeasia.org/ref_library/others/Global%20market%20performance_13Sep07.pdf, accessed 5 December 2008).

Dean, R.W. (1942) Emergence, migration, and control of the apple maggot. *Proceedings of the 87th Annual Meeting of the New York State Horticultural Society*. New York State Horticultural Society, Geneva, NY, pp. 291–295.

Deutsch, A. (ed.) (2004) IPM loses a giant. *IPMnet News* 126.

Duan, J.J. and Prokopy, R.J. (1992) Visual and odor stimuli influencing effectiveness of sticky spheres for trapping apple maggot flies. *Journal of Applied Entomology* 113, 271–279.

Duan, J.J. and Prokopy, R.J. (1993) Toward development of pesticide-treated spheres for apple maggot flies: carbohydrates and amino acids as feeding stimulants. *Journal of Applied Entomology* 115, 176–184.

Duan, J.J. and Prokopy, R.J. (1995) Development of pesticide-treated spheres for controlling apple maggot flies: pesticides and residue-extending agents. *Journal of Economic Entomology* 88, 117–126.

EcoApple Working Group (2008) Research priorities: results January 2008 (neipmc.org/priority/2008/eco-apple.pdf, accessed 23 June 2008).

Feder, J.L. (1998) The apple maggot, *Rhagoletis pomonella*: flies in the face of conventional wisdom about speciation? In: Howard, D.J. and Berlocher, S.H. (eds) *Endless Forms: Species and Speciation*. Oxford University Press, New York, pp. 130–144.

Feder, J.L. and Filchak, K.E. (1999) It's about time: the evidence for host-plant mediated selection in the apple maggot fly, *Rhagoletis pomonella*, and its implications for fitness trade-offs in phytophagous insects. *Entomologia Experimentalis et Applicata* 91, 211–225.

Feder, J.L., Reynolds, K., Go, W. and Wang, E.C. (1995) Intra- and interspecific competition and host race formation in the apple maggot fly, *Rhagoletis pomonella* (Diptera: Tephritidae). *Oecologia* 101, 416–425.

Fein, B.L., Reissig, W.H. and Roelofs, W.L. (1982) Identification of apple volatiles attractive to the apple maggot. *Journal of Chemical Ecology* 8, 1473–1487.

Foster, S.P. and Harris, M.O. (1997) Behavioural manipulation methods for insect pest management. *Annual Review of Entomology* 42, 123–146.

French, S. and Rogers, G. (2005) *LOHAS Market Research Review: Marketplace Opportunities Abound*. Natural Marketing Institute, Harleysville, Pennsylvania.

Gilliom, R.J., Barbash, J.E., Crawford, C.G., Hamilton, P.A., Martin, J.D., Nakagaki, N., Nowell, L.H., Scott, J.C., Stackelberg, P.E., Thelin, G.P. and Wolock, D.M. (2007) *Pesticides in the Nation's Streams and Ground Water, 1992–2001*. United States Geological Survey, United States Department of the Interior, Washington, DC.

Green Shield Certified (2007) Green Shield Certified. Effective pest control. Peace of mind (www.greenshieldcertified.org, accessed 23 June 2008).

Green, T.A. (2005) Ronald J. Prokopy. *Insect Pest Control Newsletter* 65, 25–26.

Green, T.A. (2006) Successes and failures with eco-labels. *Proceedings of the New York State Vegetable Growers Conference*, 13 February 2006, Cornell University, Syracuse, New York. New York State Vegetable Growers Association, Kirkville, New York.

Green, T.A. (2008) Links to IPM certification and recognition programs (www.ipminstitute.org/links.htm, accessed 23 June 2008).

Green, T.A. and Prokopy, R.J. (1986) Visual monitoring trap for the apple blotch leafminer, *Phyllonorycter crataegella* (Lepidoptera: Gracillariidae). *Environmental Entomology* 15, 562–566.

Green, T.A. and Prokopy, R.J. (1998) Diurnal behavior of the apple blotch leafminer moth, *Phyllonorycter crataegella* (Lepidoptera: Gracillariidae). *Canadian Entomologist* 130, 415–425.

Gruchalla, R.S., Pongracic, J., Plaut, M., Evans, R., Visness, C.M., Walter, M., Crain, E.F., Kattan, M., Morgan, W.J., Steinbach, S., Stout, J., Malindzak, G., Smartt, E. and Mitchell, H. (2005) Inner city asthma study: relationships among sensitivity, allergen exposure, and asthma morbidity. *Journal of Allergy and Clinical Immunology* 115, 478–485.

Harris Interactive (2007) 'Green' issues and professional pest services. Prepared for the National Pest Management Association (NPMA) and presented at Going Green: Marketing to the 21st Century Consumer, Denver, Colorado.

Hermanson, M.H., Moss, D.J., Monosmith, C.L. and Keeler, G.J. (2007) Spatial and temporal trends of gas and particle phase atmospheric DDT and metabolites in Michigan: evidence of long-term persistence and atmospheric emission in a high-DDT-use fruit orchard. *Journal of Geophysical Research* 112. DOI:10.1029/2006JD007346, 2007.

Hodson, A.C. (1943) Lures attractive to the apple maggot. *Journal of Economic Entomology* 36, 545–548.

Hoffman, S., Mittenthal, R., Chandler, B., Lafleur, G., Wright, S., Appleton, P., Becker, M., Dynock, S. and Prokopy, R. (2002) Influence of odor bait, cultivar type, and adjacent habitat composition on performance of perimeter traps for controlling apple maggot flies. *Fruit Notes* 67(2), 20–24.

Hollingsworth, C.S. (1994) Integrated pest management certification: a sign by the road. *American Entomologist* 40, 74–75.

Hollingsworth, C.S. and Coli, W.M. (2000) Massachusetts integrated pest management guidelines: crop-specific definitions. *Journal of Extension* 38(3), www.joe.org/joe/2000june/tt1.html.

Hollingsworth, C.S., Pascall, M.J., Cohen, N.L. and Coli, W.M. (1993) Support in New England for certification and labeling of produce grown using integrated pest management. *American Journal of Alternative Agriculture* 8, 78–84.

Hood, E. (2006) The apple bites back. *Environmental Health Perspectives* 114(8), A470–476.

Houlihan, J., Wiles, R., Thayer, K. and Gray, S. (2003) *Body Burden: The Pollution in People*. Environmental Working Group Washington, DC.

Hu, X.P. and Prokopy, R.J. (1996) Progress in 1995 toward development of toxicant-treated spheres for controlling apple maggot flies. *Fruit Notes* 61(2), 10–13.

Hu, X.P. and Prokopy, R.J. (1998) Lethal and sublethal effects of imidacloprid on apple maggot fly, *Rhagoletis pomonella* Walsh (Diptera:Tephritidae). *Journal of Applied Entomology* 122, 37–42.

Hu, X.P., Kaknes, A. and Prokopy, R. (1997) Improved pesticide-treated wooden spheres for controlling apple maggot flies. *Fruit Notes* 62(2), 3–5.

Hu, X.P., Shasha, B., McGuire, M. and Prokopy, R. (1998) Controlled release of sugar and toxicant from a novel device for controlling pest insects. *Journal of Controlled Release* 50, 257–265.

Illingsworth, J.F. (1912) A Study of the Biology of the Apple Maggot *Rhagoletis pomonella* Together with an Investigation of Methods of Control. *New York Agricultural Experiment Station Bulletin 324*, Geneva, New York.

IOBC (2007) Commission on IP-guidelines and endorsement. International Organization for Biological Control (www.iobc.ch/, accessed 23 June 2008).

IPM Institute (2008a) IPM and sustainable ag certifiers experience continued growth in 2007. *IPM in the Marketplace* 9(1) (www.ipminstitute.org/newsletter/newsletter_v9i1.htm, accessed 23 June 2008).

IPM Institute (2008b) *EcoApple Protocol and Grower Self-Assessment*. IPM Institute of North America, Madison, Wisconsin.

Kiely, T., Donaldson, D. and Grube, A. (2004) Pesticides industry sales and usage: 2000 and 2001 market estimates. US EPA Biological and Economic Analysis Division, Office of Pesticide Programs and Office of Prevention, Pesticides, and Toxic Substances, Washington, DC.

Kring, J.B. (1970) Red spheres and yellow panels combined to attract apple maggot flies. *Journal of Economic Entomology* 63, 466–469.

Lafleur, G., Rull, J. and Prokopy, R. (2001) Do sugar caps atop pesticide-treated spheres affect apple maggot fly attraction to the spheres? *Fruit Notes* 66, 37–39.

Linn, C.E. Jr, Feder, J.L., Nojima, S., Dambrowski, H.R., Berlocher, S.H. and Roelofs, W.L. (2003) Fruit odor discrimination and sympatric host race formation in *Rhagoletis*. *Proceedings of the National Academy of Sciences of the United States of America* 100, 11490–1493.

Lu, C., Toepel, K., Irish, R., Fenske, R.A., Barr, D.B. and Bravo, R. (2005) Organic diets significantly lower children's exposure to commonly found organophosphorus pesticides in food. *Environmental Health Perspectives* 114, 260–263.

Luckman, W.H. and Metcalf, R.L. (1994) The pest management concept. In: Metcalf, R.L. and Luckman, W.H. (eds) *Introduction to Pest Management*. John Wiley and Sons, New York, pp. 1–32.

NEPMA (2007) *NEPMA IPM Registry*. New England Pest Management Association Concord, New Hampshire.

Organic Trade Association (2007) Executive summary: Organic Trade Association's 2007 Manufacturer Survey (www.ota.com/pics/documents/2007ExecutiveSummary.pdf, accessed 23 June 2008).

Papaj, D.R. and Prokopy, R.J. (1989) Ecology and evolutionary aspects of learning in phytophagous insects. *Annual Review of Entomology* 34, 315–350.

Pool, W.M. (1999) IPM label products on supermarket shelves. Presentation at 'Environmental Marketing: Opportunities for Producers, Distributors and Retailers', IPM Institute of North America, Madison Wisconsin.

Porter, B.A. (1928) *The Apple Maggot*. USDA Technical Bulletin 66, 1–47.

Pottinger, R.P. and LeRoux, E.J. (1971) The biology and dynamics of *Lithocolletis blancardella* (Lepidoptera: Gracillariidae) on apple in Quebec. *Memoirs of the Entomological Society of Canada* 77, 437 pp.

Prokopy, R.J. (1968a) Sticky spheres for estimating apple maggot adult abundance. *Journal of Economic Entomology* 61, 1082–1085.

Prokopy, R.J. (1968b) Visual responses of apple maggot flies, *Rhagoletis pomonella*: orchard studies. *Entomologia Experimentalis et Applicata* 11, 403–422.

Prokopy, R.J. (1972) Evidence for a marking pheromone deterring repeated oviposition in apple maggot flies. *Environmental Entomology* 1, 326–332.

Prokopy, R.J. (1975) Apple maggot fly control by sticky red spheres. *Journal of Economic Entomology* 68, 197–198.

Prokopy, R.J. (1986) *Toward a World of Less Pesticide*. Massachusetts Agricultural Experiment Station Research Bulletin 710, University of Massachusetts, Amherst, Massachusetts.

Prokopy, R.J. (1998) Review: the case of people versus pests. (Review of *Nature Wars*, M.L. Winston.) *BioScience* 48, 959–961.

Prokopy, R.J. (2003) Two decades of bottom-up, ecologically based pest management in a small commercial orchard in Massachusetts. *Agriculture, Ecosystems, and Environment* 94, 299–309.

Prokopy, R.J. and Kogan, M. (2003) Integrated pest management. In: Resh, V.H. and Cardé, R.T. (eds) *Encyclopedia of Insects*. Academic Press, San Diego, California, pp. 589–595.

Prokopy, R.J. and Mason, J.L. (1996) Behavioral control of apple maggot flies. In: McPheron, B. and Steck, G. (eds) *Fruit Flies*. St Lucia Press, Florida, pp. 555–559.

Prokopy, R.J. and Owens, E.D. (1978) Visual monitoring trap for European apple sawfly. *Entomologia Experimentalis et Applicata* 24, 609–620.

Prokopy, R.J. and Owens, E.D. (1983) Visual detection of plants by herbivorous insects. *Annual Review of Entomology* 28, 337–364.

Prokopy, R.J. and Roitberg, B.D. (2007) Arthropod pest behavior and IPM. In: Kogan, M. and Jepson, P. (eds) *Ecological Approaches to Pest Management*. Cambridge University Press, Cambridge, UK, pp. 87–121.

Prokopy, R.J., Moericke, V. and Bush, G.L. (1973) Attraction of apple maggot flies to odor of apples. *Environmental Entomology* 2, 743–749.

Prokopy, R.J., Hubbell, G.L., Adams, R.G. and Hauschild, K.L. (1982) Visual monitoring trap for tarnished plant bug adults on apple. *Environmental Entomology* 11, 200–203.

Prokopy, R.J., Diehl, S.R. and Cooley, S.S. (1988) Behavioral evidence for host races in *Rhagoletis pomonella* flies. *Oecologia* 76, 138–147.

Prokopy, R.J., Christie, M., Johnson, S.A. and O'Brien, M. (1990) Transitional steps toward second-stage integrated management of arthropod pests of apple in Massachusetts orchards. *Journal of Economic Entomology* 83, 2405–2410.

Prokopy, R.J., Mason, J.L., Christie, M. and Wright, S.E. (1996) Arthropod pest and natural enemy abundance under second-level versus first-level integrated pest management practices in apple orchards: a four-year study. *Agriculture, Ecosystems and Environment* 57, 35–47.

Prokopy, R.J., Mason, J.L. and Wright, S.E. (1997) Action thresholds for controlling *Phyllonorycter crataegella* leafminers based on prebloom captures of adults on visual traps. *Journal of Agricultural Entomology* 14, 163–169.

Prokopy, R.J., Wright, S.E., Chandler, B.W. and Hu, X.P. (1998) Evaluation of varying doses of different toxicants for use on spheres to control apple maggot flies. *Fruit Notes* 63(4), 6–10.

Prokopy, R., Wright, S. and Black, J. (1999) Commercial-orchard trials of attracticidal spheres for controlling apple maggot flies. *Fruit Notes* 64(4), 14–17.

Prokopy, R.J., Wright, S.E., Black, J.L., Hu, X.P. and McGuire, M.R. (2000) Attracticidal spheres for controlling apple maggot flies: commercial-orchard trials. *Entomologia Experimentalis et Applicata* 97, 293–299.

Prokopy, R., Chandler, B., Dynok, S., Appleton, P. and Becker, S. (2001) Commercial-orchard evaluation of pesticide-treated spheres for apple maggot control in 2001. *Fruit Notes* 66, 30–34.

Prokopy, R.J., Chandler, B.W. and Wright, S.E. (2003) Improved sugar delivery onto pesticide-treated spheres for controlling *Rhagoletis pomonella* flies (Diptera: Tephritidae). *Canadian Entomologist* 135, 1–10.

Prokopy, R.J., Wright, S.E. and Chandler, B.W. (2004a) Feeding stimulants for pest control. World Intellectual Property Organization, International Publication Number WO 2004/071997.

Prokopy, R., Jácome, I., Bigurra, E., Blanco, M., Chandler, B. and Wright, S. (2004b) Evaluation of pesticide-treated spheres for control of apple maggot flies in 2003. *Fruit Notes* 69(2), 21–25.

Prokopy, R.J., Jácome, I. and Bigurra, E. (2005) An index for assigning distances between odor-baited spheres on perimeter trees of orchards for control of apple maggot flies. *Entomologia Experimentalis et Applicata* 115, 371–377.

Reissig, W.H. (1975) Performance of apple maggot traps in various apple tree canopy positions. *Journal of Economic Entomology* 68, 534–538.

Reissig, W.H., Fein, B.L. and Roelofs, W.L. (1982) Field tests of synthetic apple volatiles as apple maggot attractants. *Environmental Entomology* 11, 1294–1298.

Reynolds, A., Prokopy, R., Green, T. and Wright, S. (1996) Apple maggot fly (Diptera: Tephritidae) responses to perforated red spheres. *Florida Entomologist* 79, 173–179.

Salam, M.T., Li, Y.F., Langholz, B. and Gilliland, F.D. (2004) Early-life environmental risk factors for asthma: findings from the children's health study. *Environmental Health Perspectives* 112, 760–765.

Suckling, K. and Hodges, W. (2007) *Status of the Bald Eagle in the Lower 48 States and the District of Columbia: 1963–2007*. Center for Biological Diversity, Tucson, Arizona.

Taylor, C.R. (1980) The nature of benefits and costs of use of pest control methods. *American Journal of Agricultural Economics* 62, 1007–1011.

Thornton, J.W., McCally, M. and Houlihan, J. (2002) Biomonitoring of industrial pollutants: health and policy implications of the chemical body burden. *Public Health Reports* 117, 315–323.

UMass (1998) Labeling for growers and consumers. In: *Vigorous Farms in a Safe Environment: Solutions for the Next Century*. UMass Extension IPM, Amherst, Massachusetts.

US Census Bureau (2007) International Data Base (http://www.census.gov/ipc/www/idb/worldpopinfo.html, accessed 5 December 2008).

USDA (2002) Organic food standards and labels: the facts. National Organic Program, USDA, Washington, DC, 2 pp.

USDA (2004) National road map for integrated pest management (http://www.ipmcenters.org/IPMRoadMap.pdf, accessed 5 December 2008).

USDA (2006) USDA roles in market-based environmental stewardship (http://www.ocio.usda.gov/directives/doc/DR5600-003.htm, accessed 5 December 2008).

US EPA (2007) *Office of Pesticide Programs Annual Reports* (http://www.epa.gov/oppfead1/annual/, accessed 5 December 2008).

US GAO (2001) *Agricultural Pesticides: Management Improvements Needed to Further Promote Integrated Pest Management.* United States General Accountability Office, Washington, DC.

USGBC (2007) *LEED for High Performance Operations, v. 2008. Draft Credits for 2nd Comment Period.* United States Green Building Council, Washington, DC.

Van Driesche, R.G. and Taub, G. (1983) Impact of parasitoids on *Phyllonorycter* leafminers infesting apple in Massachusetts, USA. *Protection Ecology* 5, 303–317.

Walsh, B.D. (1867) The apple-worm and the apple maggot. *Journal of Horticulture* 2, 338–343.

Walsh, B.D. and Riley, C.V. (1868) *American Entomologist, Volume 1*. R.P. Studley and Company, St Louis, Missouri.

WHO (2006) WHO gives indoor use of DDT a clean bill of health for controlling malaria: WHO promotes indoor spraying with insecticides as one of three main interventions to fight malaria (www.who.int/mediacentre/news/releases/2006/pr50/en/, accessed 23 June 2008).

Whyatt, R.M., Rauh, V., Barr, D.B., Camann, D.E., Andrews, H.F., Garfinkel, R., Hoepner, L.A., Diaz, D., Dietrich, J., Reyes, A., Tang, D., Kinney, P.L. and Perera, F.P. (2004) Prenatal insecticide exposures and birth weight and length among an urban minority cohort. *Environmental Health Perspectives* 112, 1125–1132.

Wright, S., Hu, X. and Prokopy, R. (1997) Tests of imidacloprid spheres for controlling apple maggot fly. *Fruit Notes* 62(4), 1–4.

Wright, S., Chandler, B., Fleury, R. and Prokopy, R. (2000) Pursuit of effective pesticide-treated spheres for controlling apple maggot. *Fruit Notes* 65, 45–49.

Wright, S., Chandler, B. and Prokopy, R. (2001) Improvement of sugar delivery systems for rain-activated, pesticide-treated spheres. *Fruit Notes* 66, 25–29.

Index

Index of Common Names of Organisms

Index of Latin Names